Skin Photoaging (Second Edition)

Online at: https://doi.org/10.1088/978-0-7503-5112-6

Skin Photoaging (Second Edition)

Edited by
Rui Yin
Department of Dermatology, Southwest Hospital, Third Military Medical University (Army Medical University), Chongqing, People's Republic of China

Yang Xu
Department of Dermatology, First Affiliated Hospital of Nanjing Medical University, Nanjing, Jiangsu Province, People's Republic of China

Chengfeng Zhang
Department of Dermatology, Huashan Hospital, Fudan University, Shanghai, People's Republic of China

IOP Publishing, Bristol, UK

ISBN 978-0-7503-5112-6 (ebook)
ISBN 978-0-7503-5110-2 (print)
ISBN 978-0-7503-5113-3 (myPrint)
ISBN 978-0-7503-5111-9 (mobi)

DOI 10.1088/978-0-7503-5112-6

Version: 20241201

IOP ebooks

British Library Cataloguing-in-Publication Data: A catalogue record for this book is available from the British Library.

Published by IOP Publishing, wholly owned by The Institute of Physics, London

IOP Publishing, No.2 The Distillery, Glassfields, Avon Street, Bristol, BS2 0GR, UK

US Office: IOP Publishing, Inc., 190 North Independence Mall West, Suite 601, Philadelphia, PA 19106, USA

Contents

Preface

As a medical professional dedicated to the field of dermatology, I am delighted to present this comprehensive book on photoaging. Over the years, I have witnessed first-hand the impact of environmental factors, particularly sunlight, on the skin ageing process. Photoaging has become an important topic in dermatology, prompting us to explore the mechanisms of photoaging, preventive measures, and the latest rejuvenation therapies.

In this book, we present the complex mechanisms of photoaging and show how UV radiation and other environmental factors accelerate skin ageing. Understanding these processes is essential for developing effective skin protection and rejuvenation strategies.

We also emphasize the importance of sun protection as a cornerstone of anti-aging skincare. Through the latest research and advances in sun protection technology, we provide insight into how individuals can protect their skin from UV radiation.

In addition, the book details some of the most satisfying facial rejuvenation treatments performed in the clinic today. From popular treatments such as Botox injections and dermal fillers to cutting-edge technologies such as microneedling, chemical peels, low-energy light therapy, lasers, photodynamic therapy, radio-frequency, ultrasound, and even at-home devices, these methods offer a wide range of clinical treatment options for treating photoaging and restoring youthful skin.

In 2014, the first edition of *Photoaging* was published, which was very well received by our readers. Over the past decade, there has been a sea-change in research into the mechanisms and clinical treatments targeting photoaging of the skin. New rejuvenation modalities, devices and rejuvenating therapeutic materials have emerged with satisfactory clinical results. As a result, this second edition of *Photoaging* has been significantly improved and enriched.

I am very grateful to the contributing authors of this book, who are medical experts in the fields of dermatology, medical aesthetics and micropigmentation, for selflessly sharing their knowledge and insights, and we are confident that the second edition will serve as a source of reference information for dermatologists and healthcare practitioners in the field of medical aesthetics to provide better care to their patients and individuals who need to find ways to counteract the effects of photoaging.

Rui Yin MD, PhD

Acknowledgments

When we were invited to publish the second edition of this book, we discussed and updated the writing framework together. We realized that writing this book well would not be an easy task. Just as everyone was preparing to begin writing, the sudden outbreak of COVID-19 had an enormous impact and brought great difficulty to our lives and work. In the last three years we have experienced unprecedented difficulties. Without the efforts and experience of our invited contributors, this book could not have been completed. They deserve our recognition and deep appreciation, and we thank all of our collaborators. We are also grateful to the publishers for their help and understanding.

Rui Yin MD, PhD
Yang Xu MD, PhD
Chengfeng Zhang MD, PhD

Editor biographies

Rui Yin

Rui Yin, MD, PhD, is a professor of dermatology at Southwest Hospital, affiliated with Amy Medical University, and a visiting associate professor at the Wellman Center for Photomedicine at Massachusetts General Hospital. Her research interests are focused on photodynamic therapy and skin laser treatments for photoaging and skin aging. Dr. Yin has published 40 peer-reviewed articles in English and over 40 peer-reviewed articles in Chinese. She has contributed to over 50 conference proceedings and has authored or edited more than 10 books. In 2012, she was recognized as one of the Top Ten National Outstanding Young Dermatologists by the China Dermatological Association.

Yang Xu

Yang Xu, MD, PhD, is an associate professor in the Department of Dermatology at the First Affiliated Hospital of Nanjing Medical University. From 2013 to 2015, she served as a visiting scholar in the Department of Dermatology at the University of Michigan. Her research focuses on photo-related damage to human skin. Dr. Xu has authored over 30 scholarly articles, reviews, abstracts, and textbook chapters, and has also authored or edited five books. In 2017, she was honored as one of the Top Ten National Outstanding Young Dermatologists by the China Dermatological Association.

Chengfeng Zhang

Chengfeng Zhang, MD, PhD, is a professor of dermatology at Huashan Hospital, affiliated with Shanghai Medical College, Fudan University, in China. He is also a visiting scholar in the Division of Biology and Pathobiology of the Skin, Department of Dermatology, at the Medical University of Vienna in Austria. His research interests span pigmentation disorders such as melasma and vitiligo, acne, and skin-barrier-related skin conditions. Dr. Zhang has published over 50 international peer-reviewed articles in top dermatology journals, including the Journal of the American Academy of Dermatology, JAMA Dermatology, Journal of Investigative Dermatology, and the British Journal of Dermatology. Additionally, he serves as an associate editor and social media editor for the Journal of Investigative Dermatology.

List of contributors

Zhuanli Bai
Department of Plastic and Aesthetic Maxillofacial Surgery, The First Affiliated Hospital of Xi'an Jiaotong University, Xi'an, People's Republic of China

Xu Chen
Jiangsu Key Laboratory of Molecular Biology for Skin Diseases and STIs, Institute of Dermatology, Chinese Academy of Medical Sciences and Peking Union Medical College, Nanjing, People's Republic of China

Daniel Meng-Yen Hsieh
Department of Dermatology, Peking University First Hospital, Beijing, People's Republic of China

Min Jiang
Department of Dermatology, Huashan Hospital, Fudan University, Shanghai, People's Republic of China

Xian Jiang
Department of Dermatology, West China Hospital, Sichuan University, Chengdu, People's Republic of China

Shanglin Jin
Department of Dermatology, Huashan Hospital, Fudan University, Shanghai, People's Republic of China

Lei Li
Department of Dermatology, Southwest Hospital, Third Military Medical University (Army Medical University), Chongqing, People's Republic of China

Xiaoxue Li
Department of Dermatology,West China Hospital, Sichuan University, Chengdu, People's Republic of China

Huaxu Liu
Hospital for Skin Diseases, Shandong First Medical University, Jinan, People's Republic of China; Shandong Provincial Institute of Dermatology and Venereology, Shandong Academy of Medical Sciences, Jinan, People's Republic of China

Ziqi Liu
Department of Dermatology, Huashan Hospital, Fudan University, Shanghai, People's Republic of China

Qi Luan
Department of Dermatology, The Second Affiliated Hospital of Xi'an Medical University, Xi'an, People's Republic of China

Xiuxiu Wang
Department of Dermatology, Huashan Hospital, Fudan University, Shanghai, People's Republic of China

Xiang Wen
Department of Dermatology, West China Hospital, Sichuan University, Chengdu, People's Republic of China

Xiaojin Wu
Department of Dermatology, Shanghai Ninth People's Hospital affiliated to Shanghai Jiaotong University School of Medicine, Shanghai, People's Republic of China

Yan Wu
Department of Dermatology, Peking University First Hospital, Beijing, People's Republic of China

Li Xie
Department of Dermatology, West China Hospital, Sichuan University, Chengdu, People's Republic of China

Xiaoxue Xing
Department of Dermatology, Huashan Hospital, Fudan University, Shanghai, People's Republic of China

Zhongyi Xu
Department of Dermatology, Huashan Hospital, Fudan University, Shanghai, People's Republic of China

Zhen Zhang
Department of Dermatology, Shanghai Ninth People's Hospital affiliated to Shanghai Jiaotong University School of Medicine, Shanghai, People's Republic of China

Yue Zheng
Department of Dermatology, Nanfang Hospital, Southern Medical University, Guangzhou, People's Republic of China

Shaomin Zhong
Department of Dermatology, Peking University First Hospital, Beijing, People's Republic of China

Introduction

Cutaneous aging is an accumulated process affected by the superimposed additional changes or 'accelerated aging' due to environmental damage, including UV irradiation. Cutaneous aging, as for all other organs, undergoes chronological aging, involving slow deformation of appearance and deterioration in tissue function. It can be divided into two groups: intrinsic aging and photoaging. This latter process mainly results from UV exposure, which depends primarily on the degree of sun exposure and skin pigment. Rejuvenation or slowing the aging process has always been one of the major goals of medicine. Recently, due to increased life expectancy and, in turn, an increase in the aging population, preventive and regenerative approaches, in particular in the field of dermatology, are in high demand. Relaxation of facial rhytides and restoration of lost volume with non-surgical treatments has been the most significant change in facial rejuvenation in the last ten years. Prevention of UV-induced damage is currently achieved by the application of sunscreens. The development of cosmetically pleasing sunscreens that protect against both UVA and UVB irradiation as well as cosmeceuticals that resist the UV signaling pathways leading to photoaging are major steps forward in preventing and reversing photoaging. Many skincare products containing antioxidant ingredients have been introduced. These ingredients can help reduce skin damage from ultraviolet and blue light, thereby slowing down the skin aging process. For skin rhytides and laxity, minimally invasive procedures have gained popularity as they are less invasive and require less downtime, and thus are superior to traditional cosmetic interventions. They include cosmetological injections and filling, chemical peeling, laser, radiofrequency, and photodynamic therapy, and so on. Although these novel modalities demonstrate significant improvements in clinical signs, questions remain regarding the ideal treatment parameters that should be employed for optimal results. More controlled randomized comparative clinical trials are necessary to elucidate the most effective way to use these devices for their various clinical applications. Combination therapy and individualized treatment for some special subjects are also recommendations.

In recent years, the treatment of skin photoaging has been developing with each passing day, and new technologies and equipment are being used widely in clinical practice. Therefore, we have revised the manuscript on the basis of the first edition to make it more conducive to clinical use.

IOP Publishing

Skin Photoaging (Second Edition)

Rui Yin, Yang Xu and Chengfeng Zhang

Chapter 1

Skin photoaging

Yang Xu, Xu Chen and Yue Zheng

Skin aging encompasses both endogenous and photoaging factors, influenced by ultraviolet radiation (UVA, UVB), infrared A (IRA), and visible light. This complex process results in a spectrum of distinct histopathological and clinical presentations, necessitating targeted interventions. Photoaging, characterized by chronic inflammation, compromised skin barrier function, and diminished skin-related functions, underscores the paramount importance of preventive measures. Therefore, emphasizing proactive strategies against photoaging becomes imperative in medical practice.

1.1 Introduction

Aging is an important scientific problem. Skin aging is closely related to the occurrence of a wide range of skin diseases. According to the different pathological mechanisms of aging, skin aging can be divided into endogenous aging and photoaging. The former is the natural process of systemic senescence caused by aging. The latter is based on the endogenous aging of skin tissue, due to the superimposed effect of chronic ultraviolet (UV) exposure.

The typical manifestations of endogenous skin aging include loose skin, dry rough skin, wrinkle formation, cherry-like hemangioma, and seborrheic keratosis. The typical manifestation of photoaging is more significant wrinkle formation (especially deep wrinkle increase). In addition, it can appear as papular elastic fiber degeneration, irregular pigmentation, punctate hypopigmentation, open acne, telangiectasia, venous lakes, sebaceous gland hyperplasia, stellate pseudoscar and other manifestations. Skin photoaging is closely related to precancerous lesions such as actinic keratosis and malignant tumors such as cutaneous squamous cell carcinoma.

Complex factors such as different locations, different living and working habits (leading to differences in UV exposure intensity), and differences in the intensity and frequency of sunscreen measures applied make it difficult to accurately assess the UV exposure of different individuals. The UV exposure of the population increases

doi:10.1088/978-0-7503-5112-6ch1

gradually with the increase of altitude and the decrease of latitude of a location. Studies have shown that the annual UV exposure of Europeans, Americans and Australians in indoor work ranges from 10 000 to 20 000 J m^{-2}, 20 000 to 30 000 J m^{-2}, and 20 000 to 50 000 J m^{-2} (Godar 2005). However, when holidays are taken into account, UV exposure can increase by 30% or more (Godar 2005). Even in environments with similar UV exposure levels, sunlight-induced skin changes vary significantly among individuals, possibly due to differences in skin response and repair ability to UV-caused skin damage from different genetic backgrounds. In the same Caucasian population, skin types I and II have different clinical manifestations of photodamage from skin types III and IV (Lim *et al* 2007). Type I and type II skin mainly show atrophic skin lesions, but wrinkle formation is not significant, and can be accompanied by focal hypopigmentation, solar keratosis and skin malignancy. Type III and IV skin often show deep wrinkle formation, rough skin, and moles (Lim *et al* 2007).

At present, the spectrum of action causing skin photoaging is still unclear. Current *in vitro* and *in vivo* evidence cannot elucidate well the exact relationship between UVB and UVA and the occurrence of photoaging. There are significant differences in the optical characteristics of UVB and UVA reaching the Earth's surface: the energy intensity of UVB radiation is significantly higher than UVA, the amount of UVA radiation on the Earth's surface is far greater than that of UVB. The penetration ability of UVA in skin tissue is higher than UVB (current studies indicated that the influence of the latter is more limited to the epidermis). Therefore, it can be speculated that UVB radiation has a more significant effect on the epidermis, and UVA may play a more critical role in the dermal tissue changes caused by photoaging. In a simulated skin tissue model constructed *in vitro*, the cells in the epidermis were mainly affected by UVB radiation, producing sunburned cells (Rahman *et al* 2021) and elevated cyclobutane pyrimidine dimer (Garcia-Ruiz *et al* 2022). Apoptosis of upper dermal fibroblasts was induced by UVA producing increased matrix metalloproteinases (Seo *et al* 2018, Lan *et al* 2019, Park and Park 2019, Wu *et al* 2019, Xue *et al* 2022). In addition to UV rays, the infrared and visible components of sunlight may also contribute to the generation of skin photoaging. Infrared rays can be divided into IRA (700–1400 nm), IRB (1400–3000 nm) and IRC (3000 nm^{-1} mm) according to their wavelength. The latter two have a weak influence on photoaging because they are mainly blocked and absorbed by the skin surface and upper epidermis (Krutmann *et al* 2021). IRA's high penetration ability allows it to reach the deep dermis (Schieke *et al* 2003). It has been found in both human and mouse studies that IRA can cause skin wrinkle formation (Barolet *et al* 2016). IRA caused increased levels of matrix metalloproteinase-1 (MMP-1), which was inhibited by the combination of sunscreen and antioxidants, but not by sunscreen alone (Grether-Beck *et al* 2015). Previous studies in guinea pig models have found that infrared rays can induce an increase in the synthesis of dermal matrix, and infrared rays can further aggravate the damaging effect of UVA on dermis (such as elastic fiber degeneration) (Schieke *et al* 2003). The link between visible light and skin photoaging remains unclear. It has been found that visible light itself or visible light combined with UVA1 can cause persistent skin pigmentation in dark skin types. However, this effect was not significant in light skin types (Mahmoud *et al* 2010, Kohli *et al* 2018). Melasma is currently considered as a skin disease associated with photoaging (Passeron and Picardo 2018). In visible light, the short-wavelength components blue violet light and blue light can promote melanocyte to increase melanin synthesis through the mediation of the opsin 3 receptor (Duteil *et al* 2014, Regazzetti *et al* 2018). Importantly, the combination of shortwave visible light protectants (e.g. iron oxide) and UVA/UVB sunscreen produced a more significant intervention effect in reducing melasma recurrence than sunscreen alone (Boukari *et al* 2015). Therefore, visible light may also play a role in the occurrence and progression of skin photoaging.

1.2 Mechanism of photoaging

1.2.1 Sunlight components that may be associated with skin photoaging

UV rays in sunlight can be divided into UVA (320–400 nm), UVB (280–320 nm) and UVC (100–280 nm) based on wavelength differences. The ozone layer in Earth's atmosphere absorbs most of the UVC in sunlight, so it is mostly UVA and UVB that reach the Earth's surface. As mentioned above, part of the visible light component of sunlight has also been proved to be related to the occurrence of photoaging, for example, the infrared component at wavelengths ranging from 760 nm to 1 mm, and the blue violet component at wavelengths ranging from 400 to 480 nm.

1.2.2 DNA damage and DNA repair

Due to the low penetration capacity of UVB, almost most of UVB is absorbed by the epidermis after exposure to UV radiation. UVB can cause cells to produce a wide range of cell biological changes. First, UVB radiation causes significant DNA damage to keratinocytes and melanocytes in the epidermis, as indicated by an increase in 8-oxoguanine, cyclobutene pyrimidine dime, pyrimidine (6-4) pyrimidone and other photodamage products (Rahman *et al* 2021, Tanaka *et al* 2019). DNA photodamage products produced during chronic UV damage of skin are closely associated with skin photoaging and possible skin photocarcinogenesis (Yang *et al* 2019, Pollet *et al* 2018, Carrara *et al* 2019), for example, p53 mutations detected in skin tumors characterized by cytosine to thymine translocation and cytosine cytosine–thymine thymine replacement at the bipyrimidine site (Pfeifer and Besaratinia 2012).

Although the radiation intensity is lower than UVB, UVA also has a strong ability to cause DNA damage to tissues due to its higher amount in sunlight. UVA mainly causes oxidative DNA damage products, including 8-oxo-7 and 8-dihydro-2′-deoxyguanosine (De Gruijl and Rebel 2008, Beani 2014). In addition, UVA can also induce the production of a large number of cyclobutane pyrimidine dimers. But unlike UVB-induced DNA photoproducts, UVA preferentially leads to the production of TT site cyclobutane pyrimidine dimers without the production of pyrimidine (6-4) pyrimidone (Mouret *et al* 2006).

In order to antagonize the adverse effects of UV on skin tissue, the body will start the DNA repair mechanism to repair the damaged DNA and avoid the accumulation of genotoxic substances (in particular the UV-induced DNA mutation (Cadet *et al* 2015)) to maintain the stability of the internal environment. Nucleotide excision repair (NER) is an effective mechanism for repairing DNA photoproducts caused by UV (Hoeijmakers 2001, de Laat *et al* 1999), however, this mechanism is more effective at recognizing and repairing pyrimidine dimers and is less capable of repairing cyclobutane pyrimidine dimers. NER includes two forms: whole-genome NER and transcription-coupled NER. Transcription coupling NER is a specific repair mode that removes photodimers that block transcription while restoring UV-suppressed gene transcription (Mullenders 2018). It is of concern that microorganisms that are continuously exposed to UV can produce photolysis enzymes, such as cyclobutane pyrimidine dimer photolysis enzyme and pyrimidine(6-4)pyrimidone photolysis enzyme, which have excellent ability to repair DNA damage photoproducts (Ramirez *et al* 2021). Although the human body cannot synthesize the above-mentioned photolysis substances, some topical preparations containing photolysis enzymes of microbial origin have been explored for human use (Ramirez *et al* 2021). When cells are under stress, the normally functioning p53 protein leads to cell cycle arrest, and promotes DNA repair or initiates apoptosis to remove cells with high levels of DNA damage products (Ianni *et al* 2021). Therefore, p53 plays a key role in avoiding the accumulation of genotoxicity caused by UV-induced DNA damage products (Ianni *et al* 2021). When the p53 protein is dysfunctional and its protective mechanism fails to perform, more mutations and chromatin instability occur, leading to abnormal cell proliferation and progressive development to actinic keratosis and squamous cell carcinoma (Hussein *et al* 2004). DNA damage that is not repaired properly leads to faulty DNA replication, which causes gene mutations. Some of these mutations may cause changes in key cell biological functions, thus participate in the occurrence of cancer and aging (Martincorena and Campbell 2015). However, it has been found that UVA-induced DNA photodamage products are more likely to accumulate in the skin tissue, because the clearance rate of UVA-induced cyclobutane pyrimidine dimers is lower than that of UVB-induced cyclobutane pyrimidine dimers (Mouret *et al* 2011).

1.2.3 UV induced autophagy and cell death

After suffering injury stress, cells perform cell fate selection, cell survival or cell death, in order to maintain the stability of the internal environment (Gudipaty *et al* 2018). Autophagy is a key cell biological mechanism that degrades intracellular substances and organelles (Thorburn 2018). Under stress, the initiation of autophagy is a key way to maintain the stability of the internal environment (Perrotta *et al* 2020). Many studies in invertebrates and vertebrates have found an association between autophagy dysfunction and aging (Wong *et al* 2020). How is autophagy involved in the aging process? The current explanation is that autophagy dysfunction occurs with age, resulting in abnormal protein homeostasis (Kaushik *et al* 2021). Research evidence has suggested that the autophagy mechanism is involved in the occurrence and progression of skin photoaging. UV-induced elevated lipid peroxidation (Indirapriyadarshini *et al* 2022) is a very important factor in the generation of photoaging. The autophagic pathway plays an important role in the degradation of these photoage-inducing substances (Wang *et al* 2019a). For example, increased levels of 25-hydroxycholesterol (an oxidized cholesterol product) in keratinocytes following UV irradiation can induce increased levels of intracellular autophagy (Olivier *et al* 2017). UVA radiation and phospholipid oxides produced by UVA radiation can induce autophagy in epidermal keratinocytes, and the level of active phospholipid oxide in cells with autophagy defects is significantly increased (Zhao *et al* 2013). The molecule Bach2, which has an antagonistic effect on skin photoaging, has been identified in skin fibroblasts and can inhibit the expression of aging-related genes. The autophagy mechanism is involved in the inhibition of Bach2 on the photoaging effect (Wang *et al* 2021b). It should be noted that different autophagy regulatory effects were observed in keratinocytes exposed to different doses of UVB. For example, a relatively low dose of UVB radiation induced an increase in the autophagic flux of keratinocytes (Bahamondes Lorca and Wu 2020). However, relatively high dose of UVB irradiation showed an inhibition effect on autophagy flux (Chen *et al* 2018). Considering the complexity of human exposure under different living conditions, there may be a mixed regulatory state of skin autophagy induction and autophagy inhibition after different levels of UV exposure. Therefore, the exact biological role of complicated regulation of autophagy in the occurrence and development of skin photoaging is still a great concern.

UV radiation, on the other hand, induces cell death in skin cells. For example, as described above, cells perform apoptosis mechanisms to eliminate cells with severe DNA damage and avoid photocarcinogenesis. Apoptosis of keratinocytes could be induced by UVB (Noh *et al* 2019), while apoptosis of skin fibroblasts was found to be induced by UVA (Xue *et al* 2022). Based on the differences in the molecular characteristics of cell death, the International Nomenclature Committee on Cell Death has made recommendations for naming different types of cell death, including endogenous apoptosis, exogenous apoptosis, necroptosis, ferroptosis, pyroptosis, parthanatos, autophagy dependent cell death, cellular senescence, etc (Galluzzi *et al* 2018). It was recently confirmed that UVB can induce gasdermin E-mediated pyroptosis (Chen *et al* 2020, Liu *et al* 2021) and ferroptosis (Vats *et al* 2021) in keratinocytes. These studies suggest the complexity of how cell death is performed after UV damage. UV radiation can induce premature cellular senescence of skin cells, and the resulting senescent cell accumulation and senescence-related secretory phenotype are important factors driving the occurrence of skin photoaging (Fitsiou *et al* 2021). However, it is still unclear whether and how execution or dysfunction of regulated cell death such as apoptosis, pyroptosis, and ferroptosis are involved in the occurrence of photoaging.

1.2.4 Oxidative stress

Reactive oxygen products are a problem that the organism needs to face all the time. Reactive oxygen products can be produced from endogenous metabolic processes or from the external environment. Moderate levels of oxidative products have certain physiological effects, such as regulating signal transduction and cell biological functions (Zhang *et al* 2020, Valko *et al* 2006). To counter the threat posed by overreactive oxygen products, the body has evolved complex antioxidant mechanisms to maintain homeostasis in the internal environment. Known antioxidant mechanisms that can be used and utilized by the body include antioxidant enzymes (such as superoxide dismutase, catalase, etc) and low molecular weight antioxidant molecules (carotenoids, ascorbic acid, vitamin E, lipoic acid and glutathione, etc) (Zhang *et al* 2020). The impaired balance of reactive oxygen products and antioxidant mechanisms in body tissues is closely related to the occurrence of various diseases (Zhang *et al* 2020, Forman and Zhang 2021, Butterfield and Halliwell 2019, de Bhailis *et al* 2021, Munzel and Daiber 2018, Perez *et al* 2019). UV radiation can cause keratinocytes and fibroblasts in skin to produce reactive oxygen products (Terra *et al* 2012, Samivel *et al* 2020, Lone *et al* 2020, Chung *et al* 2022). Excessive accumulation of reactive oxygen products causes direct damage to cells and participates in the occurrence of aging, such as mitochondrial DNA damage, protein carboxylation, 4-hydroxynonaldehyde production, etc (Gu *et al* 2020). Reactive oxygen products can also promote the production of matrix metalloproteinases through the activation of AP-1 mediated by mitogen activated protein kinase and c-Fos/c-Jun signal, and participate in the process of photoaging (Chen *et al* 2016). The production of reactive oxygen products is also involved in the activation of NF-κB signal, which mediates the expression regulation of a variety of inflammatory cytokines (Wang *et al* 2019b). It is worth noting that keratinocytes with autophagy defects show more significant DNA damage and aging when faced with oxidative stress (Song *et al* 2017). This evidence suggest that the role of oxidative stress in skin photoaging should be closely explored in combination with cellular biological regulatory mechanisms.

Exposure to UV radiation alone is rare under everyday conditions, and exposure to sunlight containing UV rays is more common. Interestingly, Hudson *et al* found that the synergistic irradiation of ultraviolet, visible and infrared radiation increased the production of reactive oxygen products in fibroblasts, but no similar effect was observed in keratinocytes (Hudson *et al* 2020). This suggests that there may be very complex synergistic or antagonistic effects of sunlight components other than UV rays on the radiative effects of UV rays. In the study of skin photoaging, especially in the construction of research models, it may be necessary to pay attention to the application of sunlight components other than UV rays.

1.2.5 Signaling pathways associated with photoaging

MAPK plays an important role in the skin aging process (Gu *et al* 2020). MAPK is a serine-threonine protein kinase that is activated under the induction of cellular stress, hormones, neurotransmitters, cytokines and other factors (Gaestel 2015). The MAPK pathway includes the extracellular signal-regulated protein kinase pathway (ERK), c-Jun NH2-terminal kinase pathway (JNK), and p38 MAPK pathway. Activation of the ERK pathway mediated by TLR4 is involved in cytokine secretion and matrix metalloproteinase (MMP) expression in skin fibroblasts of premature aging induced by UVA radiation (Seo *et al* 2018). The role of JNK and p38 MAPK pathways in photoaging was further confirmed in the photoaging model constructed using *Caenorhabditis elegans* (Prasanth *et al* 2020).

The AP-1 transcription factor family consists of Jun protein and Fos protein (Hesari *et al* 2018). Since c-Jun is regulated by the JNK pathway, the JNK pathway actively participates in the regulation of AP-1 signaling (Tian *et al* 2020, Lee *et al* 2019, Wang *et al* 2021a, Badarni *et al* 2019). UV radiation activates AP-1 signaling in skin cells (Kim *et al* 2018, Oh *et al* 2020a, Gao *et al* 2018a). Activation of AP-1 has an inhibitory effect on transforming growth factor-β, which interferes with dermal tissue's ability to synthesize collagen properly (Fisher *et al* 2002). The strategy of targeting inhibition of the abnormal activation of AP-1 and restoration of transforming growth factor-β/Smad signaling has obtained effective observational data in the recovery of collagen synthesis in UV-damaged skin fibroblasts (Gao *et al* 2018b, Park *et al* 2018, Gao *et al* 2018a, Oh *et al* 2021). The activation of AP-1 is involved in the abnormal secretion of MMP-1, MMP-3 and MMP-9 in UV-damaged skin (Kim *et al* 2018, Choi *et al* 2020, Oh *et al* 2020b, Cheong *et al* 2018).

1.2.6 Matrix metalloproteinase and tissue inhibitor of metalloproteinases

The main biological function of MMP protein family is to degrade extracellular matrix proteins, including collagen, elastin, fibronectin, proteoglycan, etc (Pittayapruek *et al* 2016). MMPs can be secreted by keratinocytes and fibroblasts in skin (Kumar and Mandal 2019, Nakyai *et al* 2017). The occurrence and progression of MMP protein family in skin photoaging has been widely understood, and its abnormal expression has been used as a molecular marker for photoaging (Shin *et al* 2019). In short (Pittayapruek *et al* 2016), MMP-1 (collagenase) is involved in the degradation of type I collagen and type III collagen in skin photoaging, MMP-2 and MMP-9 (gelatinases) are involved in the degradation of type IV collagen in skin photoaging, MMP-3 (matrix lysine) is involved in the degradation of type I collagen and the activation of MMP-1, MMP-7 and MMP-9 in skin photoaging, MMP-10 (matrix lysine) activates a variety of pre-MMP in skin photoaging, and MMP-7 (matrix soluble factor) and MMP-12 are involved in the degradation of elastic fibers in skin photoaging. Tissue inhibitor of metalloproteinases (TIMP) is a natural inhibitor of the MMP protein family in tissues. In the study of skin photoaging of hairless mice (Han *et al* 2019), it was found that the transcription level of TIMP-1 in the skin tissue of photoaging mice was reduced, and specific anti-photoaging intervention could restore and increase the level of TIMP-1.

1.3 Clinical features and histological manifestation of photoaged skin

1.3.1 Clinical features of photoaged skin

Natural light contains ultraviolet radiation (UVR), visible light, and infrared radiation (IR). IR and different types of UVR have different impacts on human skin and are responsible for various clinical manifestations. UVR from sunlight can be divided into three types according to wavelength: UVA (320–400 nm), UVB

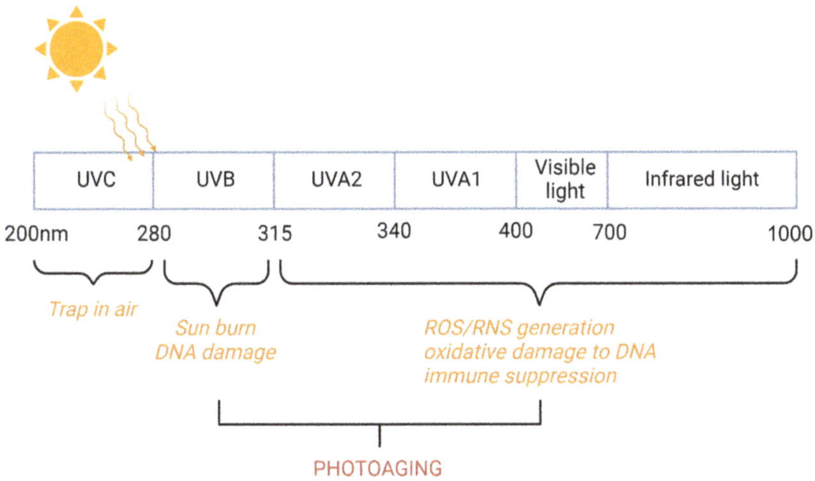

Figure 1.1. The solar radiation spectrum and its effects on skin.

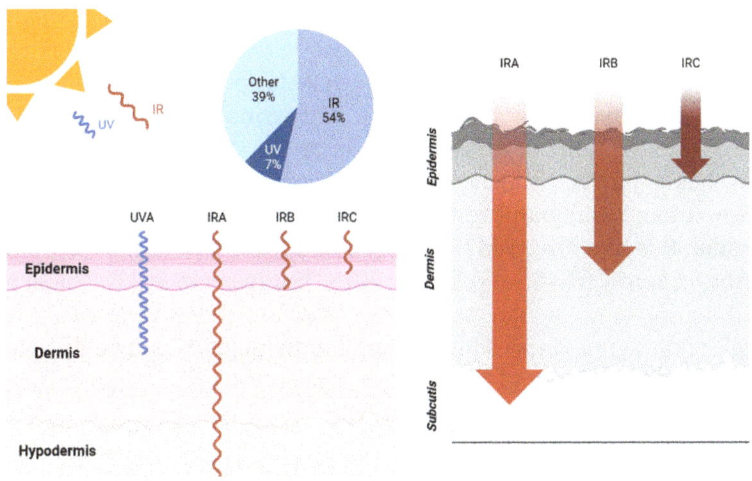

Figure 1.2. The penetration potential of UVR and IR in skin.

(280–320 nm), and UVC (200–280 nm) (figure 1.1). UVC cannot reach the Earth's surface and is absorbed by the stratospheric ozone layer.

Some UVB are absorbed by the ozone layer. It constitutes only about 5% of the total UVR and is highly destructive to DNA and epidermal keratinocytes. Biologically active UVB penetrates the superficial layers of skin to the basal layer of the epidermis (figure 1.2), where it produces reactive oxygen species (ROS) and reactive nitrogen species (RNS), leading to inflammation, erythema (sunburn), skin aging, and in some cases non-melanoma skin cancer.

UVA can invade the dermis and is about 98% responsible for major skin aging. UVA is less energetic than UVB, but it accounts for (90%–95%) of total UVR. UVA radiation can penetrate deeper into the epidermis and dermis of the skin than UVB (figure 1.2). It can activate melanin pigmentation in the top layer of the skin and creates a tanning appearance in a short time. However, it can penetrate into and damage the connective tissues and blood vessels with other major skin components including the extracellular matrix (ECM). As a result, the skin gradually loses its elasticity and tends to wrinkle. UVA damages the DNA in keratinocytes, making keratinocytes a major site for most skin cancers.

Visible light penetrates deeply into biological tissues and seems inoffensive.

IR has the lowest energy and accounts for about 45% of the solar spectrum reaching human skin. IR includes IRA (700–1400 nm), IRB (1400–3000 nm), and IRC (3000 nm^{-1} mm), of which only IRA can penetrate into the skin. It represents about 30% of IR radiation, of which 65% reaches the dermis and 10% reaches the subcutaneous tissue (figure 1.2). Upon IR radiation, abnormally high levels of MMP-1 are evident, resulting in excessive destruction of collagen, weakening of ECM, wrinkle formation and dermal thinning. IRA can also induce excessive growth of blood vessels resulting in continuous redness of the skin (Worrede *et al* 2021, Farage *et al* 2017, Ansary *et al* 2021).

The changes caused by photoaging are superimposed on the changes of intrinsic aging and are responsible for most of the age-associated features of skin appearance and cause consistent changes to the integumentary system. Altogether, photoaging is characterized by erythema, elastosis, telangiectasia, pigmentation or hypopigmentation, laxity, atrophy, rough skin textures and wrinkles. Wrinkles might be fine, coarse, or both. Chronic dryness and itching are also prevalent. Importantly, photoaging has a higher incidence of precancerosa such as actinic keratosis and Morbus Bowen, and malignant tumor, such as basal cell carcinoma and squamous cell carcinoma. It is best observed on sun-exposed sites, such as the face, lateral neck, and extensor forearms (Rittié and Fisher 2015, Farage *et al* 2017, Zhang and Duan 2018).

Although photoaging varies widely among individuals, it can be divided into two broad categories, atrophic and hypertrophic, according to its clinical, histological and molecular differences (Sachs *et al* 2019).

In atrophic photoaging (AP), skin appears thin and shiny with erythema, dyspigmentation and telangiectasia. Compared with hypertrophic photoaging, skin cancers such as keratinocyte carcinoma and melanoma are significantly increased (Sachs *et al* 2019).

Hypertrophic photoaging (HP) is characterized by thickened skin with coarse wrinkles. In addition, HP shows more sallowness, a higher global photoaging score and less erythema (figure 1.3) (Sachs *et al* 2019, Farage *et al* 2017).

One of the typical signs in photoaging presents as solar elastosis, which occurs in both HP and AP to different histological degrees (see section 1.3.2.3.1). Solar elastosis is a disease of dermal elastic tissue degeneration caused by long-term sun exposure. It commonly presents yellow, thickened, coarsely wrinkled skin (Heng *et al* 2014). A rare clinical manifestation of solar elastosis was reported as asymptomatic, shiny, smooth, pearly telangiectatic, firm papules, 1–10 mm in diameter, that can be multiple or solitary, with various colorations, and exhibit significant solar elastosis histologically (Heng *et al* 2014, Kwittken 2000).

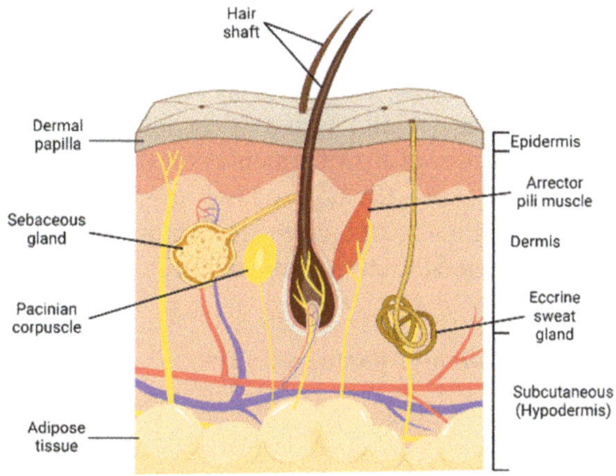

Figure 1.3. Normal skin structure showing the layers of epidermis, dermis, and hypodermis.

Solar elastosis is the most common solar elastotic syndrome, others include Favre–Racouchot disease (FRD), elastotic nodules of the ears, collagenous and elastotic plaques of the hands and colloid milia, accompanied by marked solar elastosis in histological examinations.

FRD is an occupational disorder of people with long-term sun exposure (e.g. farmers). It is characterized by thickened yellow plaques with follicular cysts surrounded by multiple open and cystically dilated comedones plugged with keratinous debris in the lumen. FRD tends to occur in older men with lighter skin, most commonly in the temporal and periorbital regions of the face, especially around the lateral canthus, and is usually bilateral. Patients with FRD are less likely to develop malignant skin neoplasms, suggesting that elastic tissue may be a beneficial tissue response (Yeh and Schwartz 2021, Heng *et al* 2014).

Elastotic nodules of the ear manifest as single or multiple bilateral, firm to hard pale nodules on the antihelix or helix (Heng *et al* 2014). Collagenous and elastotic plaques of the hands is a slowly progressive condition, mostly seen in elderly men with waxy linear plaques at the junction of the dorsal and palmar skin on the hand (Heng *et al* 2014). Colloid milium is a rare skin deposition disease that usually clinically presents as the development of yellowish translucent or flesh-colored papules or plaques on sun-exposed skin (Heng *et al* 2014, Azimi *et al* 2017).

Hyperpigmentation may present as ephelides, poikiloderma of Civatte (PC) and solar lentigines (SLs) while hypopigmentation may be characterized by idiopathic guttate hypomelanosis (IGH). Poikiloderma of Civatte usually affects Caucasians aged 40–70 years of age and typically involves sun-exposed regions of the body, particularly the face and neck. Poikiloderma of Civatte typically presents as erythematous patches covering photodamaged skin with associated hyperpigmentation and visible telangiectasias, as it is a vascular dermatosis. SLs are acquired dark brown pigmented spots that develop on sun-exposed areas of the skin, such as the face, shoulders and the back of the hands and are characterized as asymptomatic, flat, light to dark brown and well demarcated macules (Krueger *et al* 2022). Recent studies indicate that independently of UVR to a major extent, other environmental factors including particulate and possibly gaseous constituents of ambient air pollution contribute to the development of solar lentigines through the activation of aryl hydrocarbon receptor (AHR) signaling in human skin (Nakamura *et al* 2015). IGH is a benign disorder of hypopigmentation that increases with age and is rarely seen in children and young adults. It commonly occurs on the extremities as small, well-defined, hypopigmented macules. IGH may be associated with aging, UV exposure, trauma or genetic factors (Saleem *et al* 2019, Krueger *et al* 2022).

1.3.2 Histological manifestation of photoaged skin

1.3.2.1 Normal skin structure
The skin is composed of three layers: epidermis, dermis, and subcutaneous tissue (figure 1.3).

1.3.2.2 Epidermis and DEJ in photoaged skin
The UV-exposed epidermis thickens compared to the thinner epidermis of intrinsically aged skin, which is a prominent feature in photoaged epidermis. Stratum corneum (SC), the outermost layer of the epidermis, is mostly affected and thickens because of failure of degradation of corneocyte desmosomes (Zhang and Duan 2018). In a study comparing the effects of intrinsic and extrinsic aging, a histopathological examination of 83 biopsies from sun-exposed and protected skin in healthy volunteers aged 6–84 years revealed that epidermal thickness showed no differences in both sun-exposed and protected areas in different age groups, while the epidermis of sun-exposed skin is significantly thicker than that of protected skin (Farage *et al* 2017). Another prominent feature of photodamaged skin in the epidermis is apoptotic cells called 'sunburn' cells. These cells show pyknotic nuclei and a necrotic, eosinophilic cytoplasm. It was found that epidermal inclusion cysts may also be a sign of chronic UV damage. Follicular epithelial retention hyperkeratosis and comedone formation are other well-recognized features of chronic photodamage (Farage *et al* 2017).

In photoaging, the DEJ becomes flattened with shortened epidermal rete pegs and dermal rete ridges, which also occur in intrinsic skin aging to a different extent (Rittié and Fisher 2015). At the DEJ of sun-exposed skin, duplications of lamina densa are frequently observed beneath keratinocytes and anchoring fibrils are also associated with detached lamina densa, mainly on the dermis side. By contrast, in sun-protected skin of both young and old subjects, there is almost no change in the BM at the DEJ (Amano 2016). The destruction and reduplication of BM at the DEJ in sun-exposed skin may be due to the degradation of BM components: laminin 332, type IV and VII collagens, and perlecan, contributing to an increase in the level of BM-damaging enzymes, such as plasmin, MMPs, and heparanase. The structural damage of BM may correlate with the functional changes of epidermal cells and dermal cells, thus promoting the aging process by destroying the ECM of the dermis and inducing abnormal keratinocytes (figure 1.4) (Farage *et al* 2017).

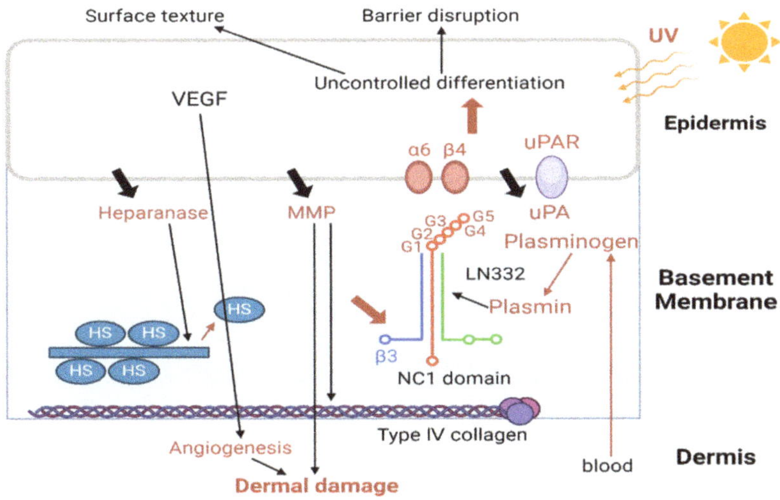

Figure 1.4. The mechanisms of destruction and reduplication of BM at the DEJ in sun-exposed skin.

1.3.2.3 Dermis in photoaged skin

1.3.2.3.1 Collagen and elastic fibril degradation (solar elastosis)

In photoaged skin, collagen fibrils and elastin fibers, which constitute the majority of the dermal extracellular matrix, appear fragmented and reduced (Sachs *et al* 2019).

During both natural and photoaging skin, quantitative and structural changes of collagen fibers and elastic fibers are the main modifications found in aged skin, while the reduction of ECM, especially collagen in dermis, is considered to be one of the main mechanisms of dermal atrophy (Shin *et al* 2019). The degradation of collagen ECM in photoaging skin is mainly mediated by a subset of matrix metalloproteinases (MMP). The effect of sunlight on the dermis leads to an increase in MMPs, such as MMP-1, MMP-3 and MMP-9, which specifically cleave collagen and other ECM proteins (Sachs *et al* 2019).

MMPs are ubiquitous endopeptidases that degrades ECM proteins and can be divided into five main subgroups: (i) collagenase (MMP-1, MMP-8 and MMP-13); (ii) gelatinase (MMP-2 and MMP-9); (iii) stromelysins (MMP-3, MMP-10 and MMP-11); (iv) matrilysins (MMP-7 and MMP-26); and (v) membrane-type (MT) MMPs (MMP-14, MMP-15 and MMP-16). MMP-1 is the main protease that initiates the fragmented collagen fibers, mainly including type I and type III in human skin. After being cut by MMP-1, collagen can be further degraded by MMP-3 and MMP-9. Physiologically, MMPs are regulated by specific endogenous tissue inhibitors of metalloproteinases (TIMPs). However, the increase of MMP level in aging skin is not accompanied by the corresponding increase of endogenous MMP inhibitor levels. The level of TIMP-1 in photoaged and intrinsic aging skin may even decline. This imbalance will accelerate the fragmentation of progressive collagen in the dermis and accelerate skin aging.

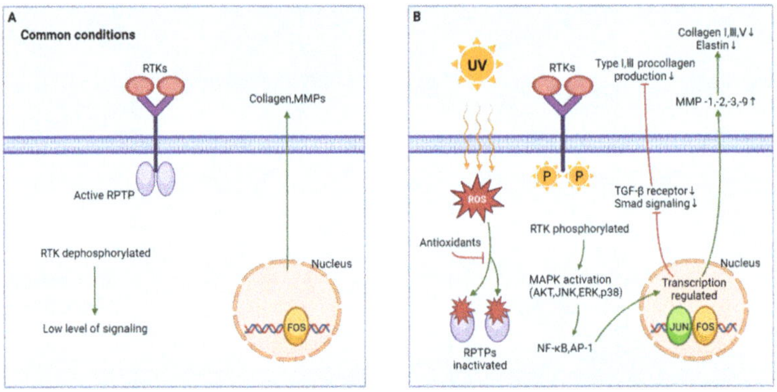

Figure 1.5. MMPs induced by ROS degrading the collagen.

The activation of MMPs-2, −3, −7, −9, −12, and −13 is capable of catabolizing elastic fibers, which is the main mechanism responsible for increased elastic fiber degradation in photoaging. MMP-12, also known as human macrophage metalloelastase, is the most active protease in elastin degradation. Other proteases such as neutrophil serine proteases cathepsin G and human leukocyte elastase are also known to decompose elastin. Therefore, solar radiation decomposes collagen and elastin faster than normal biological aging (Biskanaki *et al* 2021, Shin *et al* 2019).

ROSs are the main driving force for the increase of MMP level in aged skin (figure 1.5). ROSs, by binding to the catalytic site of RPTPs, increase the level of phosphorylated RTK and trigger downstream signaling pathways, including activation of the mitogen activated protein kinases (MAPK) family and nuclear factor-κB (NF-κB), resulting in the transcription of factor activator protein 1 (AP-1), then the AP-1 binding and histone acetylation lead to the high transcriptional activation of the MMP cluster at the 11q22.3 region in response to UVB, resulting in a high level of MMP. AP-1 also downregulates the expression of the transforming growth factor-β (TGF-β) type II receptor, thereby inhibiting collagen synthesis. Therefore, the total collagen content in photoaged skin is reduced. NF-κB activated by ROS mediates responses to ultraviolet radiation and photoaging. NF-κB activity is responsible for up-regulating MMP in dermal fibroblasts, such as MMP-1 and MMP-3. Although the main source of MMP in internal senescence is dermal fibroblasts, MMP in photoaging is also produced by epidermal keratinocytes, which accelerates the progressive fragmentation of collagen in the dermis and accelerates skin aging (Schmucker *et al* 2012, Ujfaludi *et al* 2018, Shin *et al* 2019, Zhang and Duan 2018).

In addition to MMPs, in photoaged skin, replenishment of the degraded matrix also decreases. UVB-irradiated keratinocytes can secrete factors such as inflammatory cytokines including interleukin-1 (IL-1) and tumor necrosis factor-α (TNF-α), which affect collagen synthesis in dermal fibroblasts. TNF-α stimulates the chemotaxis of inflammatory cells to the skin and downregulates procollagen mRNA, and thus may downregulate the production of type I collagen. (Biskanaki *et al* 2021) In severely photodamaged skin, fibroblasts have less interaction with intact collagen and are thus exposed to less mechanical tension, and it has been proposed that this situation might lead to decreased collagen synthesis (Farage *et al* 2017, Shin *et al* 2019, Zhang and Duan 2018).

The imbalance of the degeneration and production of the collagen and elastic fibrils and their structural changes result in the accumulation of non-functional, irregular elastic material deep in the dermis, called solar elastosis, which is a striking characteristic for photoaged skin. In photoaging skin, the formation of solar elastic tissue was observed under the skin, which replaced the collagen fibers of the skin and showed a thin-film distribution on the surface. The most severe solar elastic degeneration was observed in the limbs, followed by the back, abdomen and face (Biskanaki *et al* 2021).

The most notable histologic difference between HP and AP was the degree of solar elastotic changes. Elastotic material resides in the upper dermis and might create a wrinkled appearance by virtue of nonuniform space-filling. Histology revealed significantly less elastotic material in atrophic photoaging, which may explain the clinical phenotype of coarse wrinkling seen in HP. Collagen fibrils and gene expression of matrix metalloproteinases did not differ between the two forms of photoaging. Gene expression levels of MMP-1/3/7/9/12 enzymes, as well as type I and type III collagens, were similar in the AP and HP groups (Sachs *et al* 2019). In conclusion, in sun-exposed skin, the reduced and structurally changed collagen fibrils and elastin fibers are replaced by solar elastosis, which has a lower level in atrophic photoaging.

1.3.2.3.2 HA and proteoglycans

A recent study analyzed the amount and molecular weight of HA in paired biopsy specimens from photo-protected skin and photoaged skin of the same Japanese females, and found that both the amount and the molecular size of HA decreased in photoaged skin, which corresponded to a histological reduction of HA in the papillary dermis in the photoaged skin and suggests a link to the formation of skin wrinkles and sagging in photoaged skin (Yoshida *et al* 2018a). The main mechanism of the HA degradation is found to be mediated by HYBID (hyaluronan binding protein involved in hyaluronan depolymerization), alias KIAA1199/CEMIP, in human skin fibroblasts (Yoshida and Okada 2019).

Proteoglycans (PGs) mainly consist of decorin, biglycan and versican in human skin. In photoaged skin, versican seems to be increased, while biglycan does not change, and decorin is absent (Shin *et al* 2019).

1.3.2.3.3 The morphological changes of cells and vessels

In aged skin, the morphology of fibroblasts has a tendency to size reduction and elongation decrease, due to the impaired fibroblast attachment resulting from progressive ECM degradation while the collapsed morphology and the reduced size is a key feature of senescent fibroblasts (Shin *et al* 2019). The collapse of fibroblasts induced by the fragmentation of dermal ECM will in turn induce the increase of MMP-1 production and decrease of procollagen synthesis through AP-1/MMP-1 and TGF-β/CTGF/procollagen signaling pathways.

UVB can significantly increase the size of melanocytes and induce the accumulation of cytoplasmic brown granules (cytoplasmic melanosomes), indicating the increase of melanin synthesis in UVB-irradiated melanocytes. In addition, UVB can induce an increase in the percentage of cells with two or more nuclei. In a co-culture system, it also showed a correlation with UVB-irradiated senescent fibroblasts which could increase the synthesis and accumulation of melanosomes and the frequency of multinucleated melanocytes in melanocytes by secreting factors. In addition, it has been confirmed that true neomelanin production is caused in response to mild and repeated doses of UVB (Martic *et al* 2020).

Photoaged skin shows vascular damage while intrinsically aged skin does not. In mildly photodamaged skin, venular walls are thickened, while in severely damaged skin, the vessel walls are thinned and the supporting perivascular veil cells are reduced in number. Cross sections of photodamaged skin reveal a significant reduction in vascularity in the papillary dermis of aged skin and there are local dilations, corresponding to clinical telangiectasias (figure 1.9). Altogether, there is a marked change in the horizontal vascularization pattern with dilated and distorted vessels. Studies in humans and in the hairless skh-1 mouse model for skin aging have found that acute and chronic UVB irradiation greatly increases skin vascularization. Photoaging also showed a negative correlation to the capillary density and the pericytes to endothelial cells (PC/EC ratio) of capillaries or venules, due to the relative and absolute loss of the PC (Farage *et al* 2017). Altogether, the histological manifestation of photoaged skin is significantly seen in the epidermis and dermis. The sun-exposed epidermis thickens, especially the SC, and apoptotic cells called 'sunburn' cells can be found in photoaged epidermis, which shows pyknotic nuclei and a necrotic, eosinophilic cytoplasm. The lamina densa of the BM is frequently disrupted and reduplicated beneath keratinocytes and DEJ tends to be flattened with shortened epidermal rete pegs and dermal rete ridges. In the dermis, collagen and elastin fibers appear fragmented and reduced and are replaced by solar elastosis. Solar elastosis is a striking characteristic in photoaged skin and shows a thin-film distribution on the surface, which is significantly more prominent in HP. An increased number of atypical melanocytes can also be seen. The sizes of melanocytes and cytoplasmic brown granules significantly increase and have a higher percentage of cells with two or more nuclei. The morphology of fibroblasts also has a tendency to size reduction and elongation decrease. In addition, photoaging also showed dilation of vessels with a lower capillary density and an absolute loss of the PC.

1.4 Photoaging as a chronic inflammatory process

Photoaged skin is characterized by an increase in the numbers of mast cells, histiocytes, and monocytes. The presence of this dermal infiltrate indicates a chronic inflammatory process in photoaged skin (Farage *et al* 2017). Upon UV radiation, several inflammatory signaling pathways are activated by different surface receptors, such as epidermal growth factor (EGF) receptors, the transforming growth factor (TGF) receptor, Toll-like receptors, the interleukin-1 (IL-1) receptor, and the tumor necrosis factor (TNF) receptor. UVR can activate signaling directly through producing ROSs, or indirectly through DNA or mitochondrial damage, which causes inflammation (Ansary *et al* 2021).

This chronic inflammatory process may be complicated by several factors. UVR can induce epidermal keratinocytes and dermal fibroblasts secreting cytokines or senescence-associated secretory phenotype (SASP), and the microenvironment including fragmented collagens and elastin also plays a role in the inflammation. The structurally modified sebum may be another participant in the process. These factors are not independent, but promote each other.

Apoptosis protects the integrity of tissue by removing unfit and injured cells without initiating inflammation. Persistent types of stress, for example, oxidative, proteotoxic, and genotoxic stresses, enhance the resistance of cells to apoptosis and lead to the survival of sub-lethally damaged cells with a pro-inflammatory phenotype. The molecular basis of enhanced anti-apoptosis ability includes several mechanisms, such as: (i) functional deficiency in the p53 network, (ii) increased activity in the NF-κB-IAP/JNK axis, and (iii) changes in molecular chaperones, microRNAs, and epigenetic regulation (figure 1.6). Deficient cleansing of tissues of damaged cells results in a chronic inflammatory response. Recent studies have shown that senescent cells display a pro-inflammatory phenotype and secrete inflammatory mediators including cytokines and chemokines in order to recruit phagocytes which quickly remove apoptotic bodies or necrotic waste that appear. This pro-inflammatory state of

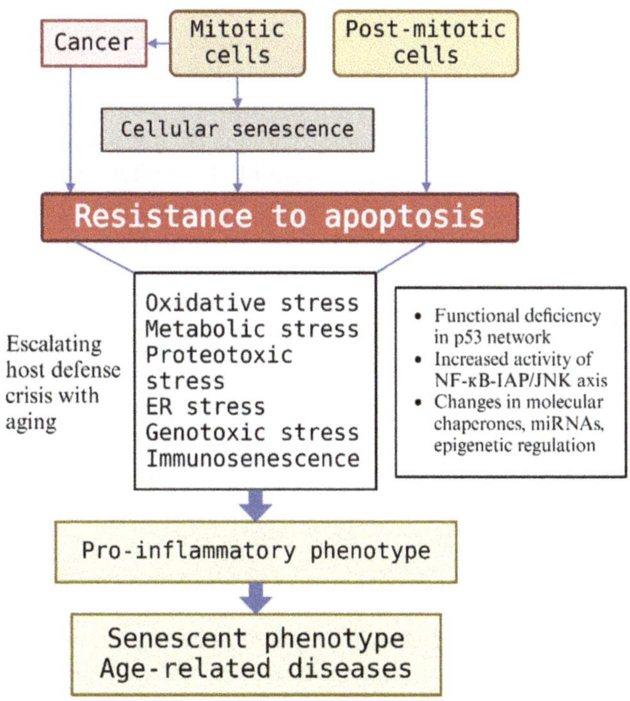

Figure 1.6. The mechanism of cellular resistance to apoptosis.

senescent cells is called the senescence-associated secretory phenotype (SASP). SASP can be induced by several senescence inducers (such as DNA damage) in fibroblasts, ECs and many other cell types (Salminen *et al* 2011). In skin, keratinocytes are the main source of cytokine production and dermal fibroblasts are the main source of MMPs. The main cytokines induced by DNA damage of keratinocytes are interleukins (IL-1, IL-3, IL-6, IL-8), colony-stimulating factors (CSFs) (granulocyte macrophage [GM]-CSF, M-CSF, G-CSF), TGF-α, TGF-β, TNF-α, and platelet-derived growth factor (PDGF). MMPs from dermal fibroblasts include MMP-1, MMP-2, MMP-3, MMP-9, MMP-11, MMP-17, and MMP-27. TNF-α is a main effector cytokine in the skin pro-inflammatory processes. TNF-α can inhibit collagen synthesis and induces the increase of MMP-9, and the epidermis can be irreversibly damaged. An increased level of TNF-α is associated with multiple pathways including NF-κB, AP-1, hypoxia-inducible factor 1-α (HIF-1α), and nuclear factor erythroid 2-related factor 2 (NRF-2), which are related to the upregulation of MMP. In both intrinsic and extrinsic aging processes, IL-1 and IL-6 can induce the activation of key transcription factors associated with inflammatory and immune responses, such as NF-κB, AP-1, c-Jun N-terminal kinase (JNK), and MAPKs. These cytokines can enhance apoptotic resistance and stimulate cancerous growth. Chronic SASP factor secretion is responsible for many age-related pathologies, for example atherosclerosis, type 2 diabetes, obesity, cardiovascular disease, sarcopenia, neurodegenerative diseases, and Alzheimer's disease. We propose that the increased resistance to apoptosis of senescent cells can aggravate the age-related host defense crisis that is enhanced by chronic inflammation (Ansary *et al* 2021, Salminen *et al* 2011)

Elastin also plays an important role in the microinflammation model of skin aging. Elastin fragments (EFs) or elastin degradation products (EDPs) can induce a variety of biological functions, including chemotaxis, gene transcription, and cell cycle regulation. Although the mature elastic fibers do not have chemotaxis, all EFs with the hydrophobic motifs GXXPG or XGXPG have chemotaxis. It is well known that solar elastosis including fragmented elastin is a prominent feature in photoaged skin. The elastin fragments induce a low-grade asymptomatic inflammation in skin aging by recruiting macrophages and neutrophils. In the inflammatory progress, elastases from macrophages and neutrophils can break intact elastic fibers into EFs which in turn can attract multiple cells, including macrophage, neutrophils, and lymphocytes into the reaction site. Human skin fibroblasts exposed to VGVAPG (EDP) have been shown to induce MMP-1 and MMP-3 expression. Therefore, all of these signaling pathways are intertwined, maintaining a vicious self-perpetuating positive feedback cycle that accelerates skin aging (figure 1.7) (Farage *et al* 2017).

A model of how damaged cells with oxidized lipids initiate the inflammatory process upon UV radiation was proposed in 2014, based on previous data (Zhuang and Lyga 2014). UV radiation induces oxidative stress in epidermal cells, resulting in damaged cells with oxidized lipids. The oxidation-specific epitopes on damaged cells and oxidized lipids are recognized by the complement system and thus cause inflammation, leading to the infiltration and activation of macrophages so as to clear damaged cells and oxidized lipids. Activated macrophages release MMPs to degrade ECM (figure 1.8(A)). Repeated UV radiation over-activates the

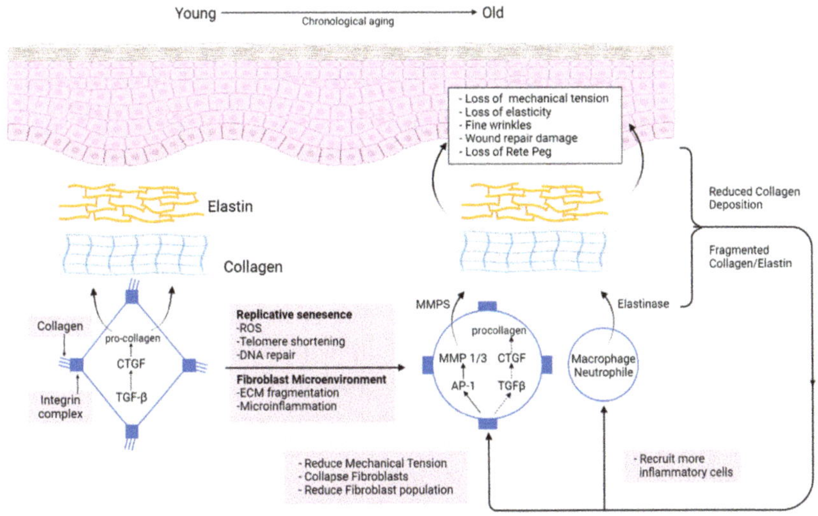

Figure 1.7. Fragmented ECM contributing to a vicious positive feedback cycle in inflammatory process.

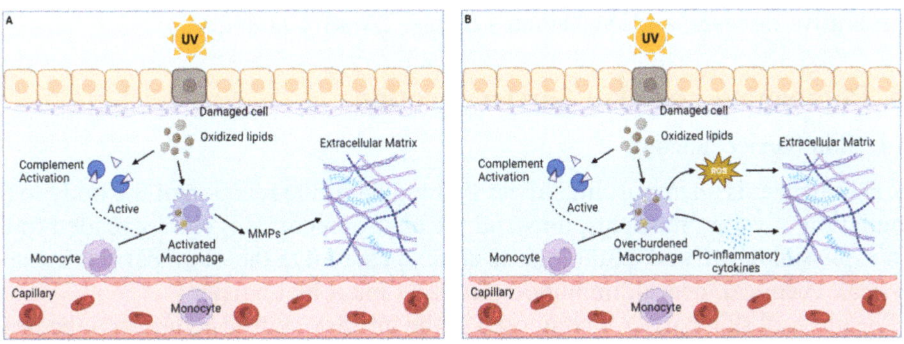

Figure 1.8. A model proposed to explain the mechanism of damaged cells with oxidized lipids initiating the inflammatory process.

complement system and causes damage to the DEJ, on which they deposit, and macrophages are overburdened with oxidized lipids. Overburdened macrophages release pro-inflammatory cytokines and ROSs. The former cause chronic inflammation and long-term damage to the dermis, while the latter triggers the damage of oxidative stress to the ECM (figure 1.8(B)). However, many details of this model require verification (Zhang and Duan 2018).

Sebum structurally modified by UVR may also participate in the chronic inflammation process. Sebum is a highly specific mixture consisting of lipids including squalene, wax esters, free fatty acids, cholesterol esters and triglycerides. When skin is photosensitized by UVR, the lipid making up sebum are easily oxidized, producing several lipid photo-oxidative products, such as squalene peroxides. Squalene peroxides can activate T cells that act as CD1a presenting autoantigens (de Jong *et al* 2014). Keratinocytes treated with squalene peroxide showed increased levels of prostaglandin E2 (PGE2) secretion. In a recent study, oxidized squalene derivatives derived from sebum (collected from healthy volunteers), that was irradiated prior to application to a cultured skin model including human primary keratinocytes, were found to significantly increase mRNA (messenger RNA) expression levels of inflammatory cytokines and receptors including IL-6, IL-8, TNF-α and IL-1R1 (Oyewole and Birch-Machin 2015). A study showed that UV exposure has dramatic effects on stratum corneum cell cohesion and mechanical integrity (Biniek *et al* 2012) and the breach of the skin barrier could lead to the penetration of lipids deeper into the skin tissue, which initiates the inflammation process. The involvement of lipid peroxidation products in UVR-induced inflammatory responses has been inadequately studied, remains highly controversial and needs further investigation.

Chronic inflammation is a body defense process but has a negative impact on human skin and other organs. Chronic inflammatory processes affect all stages of tumor development, and contributes to tumor development with the immunosuppressive effects of UVB and UVA. (Elinav *et al* 2013, Yoshida *et al* 2018b) The secretion of SASP is responsible for many age-related pathologies, for example atherosclerosis, type 2 diabetes, obesity, cardiovascular disease, sarcopenia, neurodegenerative diseases, and Alzheimer's disease (Ansary *et al* 2021).

1.5 Photoaging and skin barrier function

1.5.1 Skin barrier function

The skin barrier is an important barrier that regulates the reaction of the body to the stimulation of water, nutrients, ions and the environment. It is closely related to the skin's physiological and pathological state. Generally, the skin barrier includes physical, chemical, neural, immunological and microbial barriers.

The physical skin barrier is composed of the intrinsic skin structure, and the intact epidermis, dermis and dermal epidermal junction very important to maintain the physical barrier. The physical barrier of the epidermis is composed of stratum corneum (SC) and tight junctions (TJs) between keratinocytes under the stratum corneum. The former mainly includes keratinocytes, keratins and intercellular lipids. TJs are mainly formed in the granular layer, and are located at the upper part of the sides of two adjacent keratinocytes, like a magic glue connecting the two adjacent cells (Madison 2003). The barrier between the epidermis and dermis is mainly composed of the integral structure of the dermal epidermal junction (DEJ), namely the basement membrane (BM).

The skin chemical barrier is mainly composed of pH value, antimicrobial peptides (AMPS) and reactive oxygen species (ROSs). The pH value (4.1–5.8) of the skin physiological state plays an important role in maintaining skin structure, the skin microbiome and regulating epidermal differentiation and desquamation (Proksch 2018). AMPs are small molecular proteins with high bactericidal activity and broad-spectrum antibacterial activity, including defensins, S100, LL-37, late cornified envelope proteins, etc. ROSs produced by keratinocytes have antibacterial activity and participate in the formation of chemical barriers, but also may cause oxidative stress damage to the skin.

The skin pigment barrier is mainly maintained by melanocytes in the basal layer and the melanin they produce, which can protect the body from excessive damage from ultraviolet radiation. Melanocytes transfer melanosomes rich in melanin to surrounding keratinocytes, and then redistribute them around the nucleus. Melanin plays a protective role by absorbing and scattering ultraviolet light (Lin and Fisher 2007).

In both the epidermis and dermis, all cells and molecules closely related to innate immunity and adaptive immunity form the immune barrier of the skin. Immune cells such as Langerhans cells or T cells in the epidermis contribute to the initial defense of the host, and keratinocytes can also produce soluble immune regulatory factors. The interaction between keratinocytes and immune cells is crucial for intrinsic homeostasis and inflammatory processes (Nestle *et al* 2009).

The skin microbiome is the microbial barrier of skin, playing an important role in maintaining skin homeostasis and resisting exogenous pathogen invasion. The microbiome of healthy adult skin is highly stable and constitutes an individual's unique microbial fingerprint (Dreno *et al* 2016). 94% of the skin microorganisms are composed of actinobacteria, firmicutes and proteobacteria. The distribution of bacterial species showed obvious differences in different parts of the body. Usually, the diversity of flora in oily parts (forehead, back, nostrils, etc) is the lowest, while the diversity of flora in dry parts (forearm, palm, hip, etc) is the highest. *Propionibacterium* is dominant in oily parts, while *Staphylococcus* and *Corynebacterium* are dominant in wet parts (nostrils, armpits, finger sutures, etc).

All the skin barriers are interdependent and affect each other, influencing the physiological and pathological state of the skin in the dynamic balance. With the change of human physiological state (such as aging), the status of various skin barriers also changes.

1.5.2 Characteristics of the photoaging skin barrier

1.5.2.1 Characteristics of the skin barrier in the elderly

The clinical manifestations such as skin dryness, pruritus, dyspigmentation, and evendermatoporosis with senile purpura and stellate spontaneous pseudoscars all indicate that the skin barrier function in the elderly is significantly reduced.

The physical barrier components of aging skin change, including the keratinization and desquamation of keratinocytes, composition of intercellular lipid matrix (ILM), synthesis of natural moisturizing factors (NMFs), expression of aquaporin-3 (AQP3), skin pH, and the function of sebaceous glands and sweat, etc (Yosipovitch 2004).

Stratum corneum moisture (SCH) is also an important indicator of the skin physical barrier. A comparative study compared the skin status of elderly patients with systemic pruritus and dryness with that of normal people of the same age. It was found that the skin SCH of elderly patients with dry and itchy skin decreased significantly, the cell adhesion of keratinocytes increased, and the process of normal epidermal desquamation slowed down (Long and Marks 1992).

Sebum synthesis in elderly skin decreases and the content of lamellar bodies secreted to cells also decreases, which can lead to the reduction of the conversion of triglyceride to glycerol. The content of ceramide in SC decreases (Jensen *et al* 2005), as does the synthesis of profilagrin, leading to the reduction of NMF, all of which account for the decline of SCH. Aquaporin-3 (AQP3) is an epidermal water/glycerol transporter, which can accelerate the transport of water and glycerol through the cell membrane and maintain the hydration state of the SC. Studies have found that the expression of AQP3 in elderly skin, especially dry skin, is significantly reduced, which may be related to the skin aging process (Li *et al* 2010).

The pH value of the newborn skin surface is neutral, and then it reaches a steady state of being acid, and further gradually increases after aging. The increase of the pH value can affect the physical, chemical, microbial and immune barrier of the skin. pH value can affect the activity of some important enzymes in the SC, such as β-glucocerebrosidase, acid sphingomyelinase, lipase, phosphatase and phospholipase, etc, which in turn affect desmosome dissociation and epidermal desquamation, also can lead to a reduction in the synthesis of NMFs, ceramide, and secretion of lamellar bodies (Ali and Yosipovitch 2013), finally leading to a decline in skin barrier function.

In addition, the change of pH value leads to the change of the composition of skin microbiota. There are differences in the optimal pH for the survival of different bacteria. For example, the optimal pH for *Staphylococcus aureus* is 7.5, and for *Propionibacterium acnes* is 6.3, and also *Malassezia* can produce more allergens at pH 6.5 than at pH 5.5. In addition, pH will also affect the immune barrier. For example, increased skin pH can induce the expression of kallikree 5 and activate PAR-2, resulting in the secretion of thymic stromal lymphopoietin, and induce skin Th2 type reaction, and finally lead to an eczema-like reaction. In contrast, decreased pH can downregulate kallikree 5 activity and reduce skin inflammation.

1.5.2.2 Special barrier characteristics of photoaging skin

Compared with non-sun-exposed areas, the barrier of photoaging skin undergoes some special changes, which are mainly manifested as changes in the physical barrier caused by photo-ageing skin structure.

The literature shows that the differentiation products related to the keratinization process in the epidermis have no significant difference in the sun-exposed and non-sun-exposed parts of the skin in the elderly, but there are some characteristic changes in the basement membrane (BM), the composition of microfilaments in the dermal papilla layer (the Grenz zone), dermal collagen and elastin, and adipose tissue.

BM damage is closely related to activated matrix metalloproteinases (MMPs) and heparan sulfate proteoglycans in photoaging skin. UVB radiation can increase the synthesis of urokinase-type plasminogen activator (uPA) and MMPs (Marschall *et al* 1999). Even with MMP inhibitor, plasminogen can still promote the degradation of BM components, while the plasminogen inhibitor aprotinin could rebuild the damaged BM structure in the DEJ caused by plasminogen (Ogura *et al* 2008). Heparan sulfate proteoglycans in photoaging skin are also involved in the damage of BM.

In non-photoaging skin, there are a large number of microfilaments with a diameter of 8–11 nm in the dermal papillary layer, while microfilaments in the dermal papillary layer (Grenz zone) of photoaging skin are replaced by dense collagen fibers linearly arranged parallel to the epidermis. Duplication of lamina densa leads to a more fragile DEJ, which reduces the resistance of aging skin to shear force. Even in 30-year-old women, the lamina densa under keratinocytes might appear as duplicated, which is rarely seen in the skin of non-photoaging sites. In photoaging skin, the duplicated lamina densa was found to be increased, and it was also found to be branched in all directions.

In the reticule dermis, there are fibulin-5 and other elastic fiber components, which decrease with age, more obviously in sun-exposed areas. UVB irradiation could significantly reduce the content of fibulin-5 and elastin. However, fibulin-5 deposition was found to be increased in solar elastosis, indicating fibulin-5 as a marker of skin photoaging, while the early reduction of fibulin-5 can lead to changes in the composition of elastic fibers (Kadoya *et al* 2005).

The structure volume of adipose tissue in photoaging skin decreases, resulting in a sagging appearance. Chronic photodamage causes the depletion of human dermal white adipose tissue (dWAT) and the formation of skin fibrosis, which might be caused by adipocyte-myofibroblast transition (AMT). The process of fibrotic tissue replacing dWAT after AMT causes changes in the mechanical structure in skin and reduces the volume, which then manifests as collapsed and sagging skin (Kruglikov and Scherer 2016).

1.6 Photoaging-associated decline in skin function

1.6.1 Physiological function of skin

Skin physiological function refers to the role of skin in the process of normal physiological activities of the body, including protection, absorption, secretion and excretion, sensation, thermoregulation and immune modulation.

The physiological structure of normal skin can protect the skin from physical, chemical, biological and other harmful factors. The skin can choose to absorb some materials through skin appendages, in intercellular and cellular ways. The secretion and excretion function of the skin mainly relies on sweat glands and sebaceous glands. There are nerve ending of sensory and motor nerves in the skin, which can feel single and compound senses such as touch, pain, cold, heat, etc. In addition, the skin is a bad conductor of heat, which keeps the body temperature constant through vasoconstriction and sweat secretion. The immunocytes in the skin mainly include Langerhans cells, lymphocytes, macrophages, mast cells, etc. These cells are distributed around the superficial dermal capillaries and interact with each other, and participate in the body's immune response through the mutual regulation of the synthetic cytokines.

1.6.2 Changes of skin function in photoaging

A variety of changes in the structure of aging skin determine that its function is significantly decreased compared with that of young people, which include epidermal atrophy (mainly spinal stratum), flattened DEJ, reduced dermal thickness, ECM atrophy, and also reduction of appendage structure, subcutaneous tissue, nerve fiber endings, and microvascular structure. The atrophy of the skin structure leads to the decline of the protective function of the skin to the body. In severe cases, chronic skin functional insufficiency and fragile syndrome may occur, which is called dermatoporosis (Kaya and Saurat 2007). The typical clinical manifestations of dermatoporosis include skin atrophy, senile purpura and stellate spontaneous pseudoplaques, which generally begin to appear around the age of 70, and appear earlier if sun exposure is more frequent, and more obvious in sun-exposed areas such as the extension of the forearm and hand, sternum and anterior tibia (Kaya 2012). The fragility of aging skin increases, and the skin is prone to superficial tear wounds after blunt injury. In addition, the healing of deeper chronic wounds is delayed, and in some cases of dermatoporosis, due to the increased fragility of blood vessels, trauma can even induce massive bleeding between the subcutaneous fat layer and muscle fascia layer, resulting in a deep dissecting haematoma (DDH) (Kaya *et al* 2008). The sweat glands and sebaceous glands of senile skin are reduced and the secretion function is also significantly reduced, so the secretion function is weakened, and the reduction of nerve fiber endings in the skin determines the decline of sensory function. In addition, the reduction of the number of Langerhans cells and other factors in the epidermis makes the immune function of the skin weaker than that of young people (Zouboulis and Makrantonaki 2011).

1.7 Conclusion

In general, complex molecular mechanisms are involved in the process of skin photoaging, in which chronic inflammation plays an important role. Photoaging skin has unique histopathological and skin functional changes, distinct from chronological aging, which suggests that it is of great significance to prevent photoaging.

References

Ali S M and Yosipovitch G 2013 Skin pH: from basic science to basic skin care *Acta Derm. Venereol.* **93** 261–7

Amano S 2016 Characterization and mechanisms of photoageing-related changes in skin. Damages of basement membrane and dermal structures *Exp. Dermatol.* **25** 14–9

Ansary T M, Hossain M R, Kamiya K, Komine M and Ohtsuki M 2021 Inflammatory molecules associated with ultraviolet radiation-mediated skin aging *Int. J. Mol. Sci.* **22** 3974

Azimi S Z, Zargari O and Rudolph R I 2017 Nodular colloid milium mimicking keloid *J. Cosmet. Dermatol.* **16** e45–7

Badarni M *et al* 2019 Repression of AXL expression by AP-1/JNK blockage overcomes resistance to PI3Ka therapy *JCI Insight* **5** e125341

Bahamondes Lorca V A and Wu S 2020 Role of constitutive nitric oxide synthases in the dynamic regulation of the autophagy response of keratinocytes upon UVB exposure *Photochem. Photobiol. Sci.* **19** 1559–68

Barolet D, Christiaens F and Hamblin M R 2016 Infrared and skin: friend or foe *J. Photochem. Photobiol.* B **155** 78–85

Beani J C 2014 Ultraviolet A-induced DNA damage: role in skin cancer *Bull. Acad. Natl Med.* **198** 273–95

Biniek K, Levi K and Dauskardt R H 2012 Solar UV radiation reduces the barrier function of human skin *Proc. Natl Acad. Sci. USA* **109** 17111–6

Biskanaki F, Rallis E, Skouras G, Stofas A, Thymara E, Kavantzas N, Lazaris A C and Kefala V 2021 Impact of solar ultraviolet radiation in the expression of type I collagen in the dermis *Cosmetics* **8** 46

Boukari F, Jourdan E, Fontas E, Montaudie H, Castela E, Lacour J P and Passeron T 2015 Prevention of melasma relapses with sunscreen combining protection against UV and short wavelengths of visible light: a prospective randomized comparative trial *J. Am. Acad. Dermatol.* **72** e1

Butterfield D A and Halliwell B 2019 Oxidative stress, dysfunctional glucose metabolism and Alzheimer disease *Nat. Rev. Neurosci.* **20** 148–60

Cadet J, Douki T and Ravanat J L 2015 Oxidatively generated damage to cellular DNA by UVB and UVA radiation *Photochem. Photobiol.* **91** 140–55

Carrara I M, Melo G P, Bernardes S S, Neto F S, Ramalho L N Z, Marinello P C, Luiz R C, Cecchini R and Cecchini A L 2019 Looking beyond the skin: cutaneous and systemic oxidative stress in UVB-induced squamous cell carcinoma in hairless mice *J. Photochem. Photobiol.* B **195** 17–26

Chen T, Hou H, Fan Y, Wang S, Chen Q, Si L and Li B 2016 Protective effect of gelatin peptides from pacific cod skin against photoaging by inhibiting the expression of MMPs via MAPK signaling pathway *J. Photochem. Photobiol.* B **165** 34–41

Chen X, Li L, Xu S, Bu W, Chen K, Li M and Gu H 2018 Ultraviolet B radiation down-regulates ULK1 and ATG7 expression and impairs the autophagy response in human keratinocytes *J. Photochem. Photobiol.* B **178** 152–64

Chen Y, Xiao T, Xu S, Gu H, Li M and Chen X 2020 Ultraviolet B induces proteolytic cleavage of the pyroptosis inducer gasdermin E in keratinocytes *J. Dermatol. Sci.* **100** 160–3

Cheong Y, Kim C, Kim M B and Hwang J K 2018 The anti-photoaging and moisturizing effects of *Bouea macrophylla* extract in UVB-irradiated hairless mice *Food Sci. Biotechnol.* **27** 147–57

Choi H J, Alam M B, Baek M E, Kwon Y G, Lim J Y and Lee S H 2020 Protection against UVB-induced photoaging by *Nypa fruticans* via inhibition of MAPK/AP-1/MMP-1 signaling *Oxid. Med. Cell. Longev.* **2020** 2905362

Chung B Y, Park S H, Yun S Y, Yu D S and Lee Y B 2022 Astaxanthin protects ultraviolet B-induced oxidative stress and apoptosis in human keratinocytes via intrinsic apoptotic pathway *Ann. Dermatol.* **34** 125–31

de Bhailis A M, Chrysochou C and Kalra P A 2021 Inflammation and oxidative damage in ischaemic renal disease *Antioxidants* **10** 845

De Gruijl F R and Rebel H 2008 Early events in UV carcinogenesis-DNA damage, target cells and mutant p53 foci *Photochem. Photobiol.* **84** 382–7

de Jong A *et al* 2014 CD1a-autoreactive T cells recognize natural skin oils that function as headless antigens *Nat. Immunol.* **15** 177–85

de Laat W L, Jaspers N G and Hoeijmakers J H 1999 Molecular mechanism of nucleotide excision repair *Genes Dev.* **13** 768–85

Dreno B, Araviiskaia E, Berardesca E, Gontijo G, Sanchez Viera M, Xiang L F, Martin R and Bieber T 2016 Microbiome in healthy skin, update for dermatologists *J. Eur. Acad. Dermatol. Venereol.* **30** 2038–47

Duteil L, Cardot-Leccia N, Queille-Roussel C, Maubert Y, Harmelin Y, Boukari F, Ambrosetti D, Lacour J P and Passeron T 2014 Differences in visible light-induced pigmentation according to wavelengths: a clinical and histological study in comparison with UVB exposure *Pigment Cell Melanoma Res.* **27** 822–6

Elinav E, Nowarski R, Thaiss C A, Hu B, Jin C and Flavell R A 2013 Inflammation-induced cancer: crosstalk between tumours, immune cells and microorganisms *Nat. Rev. Cancer* **13** 759–71

2017 *The Textbook of Aging Skin* 2nd edn ed M A Farage, K W Miller and H I Maibach (Berlin: Springer)

Fisher G J, Kang S, Varani J, Bata-Csorgo Z, Wan Y, Datta S and Voorhees J J 2002 Mechanisms of photoaging and chronological skin aging *Arch. Dermatol.* **138** 1462–70

Fitsiou E, Pulido T, Campisi J, Alimirah F and Demaria M 2021 Cellular senescence and the senescence-associated secretory phenotype as drivers of skin photoaging *J. Invest. Dermatol.* **141** 1119–26

Forman H J and Zhang H 2021 Targeting oxidative stress in disease: promise and limitations of antioxidant therapy *Nat. Rev. Drug Discov.* **20** 689–709

Gaestel M 2015 MAPK-activated protein kinases (MKs): novel insights and challenges *Front. Cell Dev. Biol.* **3** 88

Galluzzi L *et al* 2018 Molecular mechanisms of cell death: recommendations of the Nomenclature Committee on Cell Death 2018 *Cell Death Differ.* **25** 486–541

Gao W, Lin P, Hwang E, Wang Y, Yan Z, Ngo H T T and Yi T H 2018a *Pterocarpus santalinus* L. regulated ultraviolet B irradiation-induced procollagen reduction and matrix metalloproteinases expression through activation of TGF-beta/Smad and Inhibition of the MAPK/AP-1 pathway in normal human dermal fibroblasts *Photochem. Photobiol.* **94** 139–49

Gao W, Wang Y S, Hwang E, Lin P, Bae J, Seo S A, Yan Z and Yi T H 2018b *Rubus idaeus* L. (red raspberry) blocks UVB-induced MMP production and promotes type I procollagen synthesis via inhibition of MAPK/AP-1, NF-$\kappa\beta$ and stimulation of TGF-β/Smad, Nrf2 in normal human dermal fibroblasts *J. Photochem. Photobiol.* B **185** 241–53

Garcia-Ruiz A, Kornacker K and Brash D E 2022 Cyclobutane pyrimidine dimer hyperhotspots as sensitive indicators of keratinocyte UV exposure(dagger) *Photochem. Photobiol.* **98** 987–97

Godar D E 2005 UV doses worldwide *Photochem. Photobiol.* **81** 736–49

Grether-Beck S, Marini A, Jaenicke T and Krutmann J 2015 Effective photoprotection of human skin against infrared A radiation by topically applied antioxidants: results from a vehicle controlled, double-blind, randomized study *Photochem. Photobiol.* **91** 248–50

Gu Y, Han J, Jiang C and Zhang Y 2020 Biomarkers, oxidative stress and autophagy in skin aging *Ageing Res. Rev.* **59** 101036

Gudipaty S A, Conner C M, Rosenblatt J and Montell D J 2018 Unconventional ways to live and die: cell death and survival in development, homeostasis, and disease *Annu. Rev. Cell Dev. Biol.* **34** 311–32

Han J H, Bang J S, Choi Y J and Choung S Y 2019 Oral administration of oyster (*Crassostrea gigas*) hydrolysates protects against wrinkle formation by regulating the MAPK pathway in UVB-irradiated hairless mice *Photochem. Photobiol. Sci.* **18** 1436–46

Heng J K, Aw D C and Tan K B 2014 Solar elastosis in its papular form: uncommon, mistakable *Case Rep. Dermatol.* **6** 124–8

Hesari A, Ghasemi F, Salarinia R, Biglari H, Tabar Molla Hassan A, Abdoli V and Mirzaei H 2018 Effects of curcumin on NF-κB, AP-1, and Wnt/β-catenin signaling pathway in hepatitis B virus infection. *J. Cell. Biochem.* **119** 7898–904

Hoeijmakers J H 2001 Genome maintenance mechanisms for preventing cancer *Nature* **411** 366–74

Hudson L, Rashdan E, Bonn C A, Chavan B, Rawlings D and Birch-Machin M A 2020 Individual and combined effects of the infrared, visible, and ultraviolet light components of solar radiation on damage biomarkers in human skin cells *FASEB J.* **34** 3874–83

Hussein M R, Al-Badaiwy Z H and Guirguis M N 2004 Analysis of p53 and bcl-2 protein expression in the non-tumorigenic, pretumorigenic, and tumorigenic keratinocytic hyper-proliferative lesions *J. Cutan. Pathol.* **31** 643–51

Ianni A *et al* 2021 SIRT7-dependent deacetylation of NPM promotes p53 stabilization following UV-induced genotoxic stress *Proc. Natl Acad. Sci. USA* **118** e2015339118

Indirapriyadarshini R, Kanimozhi G, Natarajan D and Jeevakaruniyam S J 2022 Andrographolide protects acute ultraviolet-B radiation-induced photodamages in the mouse skin *Arch. Dermatol. Res.* **315** 1197–205

Jensen J M, Forl M, Winoto-Morbach S, Seite S, Schunck M, Proksch E and Schutze S 2005 Acid and neutral sphingomyelinase, ceramide synthase, and acid ceramidase activities in cutaneous aging *Exp. Dermatol.* **14** 609–18

Kadoya K, Sasaki T, Kostka G, Timpl R, Matsuzaki K, Kumagai N, Sakai L Y, Nishiyama T and Amano S 2005 Fibulin-5 deposition in human skin: decrease with ageing and ultraviolet B exposure and increase in solar elastosis *Br. J. Dermatol.* **153** 607–12

Kaushik S, Tasset I, Arias E, Pampliega O, Wong E, Martinez-Vicente M and Cuervo A M 2021 Autophagy and the hallmarks of aging *Ageing Res. Rev.* **72** 101468

Kaya G 2012 New therapeutic targets in dermatoporosis *J. Nutr. Health Aging* **16** 285–8

Kaya G, Jacobs F, Prins C, Viero D, Kaya A and Saurat J H 2008 Deep dissecting hematoma: an emerging severe complication of dermatoporosis *Arch. Dermatol.* **144** 1303–8

Kaya G and Saurat J H 2007 Dermatoporosis: a chronic cutaneous insufficiency/fragility syndrome. Clinicopathological features, mechanisms, prevention and potential treatments *Dermatology* **215** 284–94

Kim J M, Kim S Y, Noh E M, Song H K, Lee G S, Kwon K B and Lee Y R 2018 Reversine inhibits MMP-1 and MMP-3 expressions by suppressing of ROS/MAPK/AP-1 activation in UV-stimulated human keratinocytes and dermal fibroblasts *Exp. Dermatol.* **27** 298–301

Kohli I, Chaowattanapanit S, Mohammad T F, Nicholson C L, Fatima S, Jacobsen G, Kollias N, Lim H W and Hamzavi I H 2018 Synergistic effects of long-wavelength ultraviolet A1 and visible light on pigmentation and erythema *Br. J. Dermatol.* **178** 1173–80

Krueger L, Saizan A, Stein J A and Elbuluk N 2022 Dermoscopy of acquired pigmentary disorders: a comprehensive review *Int. J. Dermatol.* **61** 7–19

Kruglikov I L and Scherer P E 2016 Skin aging: are adipocytes the next target? *Aging* **8** 1457–69

Krutmann J, Schalka S, Watson R E B, Wei L and Morita A 2021 Daily photoprotection to prevent photoaging *Photodermatol. Photoimmunol. Photomed.* **37** 482–9

Kumar J P and Mandal B B 2019 Inhibitory role of silk cocoon extract against elastase, hyaluronidase and UV radiation-induced matrix metalloproteinase expression in human dermal fibroblasts and keratinocytes *Photochem. Photobiol. Sci.* **18** 1259–74

Kwittken J 2000 Papular elastosis *Cutis* **66** 81–3

Lan C E, Hung Y T, Fang A H and Ching-Shuang W 2019 Effects of irradiance on UVA-induced skin aging *J. Dermatol. Sci.* **94** 220–8

Lee H S, Hua H S, Wang C H, Yu M C, Chen B C and Lin C H 2019 *Mycobacterium tuberculosis* induces connective tissue growth factor expression through the TLR2-JNK-AP-1 pathway in human lung fibroblasts *FASEB J.* **33** 12554–64

Li J, Tang H, Hu X, Chen M and Xie H 2010 Aquaporin-3 gene and protein expression in sun-protected human skin decreases with skin ageing *Aust. J. Dermatol.* **51** 106–12

Lim H W, Honigsmann H and Hawk J L 2007 *Photodermatology* (Boca Raton, FL: CRC Press)

Lin J Y and Fisher D E 2007 Melanocyte biology and skin pigmentation *Nature* **445** 843–50

Liu J *et al* 2021 Oncostatin M sensitizes keratinocytes to UVB-induced inflammation via GSDME-mediated pyroptosis *J. Dermatol. Sci.* **104** 95–103

Lone A N, Malik A T, Naikoo H S, Raghu R S and S A T 2020 Trigonelline, a naturally occurring alkaloidal agent protects ultraviolet-B (UV-B) irradiation induced apoptotic cell death in human skin fibroblasts via attenuation of oxidative stress, restoration of cellular calcium homeostasis and prevention of endoplasmic reticulum (ER) stress *J. Photochem. Photobiol.* B **202** 111720

Long C C and Marks R 1992 Stratum corneum changes in patients with senile pruritus *J. Am. Acad. Dermatol.* **27** 560–4

Madison K C 2003 Barrier function of the skin: 'la raison d'etre' of the epidermis *J. Invest. Dermatol.* **121** 231–41

Mahmoud B H, Ruvolo E, Hexsel C L, Liu Y, Owen M R, Kollias N, Lim H W and Hamzavi I H 2010 Impact of long-wavelength UVA and visible light on melanocompetent skin *J. Invest. Dermatol.* **130** 2092–7

Marschall C, Lengyel E, Nobutoh T, Braungart E, Douwes K, Simon A, Magdolen V, Reuning U and Degitz K 1999 UVB increases urokinase-type plasminogen activator receptor (uPAR) expression *J. Invest. Dermatol.* **113** 69–76

Martic I, Wedel S, Jansen-Duerr P and Cavinato M 2020 A new model to investigate UVB-induced cellular senescence and pigmentation in melanocytes *Mech. Ageing Dev.* **190** 111322

Martincorena I and Campbell P J 2015 Somatic mutation in cancer and normal cells *Science* **349** 1483–9

Mouret S, Baudouin C, Charveron M, Favier A, Cadet J and Douki T 2006 Cyclobutane pyrimidine dimers are predominant DNA lesions in whole human skin exposed to UVA radiation *Proc. Natl Acad. Sci. USA* **103** 13765–70

Mouret S, Leccia M T, Bourrain J L, Douki T and Beani J C 2011 Individual photosensitivity of human skin and UVA-induced pyrimidine dimers in DNA *J. Invest. Dermatol.* **131** 1539–46

Mullenders L H F 2018 Solar UV damage to cellular DNA: from mechanisms to biological effects *Photochem. Photobiol. Sci.* **17** 1842–52

Munzel T and Daiber A 2018 Environmental stressors and their impact on health and disease with focus on oxidative stress *Antioxid. Redox Signal.* **28** 735–40

Nakamura M, Morita A, Seité S, Haarmann-Stemmann T, Grether-Beck S and Krutmann J 2015 Environment-induced lentigines: formation of solar lentigines beyond ultraviolet radiation *Exp. Dermatol.* **24** 407–11

Nakyai W, Saraphanchotiwitthaya A, Viennet C, Humbert P and Viyoch J 2017 An *in vitro* model for fibroblast photoaging comparing single and repeated UVA irradiations *Photochem. Photobiol.* **93** 1462–71

Nestle F O, Di Meglio P, Qin J Z and Nickoloff B J 2009 Skin immune sentinels in health and disease *Nat. Rev. Immunol.* **9** 679–91

Noh D, Choi J G, Lee Y B, Jang Y P and Oh M S 2019 Protective effects of *Belamcandae* rhizoma against skin damage by ameliorating ultraviolet-B-induced apoptosis and collagen degradation in keratinocytes *Environ. Toxicol.* **34** 1354–62

Ogura Y, Matsunaga Y, Nishiyama T and Amano S 2008 Plasmin induces degradation and dysfunction of laminin 332 (laminin 5) and impaired assembly of basement membrane at the dermal-epidermal junction *Br. J. Dermatol.* **159** 49–60

Oh J H, Joo Y H, Karadeniz F, Ko J and Kong C S 2020a Syringaresinol inhibits UVA-induced MMP-1 expression by suppression of MAPK/AP-1 signaling in HaCaT keratinocytes and human dermal fibroblasts *Int. J. Mol. Sci.* **21** 3981

Oh J H, Karadeniz F, Kong C S and Seo Y 2020b Antiphotoaging effect of 3,5-dicaffeoyl-epi-quinic acid against UVA-induced skin damage by protecting human dermal fibroblasts *in vitro Int. J. Mol. Sci.* **21** 7756

Oh J H, Kim J, Karadeniz F, Kim H R, Park S Y, Seo Y and Kong C S 2021 Santamarine shows anti-photoaging properties via inhibition of MAPK/AP-1 and stimulation of TGF-β/Smad signaling in UVA-irradiated HDFs *Molecules* **26** 3585

Olivier E, Dutot M, Regazzetti A, Dargere D, Auzeil N, Laprevote O and Rat P 2017 Lipid deregulation in UV irradiated skin cells: role of 25-hydroxycholesterol in keratinocyte differentiation during photoaging *J. Steroid Biochem. Mol. Biol.* **169** 189–97

Oyewole A O and Birch-Machin M A 2015 Sebum, inflammasomes and the skin: current concepts and future perspective *Exp. Dermatol.* **24** 651–4

Park B, Hwang E, Seo S A, Cho J G, Yang J E and Yi T H 2018 *Eucalyptus globulus* extract protects against UVB-induced photoaging by enhancing collagen synthesis via regulation of TGF-β/Smad signals and attenuation of AP-1 *Arch. Biochem. Biophys.* **637** 31–9

Park Y M and Park S N 2019 Inhibitory effect of lupeol on MMPs expression using aged fibroblast through repeated UVA irradiation *Photochem. Photobiol.* **95** 587–94

Passeron T and Picardo M 2018 Melasma, a photoaging disorder *Pigment Cell Melanoma Res.* **31** 461–5

Perez M, Robbins M E, Revhaug C and Saugstad O D 2019 Oxygen radical disease in the newborn, revisited: oxidative stress and disease in the newborn period *Free Radic. Biol. Med.* **142** 61–72

Perrotta C, Cattaneo M G, Molteni R and De Palma C 2020 Autophagy in the regulation of tissue differentiation and homeostasis *Front. Cell Dev. Biol.* **8** 602901

Pfeifer G P and Besaratinia A 2012 UV wavelength-dependent DNA damage and human non-melanoma and melanoma skin cancer *Photochem. Photobiol. Sci.* **11** 90–7

Pittayapruek P, Meephansan J, Prapapan O, Komine M and Ohtsuki M 2016 Role of matrix metalloproteinases in photoaging and photocarcinogenesis *Int. J. Mol. Sci.* **17** 868

Pollet M *et al* 2018 The AHR represses nucleotide excision repair and apoptosis and contributes to UV-induced skin carcinogenesis *Cell Death Differ.* **25** 1823–36

Prasanth M I, Gayathri S, Bhaskar J P, Krishnan V and Balamurugan K 2020 Understanding the role of p38 and JNK mediated MAPK pathway in response to UV-A induced photoaging in *Caenorhabditis elegans J. Photochem. Photobiol.* B **205** 111844

Proksch E 2018 pH in nature, humans and skin *J. Dermatol.* **45** 1044–52

Rahman H, Kumar D, Liu T, Okwundu N, Lum D, Florell S R, Burd C E, Boucher K M, Vanbrocklin M W and Grossman D 2021 Aspirin protects melanocytes and keratinocytes against UVB-induced DNA damage *in vivo* *J. Invest. Dermatol.* **141** e3

Ramirez N, Serey M, Illanes A, Piumetti M and Ottone C 2021 Immobilization strategies of photolyases: challenges and perspectives for DNA repairing application *J. Photochem. Photobiol.* B **215** 112113

Regazzetti C, Sormani L, Debayle D, Bernerd F, Tulic M K, De Donatis G M, Chignon-Sicard B, Rocchi S and Passeron T 2018 Melanocytes sense blue light and regulate pigmentation through opsin-3 *J. Invest. Dermatol.* **138** 171–8

Rittié L and Fisher G J 2015 Natural and sun-induced aging of human skin *Cold Spring Harb. Perspect. Med.* **5** a015370

Sachs D L, Varani J, Chubb H, Fligiel S E G, Cui Y, Calderone K, Helfrich Y, Fisher G J and Voorhees J J 2019 Atrophic and hypertrophic photoaging: clinical, histologic, and molecular features of 2 distinct phenotypes of photoaged skin *J. Am. Acad. Dermatol.* **81** 480–8

Saleem M D, Oussedik E, Picardo M and Schoch J J 2019 Acquired disorders with hypopigmentation: a clinical approach to diagnosis and treatment *J. Am. Acad. Dermatol.* **80** 1233–1250.e10

Salminen A, Ojala J and Kaarniranta K 2011 Apoptosis and aging: increased resistance to apoptosis enhances the aging process *Cell. Mol. Life Sci.* **68** 1021–31

Samivel R, Nagarajan R P, Subramanian U, Khan A A, Masmali A, Almubrad T and Akhtar S 2020 Inhibitory effect of ursolic acid on ultraviolet B radiation-induced oxidative stress and proinflammatory response-mediated senescence in human skin dermal fibroblasts *Oxid. Med. Cell. Longev.* **2020** 1246510

Schieke S M, Schroeder P and Krutmann J 2003 Cutaneous effects of infrared radiation: from clinical observations to molecular response mechanisms *Photodermatol. Photoimmunol. Photomed.* **19** 228–34

Schmucker A C, Wright J B, Cole M D and Brinckerhoff C E 2012 Distal interleukin-1β (IL-1β) response element of human matrix metalloproteinase-13 (MMP-13) binds activator protein 1 (AP-1) transcription factors and regulates gene expression *J. Biol. Chem.* **287** 1189–97

Seo S W, Park S K, Oh S J and Shin O S 2018 TLR4-mediated activation of the ERK pathway following UVA irradiation contributes to increased cytokine and MMP expression in senescent human dermal fibroblasts *PLoS One* **13** e0202323

Shin J-W, Kwon S-H, Choi J-Y, Na J-I, Huh C-H, Choi H-R and Park K-C 2019 Molecular mechanisms of dermal aging and antiaging approaches *Int. J. Mol. Sci.* **20** 2126

Song X, Narzt M S, Nagelreiter I M, Hohensinner P, Terlecki-Zaniewicz L, Tschachler E, Grillari J and Gruber F 2017 Autophagy deficient keratinocytes display increased DNA damage, senescence and aberrant lipid composition after oxidative stress *in vitro* and *in vivo* *Redox Biol.* **11** 219–30

Tanaka Y, Uchi H and Furue M 2019 Antioxidant cinnamaldehyde attenuates UVB-induced photoaging *J. Dermatol. Sci.* **96** 151–8

Terra V A, Souza-Neto F P, Pereira R C, Silva T N, Costa A C, Luiz R C, Cecchini R and Cecchini A L 2012 Time-dependent reactive species formation and oxidative stress damage in the skin after UVB irradiation *J. Photochem. Photobiol.* B **109** 34–41

Thorburn A 2018 Autophagy and disease *J. Biol. Chem.* **293** 5425–30

Tian L X *et al* 2020 Ellipticine conveys protective effects to lipopolysaccharide-activated macrophages by targeting the JNK/AP-1 signaling pathway *Inflammation* **43** 231–40

Ujfaludi Z, Tuzesi A, Majoros H, Rothler B, Pankotai T and Boros I M 2018 Coordinated activation of a cluster of MMP genes in response to UVB radiation *Sci. Rep.* **8** 2660

Valko M, Rhodes C J, Moncol J, Izakovic M and Mazur M 2006 Free radicals, metals and antioxidants in oxidative stress-induced cancer *Chem. Biol. Interact.* **160** 1–40

Vats K, Kruglov O, Mizes A, Samovich S N, Amoscato A A, Tyurin V A, Tyurina Y Y, Kagan V E and Bunimovich Y L 2021 Keratinocyte death by ferroptosis initiates skin inflammation after UVB exposure *Redox Biol.* **47** 102143

Wang L, Yang Y F, Chen L, He Z Q, Bi D Y, Zhang L, Xu Y W and He J C 2021a Compound dihuang granule inhibits nigrostriatal pathway apoptosis in Parkinson's disease by suppressing the JNK/AP-1 pathway *Front. Pharmacol.* **12** 621359

Wang M, Charareh P, Lei X and Zhong J L 2019a Autophagy: multiple mechanisms to protect skin from ultraviolet radiation-driven photoaging *Oxid. Med. Cell. Longev.* **2019** 8135985

Wang M *et al* 2021b Bach2 regulates autophagy to modulate UVA-induced photoaging in skin fibroblasts *Free Radic. Biol. Med.* **169** 304–16

Wang Y, Wang L, Wen X, Hao D, Zhang N, He G and Jiang X 2019b NF-κB signaling in skin aging *Mech. Ageing Dev.* **184** 111160

Wong S Q, Kumar A V, Mills J and Lapierre L R 2020 Autophagy in aging and longevity *Hum. Genet.* **139** 277–90

Worrede A, Douglass S M and Weeraratna A T 2021 The dark side of daylight: photoaging and the tumor microenvironment in melanoma progression *J. Clin. Invest.* **131** e143763

Wu S, Hu Y, Bai W, Zhao J, Huang C, Wen C, Deng L and Lu D 2019 Cyanidin-3-o-glucoside inhibits UVA-induced human dermal fibroblast injury by upregulating autophagy *Photodermatol. Photoimmunol. Photomed.* **35** 360–8

Xue N, Liu Y, Jin J, Ji M and Chen X 2022 Chlorogenic acid prevents UVA-induced skin photoaging through regulating collagen metabolism and apoptosis in human dermal fibroblasts *Int. J. Mol. Sci.* **23** 6941

Yang H W, Kim H D and Kim J 2019 The DNA repair domain of human rpS3 protects against photoaging by removing cyclobutane pyrimidine dimers *FEBS Lett.* **593** 2060–8

Yeh C and Schwartz R A 2021 Favre–Racouchot disease: protective effect of solar elastosis *Arch. Dermatol. Res.* **314** 217–22

Yoshida H *et al* 2018a Relationship of hyaluronan and HYBID (KIAA1199) expression with roughness parameters of photoaged skin in Caucasian women *Skin Res. Technol.* **24** 562–9

Yoshida H, Nagaoka A, Komiya A, Aoki M, Nakamura S, Morikawa T, Ohtsuki R, Sayo T, Okada Y and Takahashi Y 2018b Reduction of hyaluronan and increased expression of HYBID (alias CEMIP and KIAA1199) correlate with clinical symptoms in photoaged skin *Br. J. Dermatol.* **179** 136–44

Yoshida H and Okada Y 2019 Role of HYBID (hyaluronan binding protein involved in hyaluronan depolymerization), alias KIAA1199/CEMIP, in hyaluronan degradation in normal and photoaged skin *Int. J. Mol. Sci.* **20** 5804

Yosipovitch G 2004 Dry skin and impairment of barrier function associated with itch-new insights *Int. J. Cosmet. Sci.* **26** 1–7

Zhang S and Duan E 2018 Fighting against skin aging: the way from bench to bedside *Cell Transplant.* **27** 729–38

Zhang Y, Roh Y J, Han S J, Park I, Lee H M, Ok Y S, Lee B C and Lee S R 2020 Role of selenoproteins in redox regulation of signaling and the antioxidant system: a review *Antioxidants* **9** 383

Zhao Y, Zhang C F, Rossiter H, Eckhart L, Konig U, Karner S, Mildner M, Bochkov V N, Tschachler E and Gruber F 2013 Autophagy is induced by UVA and promotes removal of oxidized phospholipids and protein aggregates in epidermal keratinocytes *J. Invest. Dermatol.* **133** 1629–37

Zhuang Y and Lyga J 2014 Inflammaging in skin and other tissues-the roles of complement system and macrophage *Inflamm. Allergy Drug Targets* **13** 153–61

Zouboulis C C and Makrantonaki E 2011 Clinical aspects and molecular diagnostics of skin aging *Clin. Dermatol.* **29** 3–14

IOP Publishing

Skin Photoaging (Second Edition)

Rui Yin, Yang Xu and Chengfeng Zhang

Chapter 2

Primary prevention for skin photoaging

Yang Xu

Preventive strategies against photoaging primarily involve comprehensive sun protection measures, including the regular application of broad-spectrum sunscreen with high SPF, UV-protective clothing, and seeking shade during peak sunlight hours. Additionally, promoting antioxidant-rich diets and topical formulations helps mitigate oxidative stress. Skincare routines incorporating retinoids, alpha hydroxy acids, and peptides enhance collagen synthesis and mitigate photoaging effects. Furthermore, advocating lifestyle modifications such as avoiding smoking and maintaining a balanced diet and adequate hydration contribute to overall skin health and resilience against photoaging.

For skin photoaging, the role of prevention is particularly important, and the relevant preventive measures include the reasonable and scientific use of sunscreen and the modification of daily sunscreen behaviors.

2.1 Sunscreen

2.1.1 Sunscreen standards

The labeling of sunscreen properties in each sunscreen product always includes UVB and UVA protection, water resistance and broad-spectrum protection. There are some differences in the standards adopted by different countries in the world, most of which are the ISO 24444 (ISO 2019) standard, FDA standard (Center for Drug Evaluation and Research 2012), Australian regulatory guidelines (Therapautic Goods Administration 2016) and Boots Star system (Diffey 1994), etc.

2.1.1.1 Sun protection factor (SPF)

Sun protection factor (SPF) is used to evaluate the ability of sunscreen to prevent sunburn and erythema. The SPF ISO 24444 standard (ISO 2019), is generally used to determine SPF, which is used by more than 60 countries around the world. First, the

Table 2.1. The relationship between SPF value and UVB protection.

SPF value	Theoretical UVB absorption
SPF30	96.7%
SPF50	98%
SPF50+	98.3%

minimal dosage of UVB irradiation is determined as the one causing skin erythema, which is called the minimum erythema dose (MED). SPF compares the dose of UV radiation required to produce 1 MED after the application of 2 mg cm^{-2} of sunscreen product divided by the dose needed to produce 1 MED on unprotected skin.

The formula for calculating SPF is

$$SPF = \frac{\text{MED with the tested sunscreen}}{\text{MED without sunscreen}}.$$

The higher the SPF value, the better the anti-sunburn erythema effect. The relationship between SPF value and theoretical UVB absorption is shown in table 2.1.

In the Australian regulatory guidelines for sunscreens the maximum allowable claim to efficacy is SPF50+, which indicates the products have a minimum SPF of 60 or higher. The range of SPF categorization is classified as low (SPF 4–10), medium (SPF 15–25), high (SPF30–50) or very high (SPF50+). In China it is required that the identification of SPF is based on the actual measured SPF value of the product. When the measured SPF value of the product is more than 50, it is marked as SPF50+.

The labeled SPF does not necessarily equate to actual real-life SPF. Factors influencing SPF results include methods and dosage of sunscreen application, inter-individual sensitivity, and variable sensitivities in different anatomical areas within the same individual.

2.1.1.2 Protection factor of UVA (PFA)

The protection factor of UVA (PFA) is a protective index to evaluate the protective effect of photo induced hyperpigmentation of sunscreen. Actually, there is no international consensus as to which method should be used to quantify UVA protection for sunscreens. Different methods are used in Australia, America, Europe and Japan.

In Australia, PFA is measured according to the International Standard ISO 24444 (ISO 2019). Sunscreen is spread over a quartz plate at 2 mg cm^{-2} and should block at least 90% of the radiation at every UVA wavelength. There is no numerical labeling of UVA-PF and therefore no grading system for UVA protection in Australia.

In America, it is recommended to measure UVA protection levels with the persistent pigment darkening (PPD) method and the protection factor UVA grading system. It is a ratio of the dose required for a minimal pigment response comparing unprotected skin with sunscreen-treated skin. When classifying UVA protection, the USA uses the low, medium, high and highest classification, which translates to one to four stars, respectively.

In the United Kingdom, the Boots Star system is used to measure the performance of sunscreen in protecting against UVA. The absorption spectrum of sunscreen is measured by a spectrophotometer, and the ratio of mean UVA absorbance and mean UVB absorbance is calculated. The closer the ratio is to 1, the higher the star grade.

In Japan, PPD is used to measure the protective ability of sunscreen against UVA, and PPD is also used in China, South Korea and the European Commission.

The formula for calculating PFA is

$$\text{PFA} = \frac{\text{MPPD with the tested sunscreen}}{\text{MPPD without sunscreen}}.$$

The PPD method is the basis for the labeling of protection of UVA (PA). A PFA less than 2 is not considered as UVA-protective, a PFA of 2–3 is given a rating of PA+, a PFA of 4–7 is given a rating of PA++, and a PFA of 8–15 is given a rating of PA+++, and 16 or more is given a rating of PA++++.

2.1.1.3 Water resistance

Water resistance is determined by the immersion of a test subject for at least 40 min in a simulated swim test device. There are again differences in the detection methods used in different countries.

In Australia, it is recommended that the subject is instructed to sit comfortably, and water jets should not have direct contact with the test sites. At the end of the test procedure, test subjects are allowed to air dry for no less than 15 min. The maximum water resistance that can be claimed is determined by the SPF after immersion: SPF 8–14 is 40 min, SPF 15–29 is 2 h and SPF30 or above is 4 h.

In America, a 'water-resistant (40 min)' claim or a 'water-resistant (80 min)' claim is obtained by the SPF test consists of alternating water-immersion and drying procedures. The water-immersion procedure immediately follows the sunscreen application. Subjects are immersed in water to cover the test area for 20 min and immersion is followed by a 15 min drying period. Then subjects repeat the cycle again, and then undergo the SPF test method. To obtain a 'water-resistant (80 min)' claim, a total of four immersion–drying sequences is needed. Consequently, the resulting SPF value represents the SPF protection retained after 40 or 80 min of water immersion.

2.1.1.4 Broad spectrum protection

The critical wavelength (CW) method evaluates the wavelength at which 90% of the cumulative area under the total absorbance curve from 290 to 400 nm occurs. In America, China and the European Union, it is regulated that only if the CW of sunscreen is $\geqslant$370 nm, can the product be marked as having a broad-spectrum sunscreen effect (Wang *et al* 2017).

2.1.1.5 UVA seal

In European Union countries, the label 'UVA seal' is also used to label the UVA protection of a sunscreen. If UVA-PF/SPF $\geqslant 1/3$, it can be labeled as 'UVA seal' (European Commission 2006).

2.1.2 Active ingredients

Avobenzone (butyl methoxydibenzoylmethane (BMDM)), which was first used as a sunscreen, is the strongest UVA filter at present. The absorption wavelength covers the entire UVA band (320–400 nm), but it has poor photostability. Without stabilization, 36% of the UV filter capacity could be lost after 1 h of sunlight irradiation. The mixture of titanium dioxide and avobenzone will also accelerate its decomposition (Sayre *et al* 2005). At present, many sunscreen formulations add other UV filters to improve its photostability, such as 2,6-diethylhexyl naphthalene dicarboxylate (Corapan® TQ), bis-ethylhexyloxyphenol methoxyphenyl triazine (Tinosorb S) or octocrylene, etc (Chatelain and Gabard 2001).

Octocrylene, also called 2-ethylhexyl 2-cyano-3,3-diphenylacrylate, is a kind of cinnamic acid UV filter. The maximum absorption wavelength of UV is 308 nm, and it has a strong absorption capacity for UV within the wavelength of 280–320 nm.

Octyl methoxycinnamate (OMC) as a type of octinoxate, is the most common and widely used UVB filter on the market. As an ultraviolet absorber, OMC can absorb UV rays in the wavelength range of 280–320 nm, and the maximum absorption occurs at 311 nm.

Benzophenone-3 (BP-3), one type of oxybenzone, is one of the best UVB filters, and also could absorb some UVA. Some sunscreen formulas add BP-3 to avobenzone to promote its photostability.

Diethylamino hydroxybenzoyl hexyl benzoate (DHHB), also known as Uvinul A Plus, is another type of UV filter (Fabian *et al* 2019). DHHB can absorb UVA1 and UVA2 bands and is oil soluble. Although the actual UVA protection is not as good as avobenzone, it is basically photostable (Shamoto *et al* 2017).

Homosalate, namely homomenthyl salicylate, is a UVB filter that can absorb the wavelength band within 295–315 nm. Octisalate can absorb UV wavelengths within 280–300 nm, and is always used as a UVB filter in sunscreens.

The main components of physical sunscreens are titanium dioxide (TiO_2) and zinc oxide (ZnO). The main principle is the reflection and refraction of light. TiO_2 can block UVB and part of UVA (wavelength range 250–340 nm), and ZnO can also block UVA and UVB (wavelength range 280–370 nm), but traditional zinc oxide or titanium dioxide particles can cause skin whitening and a thick skin feeling. At present, with the development of formulation, there are micron or even nanometer particles, but the protection principle of micronized particles is mainly the absorption of light rather than reflection and refraction (Osterwalder and Herzog 2010).

At present, many sunscreens on the market contain different UVB or UVA filters or a combination of both to achieve the sunscreen effect. Different filters have

Table 2.2. Common UVB and UVA filters.

UVB filter	Coated microfine titanium dioxide
	Diethylhexyl butamide triazone
	Ethylhexyl salicylate
	Ethylhehyl triazone
	Homosalate
	2-ethylhexyl 2-cyano-3,3-diphenylacrylate (octocrylene)
	Phenylbenzimidazole sulfonic acid
	Polysilicone-15
	Ethylhexyl methoxycinnamate
	Isoamyl p-methoxycinnamate
UVA filters	Bis-ethylhexyloxyphenol methoxyphenyl triazone
	Butyl methoxydibenzoylmethane (avobenzone)
	Diethylamino hydroxybenzoyl hexyl benzoate (DHHB)
	Disodium phenyl dibenzimidazole tetrasulfonate
	Methylene bis-benzotriazolyl tetramethylbutylphenol
	Ethylhexyl methoxycinnamate
	Isoamyl p-methoxycinnamate

different absorption wavelengths and peaks of ultraviolet rays. The common components of UVB and UVA filters are shown in table 2.2.

2.1.3 New development of sunscreens

2.1.3.1 Active photostable sunscreen ingredients

The photostability of sunscreen ingredients has a great impact on the final photo protection effect of sunscreen. In recent years, many studies have developed more active ingredients or formulas to enhance its photostability (table 2.3) (Young *et al* 2017).

2.1.3.2 Blue light protection

Blue light is emitted visible light in wavelengths between 400 and 500 nm. Blue light in daily life mainly comes from sunlight and electronic screens. Blue light has been shown to generate reactive oxygen species (ROSs) and induce photo damage and photoaging in the skin—similar to UV radiation. In addition, recent studies found that blue light could also be directly involved in the induction of pigmentation. The melanocytes in the skin could react with blue light with wavelengths of 400–415 nm through opsin 3 (OPN3), which leads to the phosphorylation of microphthalmia associated transcription factor (MITF), and finally leads to the increase of tyrosinase and the production of dopapaphrome tautomerase (DCT). A complex of tyrosinase and DCT is formed, which can continuously increase tyrosinase activity in skin melanocytes, and finally result in persistent hyperpigmentation (Setty 2018).

Table 2.3. Common sunscreen ingredients with good relative photostability and their trade names.

Ingredients	Trade name
Benzophenone-3	Neo Heliopan® BB
Bis-ethylhexyloxyphenol methoxyphenyl triazone	Neo Heliopan BMT
Coated microfine titanium dioxide	
Diethylamino hydroxybenzoyl hexyl benzoate	Uvinul A Plus
Diethylhexyl butamido triazone	Uvasorb HEB
Disodium phenyl dibenzimidazole tetrasulfonate	Neo Heliopan® AP
Ethylhexyl salicylate	Neo Heliopan® OS
Ethylhehyl triazone	Uvinul T-150
Homosalate	Neo Heliopan® HMS
Microfine zinc oxide	ZnO neutral
Octocrylene	Neo Heliopan® 303
Phenylbenzimidazole sulfonic acid	Neo Heliopan® Hydro

Table 2.4. Chemical names and trade names of active ingredients with blue light protection.

INCI name	Trade name
Methylene bis-benzotriazolyl tetramethylbutylphenol	Parsol® Max
Titanium dioxide, silica, dimethicone	Parsol® TX
Zinc oxide, triethoxycaprilylsilane	Parsol® ZX
Titanium dioxide, C12–15 alkyl benzoate (and) stearic acid (and) silica (and) alumina (and) polyhydroxystearic acid	TNP45TELR

Therefore, in some sunscreens ingredients with blue light protection have been added.

Some UV filters that have an extended protection into the visible spectrum are also popular against blue light protection. Methylene bis-benzotriazolyl tetramethylbutylphenol (Parsol® Max, DSM) is a broad-spectrum UV filter that also has protection against blue light. Zinc oxide (Parsol® ZX, DMS) and titanium dioxide (Parsol® TX) have also been found to be effective against blue light (Bernstein *et al* 2020). Recent studies have also found that iron oxides commonly used in cosmetics can contribute to blue light protection. The combination of titanium dioxide and iron oxide has been used in anti blue light formulations (Boukari *et al* 2015).

The active ingredients of sunscreen with blue light protection are shown in table 2.4 (Coats *et al* 2021).

2.1.3.3 Infrared ray protection

In vivo skin irradiated with visible light (VL) and infrared ray (IR) has shown significantly increased matrix metalloproteinases (MMP)-1 and MMP-9 expression, decreased type I procollagen expression and promoting degradation of dermal collagen, finally resulting in the acceleration of photoageing process (Cho *et al* 2008).

In basic research and related clinical studies, carnosine has been proved to significantly inhibit the induction of MMP-1 by IR, thereby alleviating skin photo damage and photoaging (Radrezza *et al* 2021). So, some sunscreens (such as Dragosine ®, Dragosine ® Plus) contain carnosine for better photo protection.

2.1.3.4 Anti-inflammation

Inflammation also plays an important role in the process of skin photoaging, so some substances with anti-inflammatory effects are also used in sunscreen formulations. In addition, 13 commercially available UV filters were found to have anti-inflammatory effects in a dose-dependent manner in mouse models, which include diethylhexyl butamido triazone, benzophenone-5 and titanium dioxide, benzophenone-3, octocryle'ne and isoamyl p-methoxycinnamate, PEG-25 PABA and homosalate, octyl triazone and phenylbenzimidazole sulfonic acid, octyl dimethyl PABA, bis-ethylhexyloxyphenol methoxyphenyl triazine and diethylamino hydroxybenzoyl hexyl benzoate (Couteau *et al* 2012).

2.1.3.5 DNA damage repair enzymes

There is direct and indirect DNA damage in the process of photo damage and photoaging. At present, some sunscreens contain two DNA damage repair enzymes, which can participate in the repair process of photo induced DNA damage, mainly including photolyases and T4-bacteriophage endonuclease V (T4NV) (Luze *et al* 2020).

2.1.3.5.1 Photolyases

Photolyase is a flavoenzyme containing the flavin adenine dinucleotide molecule. Dimer photoproducts, mainly including cyclobutane pyrimidine dimers (CPDs) and pyrimidine-pyrimidone (6-4) photoproducts (6-4PPs), are the results of UV-induced DNA damage. Two different kinds of photolyases specifically repair CPDs and 6-4PPs and thus are usually classified as CPD photolyases or 6-4 photolyases, respectively, corresponding to their different substrates.

Clinical trials found that the addition of CPD photolyases to conventional sunscreens contributes significantly to the reduction of UV-induced DNA damage and apoptosis when applied topically to human skin. The number of CPDs induced by UVB radiation decreased by 40%–45%, which demonstrates the ability of photolyases to actively repair damage (Luze *et al* 2020).

2.1.3.5.2 T4-bacteriophage endonuclease V (T4 endonuclease V, T4NV)

It has been reported that T4NV is involved in the repair process of UV-induced CPD (Zattra *et al* 2009), and reduces the effect of MMP-1, stimulates skin regeneration and skin reconstruction, and prevents the destruction of extracellular matrix components, thereby alleviating the process of photoaging (Wolf *et al* 1995).

2.1.3.6 Non-topical photo protective agents

In addition to topical sunscreens, some oral substances with photo protective effects can also be used for sunscreen. Oral photo protective agents usually include some nutritional supplements, plant extracts and some specific compounds (Yeager and Lim 2019, Krutmann *et al* 2020).

Oral intake of some nutritional supplements such as nicotinamide prevents photo-immunosuppression and the development of actinic keratoses and keratinocyte-related cancers in human (Chen *et al* 2014), and the oral intake of *Polypodium leucotomos* extract can alleviate UVB-induced erythema, and mitigate VL-induced persistent pigmentation (Kohli *et al* 2017).

Subcutaneous afamelanotide, an analog of a-melanocyte-stimulating hormone, combines with MC1-R, stimulates melanin production and has antioxidative properties, resulting in increased sunlight tolerance. It is indicated for persons with inherited cutaneous porphyrias such as erythropoietic protoporphyria (EPP) and *X*-linked protoporphyria (XLPP) (Dawe 2017).

2.1.4 Controversial issues

2.1.4.1 The impact of sunscreen on the environment

In recent years, Hawaii and Palau have legislated to ban the use of sunscreens containing octanoate and oxybenzone to protect corals. Among them, the most commonly used octinoxide is octyl methoxycinnamate (OMC), and the most commonly used oxybenzone is benzophenone-3. Studies have reported that octanoate and oxybenzone can increase coral bleaching and have certain toxic effects on corals and other marine organisms (Conway *et al* 2021). In addition, zinc oxide and titanium dioxide nanoparticles may also be ingested by coral (Tang *et al* 2017). Therefore, it is recommended to use degradable sunscreen and 'non nano' zinc oxide and titanium dioxide sunscreens, which are 'coral friendly' or 'sea friendly'.

2.1.4.2 The impact of sunscreens on human body

In recent years, the results of two clinical studies show that some sunscreen ingredients can be absorbed by the human system, including avobenzone, oxybenzone, octocrylene, homosalate, octisalate, and OMC, which are in different sunscreen formulations, and exceed the provisions of FDA that can exempt additional safety examination (Matta *et al* 2019, 2020).

However, whether such systematic absorption has an impact on the human body still needs further research, and there is still controversy in the results of some components' effects on human reproductive and endocrine systems. But in general, sunscreen ingredients with molecular weights greater than 600 Dalton, water-soluble sunscreen ingredients, or agents added by film-forming agents and cross-linking agents (to promote film-forming), can effectively reduce the penetration of sunscreen agents, which is also the one of the trends of sunscreen development going in the future.

2.2 Behavioral modification

Photo protection is the most important strategy for the prevention of photoageing, while dermatologists play a critical role in providing suitable advice. Sunscreens have been discussed in previous sections. Here we will discuss other related behavioral modifications for photo protection.

2.2.1 The reasons for behavioral modifications

It is emphasized that the important reason for the measures other than sunscreen is that the sunlight inducing photoaging also includes visible light (VL) and infrared rays (IRs) in addition to ultraviolet light (Poon *et al* 2015). Many studies *in vivo* and *in vitro* have confirmed that VL can cause the formation reactive oxygen species (ROSs) in the skin, and oxidative stress could activate multiple signal pathways, causing inflammatory reactions in the skin, DNA damage, reduced synthesis and increased degradation of matrix collagen, gene mutations, pigmentation, and, finally, skin aging (Liebel *et al* 2012). Many studies have confirmed that IR exposure could also promote an inflammatory response, vasodilation, ROSs formation, reduce the endogenous antioxidant enzyme defense system and promote aging, together with activating various heat sensitive receptors on keratinocytes, fibroblasts and melanocytes, including transient receptor potential vanilloid-1 (TRPV1). As a result, the calcium ion influx increases, collagen fiber degradation increases, and type I procollagen decreases, resulting in a variety of manifestations of photoaging (McDaniel *et al* 2018).

However, at present, most commercial sunscreens are still mainly protective against UVB and UVA. The use of sunscreen may also lead to the illusion that there is enough photo protection, which artificially increases exposure to sunlight. A survey conducted among 20 470 people in the United States showed that applying sunscreen alone did not significantly reduce the incidence of sunburn, but behavioral modifications were more effective (Linos *et al* 2011). In addition, in terms of sunscreen efficiency, behavioral modification is better than sunscreen only. Therefore, behavioral modification should also be emphasized for the prevention of photoageing.

2.2.2 Strategies for behavioral modification

In general, behavioral modification for better photo protection includes the following. (i) Try to avoid sun exposure. Pay attention to the parameters that may affect the UV intensity, including time, season, latitude, altitude and cloud cover. If possible, refer directly to the local UV index (UVI). (ii) Seek shade if possible. (iii) Wear sun protective clothing, wide-brimmed hats, and sunglasses, etc. (iv) Choose the appropriate sunscreen and use it correctly. (v) Avoid tanning practices.

2.2.2.1 General strategies

Of the four seasons of the year, ultraviolet radiation is the strongest in summer and the weakest in winter. UVB and UVA show a similar trend, and the changes in UVA radiation are greater than those of UVB (Grigalavicius *et al* 2016). Under different weather conditions, the ultraviolet radiation on sunny days is stronger than that in other weather conditions, but even on cloudy days, 80% UVR can still penetrate the clouds to reach the Earth's surface, so more attention should be paid to sun protection on summer and sunny days. During the day, the period with the strongest UV radiation is between 10 am and 4 pm, so it is better to try to avoid going out during this period.

Try to choose a sheltered place when exposed to the sun. In 2016, some researchers tested the shading effect of tree shade and solar umbrellas on UV radiation. The research was conducted in a city at a latitude of 38.6° North in the United States. The UV radiation dose under a tree and a solar umbrella was measured. It was found that the tree shade and solar umbrella can shield from most of the UV radiation. At 3:00 pm, the UV dose of tree shade was only 5.6%, while the UV dose under the solar umbrella near a swimming pool was only 17.2% (Saric-Bosanac *et al* 2019). The use of wide-brimmed hats and solar umbrellas in daily life can also help protect against the sun.

Under the same weather conditions, outdoor ultraviolet radiation is stronger than indoor ultraviolet radiation, but the window glass can generally only absorb UVB, and UVA can penetrate the glass into the room. People who are mainly engaged in indoor activities and work sitting within a certain distance (about 2~3 meters) from a window need to pay attention to sunscreen. If the window has shutters, curtains, dark glass or coated glass, the irradiation is less.

2.2.2.2 UV index forecast

The UV index (UVI) is a forecast of the probable intensity of skin damaging ultraviolet radiation reaching the surface during the solar noon hour (11:30–12:30 or 12:30–13:30 local daylight time). UVI is calculated by the National Oceanic and Atmospheric Administration (NOAA) and the Environmental Protection Agency (EPA) according to the ozone data, radiative transfer model, cloud amounts, climatic aerosol load, variable snow, surface albedo and altitude, with a data range from 0 to 15. An higher UVI indicates a greater dose rate of skin damaging UV radiation, and a shorter time before skin damage occurs.

If possible, it is a scientific method to refer to the UVI forecast for appropriate photo protection behavioral modification. UVI varies in different regions at the same time. Accordingly, more stringent sun protection strategies should be taken in places with high UVI. For the ultraviolet radiation of different UVI, EPA provides guidance on different sun protection methods (table 2.5).

Nearly 90% of the world's population lives in areas with annual peak UVI exceeding 10 (Zaratti *et al* 2014) and is highly likely to be exposed to strong ultraviolet radiation. But in the real world, the actual personal dose received is closely related to individual behavior. In previous studies, the average daily ultraviolet radiation exposure of adults and children was about 4%–5% of the ambient ultraviolet radiation dose of the day, but considerable variability exists, with a range between 1/10 and 10 times of the average value, indicating that behavior modification can significantly affect the individual's ultraviolet radiation exposure.

Table 2.5. General guidelines for photo protection advised by EPA.

Exposure	UVI	Protective actions
Minimal	0, 1, 2	Apply skin protection factor (SPF) 15 sunscreen.
Low	3, 4	SPF 15 and protective clothing (hat)
Moderate	5, 6	SPF 15, protective clothing, and UV-A&B sunglasses.
High	7, 8, 9	SPF 15, protective clothing, sunglasses and make attempts to avoid the sun between 10 am and 4 pm.
Very High	10+	SPF 15, protective clothing, sunglasses and avoid being in the sun between 10 am and 4 pm.

2.2.2.3 Sunscreen clothing

2.2.2.3.1 Ordinary clothes made of non-special sunscreen textile materials

Generally, a tighter or thicker weave fabric, a darker color, and textiles with polyester fiber or polyester material have higher sunscreen ability than the loose, thin, light colored or cotton/linen fabric.

2.2.2.3.2 Clothing made of textile materials with an anti-ultraviolet effect

Some textiles have a certain anti-ultraviolet effect after finishing with some physical or chemical UV shielding agents such as benzotriazole. However, only some textiles that meet the standards can be called 'UV protection products'.

The State Administration of Quality Supervision, Inspection and Quarantine promulgated the standard GB/T 18830 evaluation of ultraviolet protection performance of textiles, which stipulates that the ultraviolet protection factor (UPF) is used to identify the sunscreen effect of sunscreen clothing and fabrics. UPF refers to the ratio of the average effect of ultraviolet radiation calculated when the skin is not protected to the average effect of ultraviolet radiation calculated when the skin is protected by fabric. The UPF value is 50, which means that 1/50 ultraviolet light can pass through the fabric. When the textile UPF > 40 and the transmittance of UVA is lower than 5%, it is labeled as UPF40+, and the textile can be called 'anti-ultraviolet product'. If the textile UPF > 50 and the transmittance of UVA is less than 5%, it can be marked as UPF50+.

Therefore, the label of UV protection products generally contains the following contents: (i) National Standard No.: GB/T 18830-2009; (ii) UPF40+ or UPF50+. In addition, it should be noted that the sunscreen performance provided by the product may be reduced when the sunscreen clothing is used for a long time and when stretched or wet.

2.2.2.4 Sunglasses

Sunlight has both acute and chronic adverse effects on ocular tissue and periorbital skin, while sunglasses provide important protection. National standards are available in many countries. The color or darkness of the lenses does not indicate the level of UV protection—the label still need to be checked.

The industry standard of sunglasses generally adopts the international ISO standard 12312-1:2013. The European Union adopts the standard EN 1836:2005, and China adopts the standard GB 39552.1-2020. According to the visible light transmittance, the lenses of sunglasses are divided into five categories: 0, 1, 2, 3 and 4, however, these three standards do not have corresponding methods for the evaluation of sunscreen effect.

Sunglasses labeled UV 400 provide nearly 100% protection from harmful ultraviolet light rays, blocking wavelengths up to 400 nm, including UVA and UVB rays.

The Australian Sunglasses industry standard is the Australian standard (Product Safety Australia 2017) for eye protection as/AS/NZS 1067:2003, which divides the lenses into 5 categories: 0–4. Lenses with type 0–1 have no protective effect on ultraviolet rays, and the UV transmittance of lenses of type 2 is 43%–18%, which can provide medium-level UV protection. The larger the number of the lens, the higher the anti-UV effect. The transmittance of type 4 lenses is only 8%–3%. Generally, commercially available sunglasses are of type 2 or above.

In 2013, Behar-Cohen *et al* from the French National Institute of Health and Medical Research and researchers of Essilor Research and Development proposed and recommended to promote the use of an eye sun protection factor (EPF), an objective evaluation system, to evaluate the sunscreen efficacy of lenses. Based on the transmission and retroreflection of the lens, EPF can clearly show the inherent characteristics of the lens in protecting the eye and periorbital area. Similar to the SPF value of sunscreen and UPF value of sunscreen clothing, a higher EPF value indicates a better sunscreen effect of the sunglasses (Behar-Cohen *et al* 2014).

2.3 Photo protection and vitamin D

Vitamin D (VitD) is an essential lipid soluble vitamin for human body. It is also a kind of sterol hormone. It is mainly involved in maintaining calcium balance and bone health, and in regulating natural immunity and adaptive immunity. It is closely related to human health. A small amount of VitD is obtained from food and it is mainly synthesized by the skin after ultraviolet irradiation. Therefore, vitamin D deficiency caused by excessive sunscreen should be prevented in the process of anti-aging.

In 2017, an international expert group composed of 13 experts in endocrinology, dermatology, photobiology, epidemiology and anthropology reached the following consensus on whether sunscreen behavior will cause VitD deficiency. (i) The concentration of serum 25-OH VitD3 in the general population should not be less than 50 $\mathrm{nmol\,l^{-1}}$. (ii) The daily use of broad-spectrum sunscreen with high UVA protection index by healthy people will not lead to a lack of VitD, and there is no need for additional supplementation, but this possibility should be considered for dark skinned people, people who wear clothes that cover most of their bodies all year round, during pregnancy, for the elderly, and mainly-indoors workers. (iii) Some special groups, such as people with malabsorption syndrome, photosensitive diseases, skin tumors and those who need strict sunscreen after transplantation, may have a lack of VitD. Therefore, it is recommended to screen the level of VitD and make appropriate supplements for people suffering from photosensitive diseases (Passeron *et al* 2019).

2.4 Conclusion

As the light spectrum related to photoageing is not only UVA and UVB, the visible light, infrared and other bands outside the UV should also be taken into account for sun protection in daily life. The adjustment of daily living habits and sun protection behavior is very important to prevent photoageing. The key points include:

1. If possible, refer to the UV Index Forecast to select the appropriate travel time and sun protection choices.
2. Select sunscreen clothing, sunglasses, sun hats and etc.
3. Pay attention to sun protection in some specific indoor areas.
4. Select an appropriate broad-spectrum sunscreen, in particular some with new functional ingredients with anti-inflammatory, anti-oxidation, anti AHR, anti-MMP and anti-TRPV1 effects.
5. Pay attention to avoiding a lack of vitamin D caused by excessive sunscreen, and further avoid the increased risk of osteoporosis, cardiovascular disease, depression and other diseases caused by vitamin D deficiency.

References

Behar-Cohen F, Baillet G, de Ayguavives T, Garcia P O, Krutmann J, Pena-Garcia P, Reme C and Wolffsohn J S 2014 Ultraviolet damage to the eye revisited: eye-sun protection factor (E-SPF(R)), a new ultraviolet protection label for eyewear *Clin. Ophthalmol.* **8** 87–104

Bernstein E F, Sarkas H W, Boland P and Bouche D 2020 Beyond sun protection factor: an approach to environmental protection with novel mineral coatings in a vehicle containing a blend of skincare ingredients *J. Cosmet. Dermatol.* **19** 407–15

Boukari F, Jourdan E, Fontas E, Montaudie H, Castela E, Lacour J P and Passeron T 2015 Prevention of melasma relapses with sunscreen combining protection against UV and short wavelengths of visible light: a prospective randomized comparative trial *J. Am. Acad. Dermatol.* **72** 189–90.e1

Center for Drug Evaluation and Research 2012 Labeling and effectiveness testing: sunscreen drug products for over-the-counter human use—small entity compliance guide *Guidance for Industry* US Department of Health and Human Services Food and Drug Administration https://www.fda.gov/regulatory-information/search-fda-guidance-documents/labeling-and-effectiveness-testing-sunscreen-drug-products-over-counter-human-use-small-entity

Chatelain E and Gabard B 2001 Photostabilization of butyl methoxydibenzoylmethane (Avobenzone) and ethylhexyl methoxycinnamate by bis-ethylhexyloxyphenol methoxyphenyl triazine (Tinosorb S), a new UV broadband filter *Photochem. Photobiol.* **74** 401–6

Chen A C, Damian D L and Halliday G M 2014 Oral and systemic photoprotection *Photodermatol. Photoimmunol. Photomed.* **30** 102–11

Cho S, Lee M J, Kim M S, Lee S, Kim Y K, Lee D H, Lee C W, Cho K H and Chung J H 2008 Infrared plus visible light and heat from natural sunlight participate in the expression of

MMPs and type I procollagen as well as infiltration of inflammatory cell in human skin *in vivo J. Dermatol. Sci.* **50** 123–33
Coats J G, Maktabi B, Abou-Dahech M S and Baki G 2021 Blue light protection, part II-Ingredients and performance testing methods *J. Cosmet. Dermatol.* **20** 718–23
Conway A J, Gonsior M, Clark C, Heyes A and Mitchelmore C L 2021 Acute toxicity of the UV filter oxybenzone to the coral *Galaxea fascicularis Sci. Total Environ.* **796** 148666
Couteau C, Chauvet C, Paparis E and Coiffard L 2012 UV filters, ingredients with a recognized anti-inflammatory effect *PLoS One* **7** e46187
Dawe R 2017 An overview of the cutaneous porphyrias *F1000Res* **6** 1906
Diffey B L 1994 A method for broad spectrum classification of sunscreens *Int. J. Cosmet. Sci.* **16** 47–52
European Commission 2006 Commission recommendation of 22 September 2006 on the efficacy of sunscreen products and the claims made relating thereto *Document* 32006H0647 https://eur-lex.europa.eu/legal-content/EN/TXT/?uri=CELEX:32006H0647
Fabian F 2019 BASF pushes UV filter production capacities *BASF* https://basf.com/global/en/media/news-releases/2019/12/p-19-422.html
Grigalavicius M, Moan J, Dahlback A and Juzeniene A 2016 Daily, seasonal, and latitudinal variations in solar ultraviolet A and B radiation in relation to vitamin D production and risk for skin cancer *Int. J. Dermatol.* **55** e23–8
ISO 2019 Cosmetics—sun protection test methods—*in vivo* determination of the sun protection factor (SPF) ISO 24444:2019 https://www.iso.org/standard/72250.html
Kohli I *et al* 2017 The impact of oral *Polypodium leucotomos* extract on ultraviolet B response: a human clinical study *J. Am. Acad. Dermatol.* **77** 33–41.e1
Krutmann J, Passeron T, Gilaberte Y, Granger C, Leone G, Narda M, Schalka S, Trullas C, Masson P and Lim H W 2020 Photoprotection of the future: challenges and opportunities *J. Eur. Acad. Dermatol. Venereol.* **34** 447–54
Liebel F, Kaur S, Ruvolo E, Kollias N and Southall M D 2012 Irradiation of skin with visible light induces reactive oxygen species and matrix-degrading enzymes *J. Invest. Dermatol.* **132** 1901–7
Linos E, Keiser E, FU T, Colditz G, Chen S and Tang J Y 2011 Hat, shade, long sleeves, or sunscreen? Rethinking US sun protection messages based on their relative effectiveness *Cancer Causes Control* **22** 1067–71
Luze H, Nischwitz S P, Zalaudek I, Mullegger R and Kamolz L P 2020 DNA repair enzymes in sunscreens and their impact on photoageing—a systematic review *Photodermatol. Photoimmunol. Photomed.* **36** 424–32
Matta M K *et al* 2019 Effect of sunscreen application under maximal use conditions on plasma concentration of sunscreen active ingredients: a randomized clinical trial *JAMA* **321** 2082–91
Matta M K *et al* 2020 Effect of sunscreen application on plasma concentration of sunscreen active ingredients: a randomized clinical trial *JAMA* **323** 256–67
McDaniel D, Farris P and Valacchi G 2018 Atmospheric skin aging-contributors and inhibitors *J. Cosmet. Dermatol.* **17** 124–37
Osterwalder U and Herzog B 2010 The long way towards the ideal sunscreen-where we stand and what still needs to be done *Photochem. Photobiol. Sci.* **9** 470–81
Passeron T *et al* 2019 Sunscreen photoprotection and vitamin D status *Br. J. Dermatol.* **181** 916–31

Poon F, Kang S and Chien A L 2015 Mechanisms and treatments of photoaging *Photodermatol. Photoimmunol. Photomed.* **31** 65–74

Product Safety Australia 2017 Sunglasses and fashion spectacles *Mandatory Standards* Australian Competition and Consumer Commission https://www.productsafety.gov.au/product-safety-laws/safety-standards-bans/mandatory-standards/sunglasses-fashion-spectacles

Radrezza S, Carini M, Baron G, Aldini G, Negre-Salvayre A and D'Amato A 2021 Study of Carnosine's effect on nude mice skin to prevent UV-A damage *Free Radic. Biol. Med.* **173** 97–103

Saric-Bosanac S S, Clark A K, Nguyen V, Pan A, Chang F Y, LI C S and Sivamani R K 2019 Quantification of ultraviolet (UV) radiation in the shade and in direct sunlight *Dermatol. Online J.* **25** 13030/qt4wc0f6tw

Sayre R M, Dowdy J C, Gerwig A J, Shields W J and Lloyd R V 2005 Unexpected photolysis of the sunscreen octinoxate in the presence of the sunscreen avobenzone *Photochem. Photobiol.* **81** 452–6

Setty S R 2018 Opsin3-A link to visible light-induced skin pigmentation *J. Invest. Dermatol.* **138** 13–5

Shamoto Y, Yagi M, Oguchi-Fujiyama N, Miyazawa K and Kikuchi A 2017 Photophysical properties of hexyl diethylaminohydroxybenzoylbenzoate (Uvinul A Plus), a UV-A absorber *Photochem. Photobiol. Sci.* **16** 1449–57

Tang C H, Lin C Y, Lee S H and Wang W H 2017 Membrane lipid profiles of coral responded to zinc oxide nanoparticle-induced perturbations on the cellular membrane *Aquat. Toxicol.* **187** 72–81

Therapeutic Goods Administration 2016 Australian regulatory guidelines for sunscreens *Regulatory Guidance* Department of Health and Aged Care, Australian Government https://www.tga.gov.au/resources/resource/guidance/australian-regulatory-guidelines-sunscreens-args

Wang S Q, XU H, Stanfield J W, Osterwalder U and Herzog B 2017 Comparison of ultraviolet A light protection standards in the United States and European Union through *in vitro* measurements of commercially available sunscreens *J. Am. Acad. Dermatol.* **77** 42–7

Wolf P, Cox P, Yarosh D B and Kripke M L 1995 Sunscreens and T4N5 liposomes differ in their ability to protect against ultraviolet-induced sunburn cell formation, alterations of dendritic epidermal cells, and local suppression of contact hypersensitivity *J. Invest. Dermatol.* **104** 287–92

Yeager D G and Lim H W 2019 What's new in photoprotection: a review of new concepts and controversies *Dermatol. Clin.* **37** 149–57

Young A R, Claveau J and Rossi A B 2017 Ultraviolet radiation and the skin: photobiology and sunscreen photoprotection *J. Am. Acad. Dermatol.* **76** S100–9

Zaratti F, Piacentini R D, Guillen H A, Cabrera S H, Liley J B and Mckenzie R L 2014 Proposal for a modification of the UVI risk scale *Photochem. Photobiol. Sci.* **13** 980–5

Zattra E *et al* 2009 *Polypodium leucotomos* extract decreases UV-induced Cox-2 expression and inflammation, enhances DNA repair, and decreases mutagenesis in hairless mice *Am. J. Pathol.* **175** 1952–61

IOP Publishing

Skin Photoaging (Second Edition)

Rui Yin, Yang Xu and Chengfeng Zhang

Chapter 3

Anti-aging products and medical cosmeceuticals for skin photoaging

Chengfeng Zhang, Xiaoxue Xing, Min Jiang, Zhongyi Xu, Ziqi Liu and Xiuxiu Wang

The key ingredients in anti-aging products include retinoic acid, antioxidants, osmolytes, DNA repair enzymes, growth factors, and cytokines. They have proven efficacy in combating aging, but caution is advised regarding potential adverse reactions such as irritation or contact dermatitis. However, severe adverse reactions are generally rare.

3.1 Topical tretinoins in the treatment of skin photoaging

3.1.1 Overview

Retinoids are a class of substances containing vitamin A and its derivatives, which may be synthetic or natural. In 1943, retinoids were first used for dermatological diseases by Straumfjord in the treatment of acne vulgaris. Ever since then, they have been applied in various skin diseases, including acne, actinic keratosis, psoriasis and atopic dermatitis. Retinoids are now the most widely recognized drugs for the treatment of skin photoaging. The chemical structure of retinoids consists of a cyclohexene ring, side chains and polar groups. Retinoids have been developed to the fourth generation. First-generation retinoids are natural, including tretinoin (all-trans retinoic acid), viatninati and isotretinoin. The second-generation retinoids are synthetic compounds, including acitretin and etretinate. The third-generation retinoids have the advantage of receptor selectivity, including tazarotene, adapalene, arotinoid, arotinoid ethylester and arotinoid methylsulfone, etc. The fourth generation includes seletinoid G.

The receptors of retinoic acid are divided into two categories: one is the retinoic acid receptor (RAR), including three subtypes of RARα, β and γ; the other is the retinoic acid X receptor (RXR), with three subtypes of RXRα, β and γ. Retinoic acid is transported into the nucleus through the cellular retinoic acid binding protein, then binds to its nuclear receptor RAR or RXR. After its receptors combine

doi:10.1088/978-0-7503-5112-6ch3

into dimers or heteromers, retinoic acid binds to retinoic acid responsive elements (RAREs), activates the promoter regions of related genes and alters the level of gene transcription.

Skin aging is a complex biological process affected by several factors such as genetics, environmental exposure, hormonal and metabolic changes. Environmental exposure includes ultraviolet radiation (UVR), mechanical stress, infections, etc. Skin aging can be divided into intrinsic aging and photoaging. Skin photoaging mainly refers to the advanced aging caused by exposure to UVR and presents as fine or coarse wrinkles, roughness, mottled pigmentation, loss of elasticity and sometimes telangiectasias. To understand the multiple functions of topical retinoids in skin photoaging, basic knowledge of the UV-induced photoaging mechanism is needed. In the following paragraphs, we summarize specific mechanisms of topical retinoids in treating photoaged skin and their indications.

3.1.2 The mechanisms of topical retinoids in treating photoaging

3.1.2.1 Increase the production of collagen and synthesis of extracellular matrix

Collagen is an important component in offering strength to the dermis. Collagen synthesis and turnover are finely controlled and complicated processes. Activator protein-1 (AP-1) and transforming growth factor-β (TGF-β) play central roles in regulating the biological process. Studies showed that UVR can induce Smad 7, interfering with the TGF-β pathway and induce AP-1 to inhibit collagen synthesis gene expression (Yaar and Gilchrest 2007). It has been confirmed that topical retinoic acid could induce collagen type I production in photoaged skin (Griffiths *et al* 1993), and an increase in collagen composition could indirectly induce the remodeling of elastic tissue in the dermis (Berardesca *et al* 1990).

3.1.2.2 Influence the differentiation and proliferation of keratinocytes and fibroblasts

Retinoids could regulate the cell growth and differentiation of many cell types. Retinoic acid has been found to increase the proliferation of keratinocytes and fibroblasts in wound healing (Polcz and Barbul 2019). All-trans-retinoic acid (ATRA) has also been demonstrated to stimulate keratinocyte differentiation and encourage normal fibroblast differentiation (Szymański *et al* 2020). However, the results are controversial in keloid fibroblasts. In keloids and hypertrophic scars, retinoids have been shown to induce a marked reduction of fibroblast proliferation (Daly and Weston 1986).

3.1.2.3 Inhibit dermal tissue degradation

Matrix metalloproteinase (MMP) is a group of central enzymes which are responsible for the collagen degradation. Studies have shown that topical retinoic acid can effectively prevent the production of MMP and the degradation of collagen (Jurzak *et al* 2008). The mechanism is that ultraviolet light can activate c-Jun N-terminal kinase (JNK) and p38 in keratinocytes and dermal cells, inducing the production of c-Jun and transcription activator protein-1 (AP-1), thereby inducing increased expression of MMPs and stromelysin, leading to the degradation of collagen and extracellular matrix and aggravation of photoaging. Correspondingly, topical retinoic acid can inhibit the production of c-Jun protein and promote the destruction of c-Jun through the ubiquitination pathway, thereby preventing and treating skin photoaging.

3.1.2.4 Relieve dyschromia

Dyschromia is an important manifestation of facial skin photoaging in East Asian women, which seriously affects patients' social confidence. Dyschromia in photoaged skin usually manifests as the development of lentigines, ephelides and pigmentary irregularities. UVR plays a central role in pigmentation related to photoaging, by inducing the melanocytes hyperplasia and melanin production through the melanocortin-1 receptor (MC1R)-cellular cyclic adenosine monophosphate (cAMP)-microphthalmia-associated transcription factor (MITF) pathways.

Topical retinoic acid drugs, such as 0.025% all-trans retinoic acid, can be used to improve pigmentation diseases. It functions via various mechanisms such as inhibiting the melanin transportation from melanocytes to keratinocytes and inhibiting tyrosinase activity and the downstream melanin synthesis in melanocytes. Moreover, topical retinoids can help promote the penetration of other topical drugs and accelerate the turnover of keratinocytes with melanin deposition. These effects may contribute to the efficacy of topical retinoids in dyschromia caused by photoaging.

3.1.2.5 Inhibit the synthesis of abnormal elastic fibers

The accumulation of abnormal elastic fibers is an important histological marker for skin photoaging. Ultraviolet irradiation can induce the upregulation of elastin mRNA expression levels in human fibroblasts significantly, while retinoids could down-regulate the expression level of elastin induced by ultraviolet irradiation, thereby reducing abnormal elastic fiber production (Lee *et al* 1998).

3.1.2.6 Improve epidermal barrier function

In photodamaged skin, impairment of epidermal barrier function has been verified as flattening of dermal–epidermal junctions and decreased keratinocytes proliferation. Topical retinoids have been shown to increase type VII collagen and fibrillin microfibrils, thus enhancing the physical connection between the epidermis and dermis (Riahi *et al* 2016). Moreover, topical retinoids could promote epidermal proliferation, thinning of the stratum corneum and thickening of the granular layer (Sumita *et al* 2017). Studies have also shown that all-trans-retinoic acid influences the expression of genes related to barrier dysfunction, proteases and tight junctions (Li *et al* 2019).

3.1.2.7 Ameliorate DNA damage caused by UV exposure and have anti-tumor functions

Excess UVR irradiation can promote the production of free radicals and oxidative stress, thus causing DNA damage of skin cells and inducing pro-oncogene expression. The above pathological changes may further contribute to the initiation and development of skin cancers. All-trans-retinoic acid regulates more than 3000 genes in keratinocytes, including genes involved in regulating DNA repair (Lee *et al* 2009). Tazarotene, a retinoic acid derivative, was reported to be effective in treating basal cell carcinomas by reducing cell proliferation and increasing cell apoptosis (Orlandi *et al* 2004). Another retinoid, the Fenretinide (4-HPR) was found to be cytotoxic to skin tumor cells and affect tumor by ROS generation (Doldo *et al* 2015).

3.1.3 Clinical application of topical retinoids in photoaging

The currently approved topical retinoids for skin photoaging treatment by the US FDA are all-trans retinoic acid (tretinoin) cream and 0.1% tazarotene cream. However, there are several kinds of topical retinoids which may be effective in the treatment of photodamaged skin. Adapalene may also be used to treat photo-damaged skin considering its tolerability, although more data regarding its clinical use in photoaging is needed. It may take more than 4 months for topical retinoids to improve skin photoaging. The maximum improvement is usually achieved at 8–12 months after the beginning of topical treatment.

3.1.3.1 Tretinoin (all-trans retinoic acid)

Tretinoin is the biologically active form of vitamin A. It is the oxidized form of all-trans retinol. First used in the 1980s in treating photodamaged skin, tretinoin is now the most known and studied retinoid for treating photoaging. Tretinoin can bind to all subtypes of RARs and isomerize to bind RXRs. Topical tretinoin works by blocking the synthesis of collagenase, gelatinase, AP-1 and NF-κB activation induced by UVR. In the mouse model, tretinoin was found to ameliorate photoaged skin through the RAR-mediated pathway. All-trans retinoic acid was found to stimulate the expression of type I procollagen and inhibit the MMP-3, MMP-13 and c-Jun expression in the photoaged skin in mice model (Li *et al* 2017). Topical tretinoin can cause several histological changes in photodamaged skin, such as the reticulin fiber deposition, collagen formation and the angiogenesis in the dermis.

Studies revealed that tretinoin with a concentration as low as 0.02% is effective in the treatment of photoaging. Concentrations of tretinoin at 0.02%, 0.025%, 0.01% and 0.05% have already shown improvement of photoaging, while only 0.05% tretinoin cream is effective in long-term maintenance (Darlenski *et al* 2010). Another meta-analysis based on 12 clinical studies (Samuel *et al* 2005) showed that all-trans retinoic acid with a concentration of 0.01% or above can be used to improve coarse wrinkles, above 0.025% can be used to improve pigmentation, and above 0.05% for freckles. But we need to pay attention to the irritant reaction of retinoid, which is usually manifested by local erythema, a burning sensation, scaling and pruritus after the topical therapy.

In addition to topical cream, tretinoin has been used in peeling therapy for photoaging as well. Sumita *et al* (2018) compared the efficacy and safety of 0.05% tretinoin cream with 5% tretinoin peeling therapy on photoaging and concluded that 5% tretinoin in peeling is also safe and effective for moderate photoaging. More randomized and controlled clinical trials are required to further explore the effect of tretinoin in peeling therapy.

3.1.3.2 Tazarotene

Tazarotene could be used in a cream, foam or gel. This novel type of retinoid could be metabolized quickly into tazarotenic acid, which is biologically active *in vivo*. Tazarotenic acid can bind to all kinds of RAR-β and -γ, without binding to RXRs; it binds to the RAR-γ with the highest affinity. Tazarotene can be applied in the treatment of photodamaged skin by antagonizing AP-1 and activating tazarotene-inducible genes (TIG). It can also modulate the retinoid-responsive gene expression and inhibit the abnormal expression of epidermal growth factor receptor.

Tazarotene is effective in the improvement of wrinkles, pigmentation and skin texture. In a double-blind clinical study by Kang *et al* (2001), patients with photoaged skin applied tazarotene cream of different concentrations (0.025%, 0.01%, 0.05%, 0.1%) for 6 months. The authors found that tazarotene at other concentrations except at 0.025% can effectively improve coarse wrinkles. It was also found that the higher the concentration, the better the effect of improving coarse wrinkles. Furthermore, an only 0.1% concentration of tazarotene was effective in reducing pigmentation. Lowe *et al* carried out another clinical study to compare the efficacy of topical 0.1% tazarotene with 0.05% tretinoin. In this 24 week trial, they revealed a statistical significance which favors the efficacy of topical tazarotene in the improvement of fine wrinkling, mottled hyperpigmentation, and coarse wrinkling at certain time points.

3.1.3.3 Adapalene

Adapalene is a third-generation synthesized retinoid and is currently commercially available in 0.1% gel, cream, lotion and 0.3% gel. It can bind to the RAR-β and RAR-γ receptors but does not bind to RXRs. Adapalene can be used to improve wrinkles and hydration of the skin. The advantage of topical adapalene in skin photoaging is its tolerability, compared with other types of topical retinoids. It was found that after 9 months' treatment with topical adapalene, clinical features of skin photoaging were relieved compared with just its vehicle, with convincing tolerance. Bagatin *et al* (2018) conducted an investigator-blinded, parallel-group comparison study to evaluate the efficacy of topical adapalene 0.3% gel with tretinoin 0.05% cream for photoaging and they found that the adapalene 0.3% gel was comparably effective to tretinoin 0.05% cream in the treatment of mild or moderate photoaging, with similar safety properties. Further clinical trials are required to demonstrate the effects of adapalene in treating skin aging.

3.1.3.4 Alitretinoin

Alitretinoin is comprised of 9-cis retinoic acid, which can bind to all types of RARs and RXRs. Baumann *et al* revealed that topical alitretinoin can be used to treat actinic keratosis and other signs of skin photoaging (Baumann *et al* 2005). In this open-label pilot trial, twenty patients with photodamaged skin who applied 0.1% topical alitretinoin showed improvement in the symptoms of skin photoaging, with well tolerance. But specific and controlled trials with high quality are needed in the future to further judge its efficacy and safety for photodamaged skin.

3.1.3.5 All-trans retinol

All-trans retinol is a natural form of vitamin A, which can be oxidized to form all-trans retinoic acid (tretinoin) and thus be topically applied for treating skin photoaging. When the retinol is oxidized into retinoic acid, it binds to receptors and promotes the synthesis of procollagen and glycosaminoglycan. Moreover, it could inhibit the production of UV-induced MMPs. Topical retinol is believed to have similar efficacy compared to topical retinoic acid, with less risk of skin irritation than topical retinoic acid. A clinical study revealed that, after 12 weeks of topical retinol treatment, fine wrinkles in patients were improved significantly. Histological experiments showed that seven days of topical retinol treatment could readily reduce the expression of MMP, collagenase and gelatinase and increase fibroblast growth and collagen synthesis (Varani *et al* 2000).

3.1.3.6 All-trans retinol derivatives

Retinol derivatives consist of retinyl acetate, retinyl palmitate, retinyl propionate, etc. Unlike topical retinol, a cream of retinyl palmitate could not improve skin photoaging significantly. Retinoid derivatives, such as ethyl retinoate, have limited efficacy in photodamaged skin, but could exert synergistic effects in treating skin photoaging (Katz *et al* 2015). Retinyl esters (retinyl acetate and palmitate) should improve wrinkles, as they could be converted to retinol and then retinoic acid, which can stimulate the epidermal cell proliferation. Hanwkins *et al* evaluated the effect of a combination of retinyl propionate and climbazole and compared with 0.1% retinol in photodamaged facial skin. In this 16 week randomized, double-blind study, significant improvement was seen in deep wrinkles, lines and mottled pigmentation (Hawkins *et al* 2017). They concluded that the combination of propionated and climbazole delivered significant improvement in skin anti-aging which is comparable to or greater than 0.1% retinol, with well tolerance.

3.1.3.7 Isotretinoin

Isotretinoin is a type of retinoid which is synthetic and derived from vitamin A. It works by multi-functions and is recognized as highly effective in acne. Although less effective compared to tretinoin, topical isotretinoin is still recognized as effective in skin photoaging. 0.05% and 0.1% isotretinoin cream were revealed to be effective in improving fine and coarse wrinkles (Sorg *et al* 2006). Cunningham conducted a clinical study which applied topical 0.1% isotretinoin cream in photoaging for 6 months and demonstrated improvement in the symptoms of fine wrinkles and pigmentation (Cunningham 1990). Armstrong *et al* and Sendagorta *et al* conducted double-blind, vehicle-controlled clinical trials, respectively. 0.05% isotretinoin cream was applied for 12 weeks and 0.1% isotretinoin cream for the following 24 weeks. Both of them found that the topical isotretinoin cream could improve fine wrinkles, pigmentation and skin texture significantly, without severe skin irritation (Armstrong *et al* 1992, Sendagorta *et al* 1992).

3.1.3.8 Seletinoid G

Seletinoid G is a fourth generation of retinoids. Regarding receptor selectivity, it can bind to RAR-γ. Seletinoid G was shown to be effective in treating photoaged skin, with minimal irritation. Kim *et al* conducted a clinical trial to assess the efficacy and safety of seletinoid G to tretinoin (Kim *et al* 2005). In UV-irradiated skin, topical seletinoid G could block the downregulation of type I procollagen and upregulation of MMP-1 and c-Jun caused by UV irradiation. Also, no events of skin irradiation were reported in this clinical trial, suggesting better tolerance of seletinoid G compared with topical retinoids.

3.1.4 Side effects of topical tretinoids

Topical use of tretinoids may cause skin irritation, which is known as the 'retinoid reaction', usually manifested as local erythema, burning sensation, scaling and pruritus. The 'retinoid reaction' is more common in the very first days of the therapy. It is associated with the free carboxylic acid in the polar ends of the retinoids. Studies have shown that such a stimulatory response is related to the increase of mRNA expression levels of inflammatory factors in epidermal cells induced by topical use of retinoic acid, as the mRNA expression level of human monocyte chemoattracant protein-1(MCP-1) and interleukin-8 (IL-8) increased by more than three times compared with a normal control. Among the different categories of topical retinoic acid, all-trans retinoic acid and tazarotene had a greater probability of stimulatory reactions than isotretinoin, adapalene, retinol, and retinal. The irritant response to topical retinoids limits their tolerability. Patient education is extremely important in the treatment of topical tretinoids. Before use, patients should be fully informed of how to use topical tretinoids—starting from the minimal dose and gradually increasing to induce tolerance. To deal with the 'retinoid reaction' caused by topical retinoid therapy, we usually suggest first stopping the topical agent and waiting a few days for the symptoms to be relieved. Then patients can make a second attempt with reduced frequency or a lower concentration or another kind of retinoid with less irritation. A minimal amount of topical corticosteroid is also suggested sometimes to control the situation quickly. However, if in that case patients still cannot tolerate the topical therapy, topical retinoid therapy should be stopped and another approach adopted.

Photosensitization is another side effect of topical retinoid therapy. This side effect mainly happens in the initial period of topical retinoid therapy. Patients should avoid too much UV exposure or use sunscreens to prevent photosensitization events.

Retinoids are not recommended for pregnant women, due to their known embryotoxicity or teratogenicity. Although it was reported that the incidence rate of fetal malformations was actually limited to about 2% in the topical tretinoin group, it is suggested the use of topical retinoids be avoided in patients who are planning a pregnancy or are pregnant.

3.1.5 A new drug delivery system

Retinoids have poor water solubility and light stability so they are easily decomposed when exposed to light and heat. A new drug delivery system may have following advantages: reduce the side effects of topical tretinoids, improve the drug stability, especially photostability, and improve the anti-aging properties of topical retinoids by promoting the transport and distribution of the dermis. At present, a variety of new drug delivery systems are being used for the treatment of topical tretinoinds. However, further well-designed clinical studies are needed to confirm their benefits in drug delivery, efficacy and safety.

3.1.5.1 Nanoparticles: including polymer nanoparticles, solid lipid nanoparticles and inorganic nanoparticles

Studies have shown that solid lipid nanoparticles can improve the stability of topical retinol. Nanoparticle-coated all-trans retinoic acid cream is better tolerated than common formulations, and its epidermis thickness is twice that of the common formulation after 4 days of topical application, which can improve the wrinkles and texture of neck skin in a mouse model.

3.1.5.2 Liposomes, micro-sponges, nanoemulsions, inclusion complexes, etc

These new drug delivery systems have proved to be useful for the topical therapy of retinoids in acne. But to the best of our knowledge, there have been no reports of their use in topical retinoid therapy for photoaging in published data, which needs to be clarified in the future.

3.2 Antioxidants

As mentioned previously, UV exposure induces a burst of intracellular reactive oxygen species (ROSs) and free radicals in skin cells. Oxidative stress significantly obstructs cutaneous homeostasis by damaging intracellular organelles, DNA and the extracellular matrix. Subsequently, the cell cycle is arrested, and more senescence-associated secretory phenotypes (SASPs) are secreted, which finally lead to cellular senescence and skin photoaging. (Han *et al* 2014, Kammeyer and Luiten 2015). Antioxidants are substances that either directly scavenge ROS and free radicals or indirectly boost the machineries of reducing ROS generation (Khlebnikov *et al* 2007). To maintain redox homeostasis, skin cells develop an endogenous antioxidant system that includes both enzymatic antioxidants such as superoxide dismutase (SOD), catalase (CAT), and glutathione peroxidase (GSH-Px) and non-enzymatic antioxidants such as vitamin C, vitamin E, and glutathione (de Jager *et al* 2017, Valko *et al* 2006). These defenses retard the progress of photoaging by neutralizing and reducing free radicals and ROSs induced by UV exposure. However, the efficiency of an endogenous antioxidant system in skin cells gradually deteriorates with age or in harsh environments, particularly under UV exposure. Therefore, a supplement of antioxidants in cosmetics and diets is considered as a potential approach to combat skin photoaging.

Antioxidants protect the skin from photoaging by blocking UV-induced free radical reactions and regulating the endogenous antioxidant defense mechanisms (Kostyuk *et al* 2018). Some antioxidants act as potent reducing agents, scavenging ROSs from the body directly. Other antioxidants can active nuclear factor E2-related factor 2 (NRF2), a master transcriptional regulator of oxidative stress. As a response to stressors, NRF2 translocates into the nucleus after the sulfhydrylation of Kelch-like ECH-associated protein-1 (KEAP1) or phosphorylation of NRF2. Subsequently, it binds to the promoter region of downstream antioxidative response elements (AREs), engaging the expression of a bunch of phase II antioxidant enzymes, such as heme oxygenase (HO1), quinone oxidoreductase (NQO1), CAT, SOD, etc (Kobayashi and Yamamoto 2006, Ma 2013).

Recently, research has been focusing increasingly on bioactive compounds, particularly plant extracts, which are regarded as promising pharmacotherapies in photoprotection. Natural antioxidants, such as vitamin C/E, carotenoids, and phenolics have emerged recently and are receiving growing attention (Petruk *et al* 2018). Previous studies indicated that dietary intake of exogenous antioxidants and/or cosmetic application of antioxidants may serve as potential approaches to prevent skin from photoaging (Godic *et al* 2014). Although sunscreens provide some degree of photoprotection, their actual effects are often unsatisfactory when applied infrequently or people are continuously exposed to sunlight. The combination of natural antioxidants and cosmetics has demonstrated a potent synergism against photoaging (He *et al* 2021).

3.2.1 Vitamins

Vitamin C and E, two of the most well-known and effective antioxidants, play key roles in photoprotection in the human skin. The concentration of vitamin C ranges from 6 ~ 64 mg/100 g in the epidermis and 3 ~ 13 mg/100 g in the dermis, which is higher than in other human tissues. Vitamin C has been reported to alleviate skin aging by maintaining the epidermal barrier and promoting collagen synthesis in the dermis (Rhie *et al* 2001). Topical application of 8% vitamin C has shown remarkable photoprotective effects. However, higher concentrations (more than 20%) may cause skin irritation. Hence it is best to keep its concentration between 10% and 20% (Sharif *et al* 2015).

Vitamin C has been verified in several studies to suppress activating protein-1(AP-1), resulting in downregulation of MMPs and collagen damage. Pullar *et al* found that vitamin C protected keratinocytes from apoptosis and inhibited lipid peroxidation induced by UV exposure (Pullar *et al* 2017). Topical application of 5% vitamin C was proven to prevent skin wrinkling in animal experiments (Bissett *et al* 1990). Moreover, several clinical trials have unraveled the photoprotective properties of vitamin C. Topical application of 10% vitamin C for 12 weeks or 5% vitamin C for 6 months significantly improves photoaging scores and skin wrinkles (Fitzpatrick and Rostan 2002, Humbert *et al* 2003).

The pharmacotherapy of vitamin E is primarily represented in the protection from UV-induced pigmentation, associated with the reduction of free radical concentration and the restoration of cell lipid oxidation (Burke *et al* 2000, Masaki 2010). Clinical studies suggest that the topical application of vitamin E alleviates skin wrinkles and sunspots caused by photoaging. Further, Ichihashi *et al* reported that vitamin E reduced the activity of tyrosine hydroxylase and the synthesis of melanin in both cultured human melanoma cells and normal melanocytes (Ichihashi *et al* 1999).

Intriguingly, numerous studies reveal that vitamins C and E have synergistic effects on photoprotection. Although vitamin C is water-soluble and rich in the aqueous phase, it can protect liposoluble vitamin E by decreasing lipid oxidation (Halpner *et al* 1998, Jones *et al* 1995). After oral administration of both vitamins C and E, cutaneous resistance to UV irradiation is dramatically improved (Eberlein-Konig *et al* 1998). For topical application, the combination of 15% vitamin C and 1% vitamin E minimizes skin erythema by reducing the generation of pyrimidine dimers and UV-damaged cells (Lin *et al* 2003). Synergism effects can also be observed in cosmetics containing both vitamin C and vitamin E, which eliminate facial wrinkles and improve skin smoothness and brightness (Grether-Beck *et al* 2015, Jagdeo *et al* 2021, Lintner *et al* 2020).

3.2.2 Carotenoids

Carotenoids are diet-derived photoprotectants, such as β-carotene from carrots, lycopene from tomatoes, and astaxanthin from shrimp and lycophytes.

3.2.2.1 β-carotene

β-carotene was reported to reduce the expression of MMPs in both HaCaT keratinocytes and hairless mouse skin (Wertz *et al* 2004, Minami *et al* 2009). Furthermore, Cho *et al* observed that a daily intake of 30 mg of β-carotene improved human facial wrinkles and elasticity, and upregulated the expression of procollagen and reduced UV-induced DNA damage (Cho *et al* 2010). However, the efficacy of β-carotene against photoaging is still debatable, and clinical research findings are inconsistent. According to Hughes' long-term follow-up, dietary supplementation of β-carotene did not appear to impact skin aging (Hughes *et al* 2013).

3.2.2.2 Lycopene

Lycopene is an intermediate of carotenoid synthesis in plants. It cannot be produced by the human body and must be obtained through dietary supplementation. Lycopene is an important ROSs passivator. It can, for example, eliminate singlet oxygen twice as effectively as β-carotene and ten times as completely as vitamin E (Mozos *et al* 2018, Przybylska 2020). It has been reported that lycopene protects keratinocytes from UVB radiation. Oral supplementation of lycopene-rich diet inhibits the UVR-induced synthesis of MMP-1 and protected the dermis from photodamage (Grether-Beck *et al* 2017).

3.2.2.3 Astaxanthin

Astaxanthin is one of the most conspicuous carotenoids, with the strongest capability to eliminate free radicals in both the inner and exterior regions of the cell membrane at the same time (Kurashige *et al* 1990). It is estimated that its antioxidant capacity is 100–500 times more effective than vitamin E. Astaxanthin significantly decreases the expression of MMP-1 in UVB-treated fibroblasts and reduces collagen degradation in dermis (Tominaga *et al* 2017). Furthermore, astaxanthin has been verified to rejuvenate photoaged skin in both oral and topical clinical trials. According to a 10 week placebo-controlled double-blind clinical trial, oral administration of astaxanthin enhanced the minimum erythema dose (MED) of UV radiation and reduced transepidermal water loss (TEWL) in the irradiated skin area. In addition, a 6 week randomized, double-blind, placebo-controlled research trial with 36 healthy male volunteers was also conducted. Oral administration of 6 $mg\,d^{-1}$ astaxanthin enhanced TEWL and showed a substantial trend of improvement of cheek wrinkles and elasticity (Tominaga *et al* 2012). Likewise, 28 healthy middle-aged women took part in a 6 week double-blind, placebo-controlled clinical trial. Substantial improvements in skin moisture content and elasticity were observed after oral administration of $4\,mg\,d^{-1}$ anthocyanin (Yamashita 2015). However, a 16 week, double-blind, placebo-controlled research trial of 65 healthy women revealed that skin wrinkles, skin moisture content, and interleukin 1 levels did not differ significantly before and after taking astaxanthin orally at 6 or 12 mg d^{-1} but worsened dramatically in the placebo group (Tominaga *et al* 2017). The impact of combined application of astaxanthin and collagen is also extremely noticeable. A 12 week double-blind placebo-controlled trial included 44 healthy volunteers. Oral astaxanthin ($2\,mg\,d^{-1}$) combined with collagen hydrolysate (3 g d^{-1}) boosted type I procollagen expression while decreasing MMP-1 and MMP-12 expression in photoaged facial skin, resulting in significant improvements in skin elasticity and TEWL (Yoon *et al* 2014). For topical application, astaxanthin also showed considerable protective effects. 11 women who received astaxanthin 0.7 mg g^{-1} twice a day experienced significant improvements in skin dryness and skin moisture content (Seki *et al* 2001). 30 healthy women were prescribed $6\,mg\,d^{-1}$ astaxanthin for oral application and 78.9 μM astaxanthin solution for topical application. After 8 weeks, skin elasticity was dramatically increased, while age spots shrank (Tominaga *et al* 2012). These findings indicate that astaxanthin can efficiently improve the elasticity and epidermal barrier of human skin.

3.2.3 Phenols

Phenols are abundant in vegetables, fruits, red wine and green tea. They can be separated into three classes—flavonoids, phenolic acids, and astragalus—according to their chemical features. Most of them exhibit potent anti-photoaging properties, although the results are mainly based on basic experiments.

3.2.3.1 Flavonoids

Flavonoid chemicals include catechins, isoflavones, procyanidins and anthocyanins, silymarin, and so on. Green tea polyphenols are derivatives from catechin and possess an even stronger antioxidant potential than vitamins C and E (Rice-Evans 1999). They could ameliorate UV-induced pigmentation, the generation of intracellular ROSs and MMPs (Roh *et al* 2017). Moreover, tea polyphenols protect DNA from UV and hydrogen peroxide-induced damages in skin fibroblasts (Parshad *et al* 1998). Topic application of green tea polyphenols reduced the production of UVB-induced pyrimidine dimer in the epidermal and dermal cells (Katiyar *et al* 2000). However, the benefit of the oral administration of tea polyphenols on photoaging has not been confirmed in an eight week randomized controlled double-blind trial, which included 40 women with moderate photoaging. Although oral administration of 300 mg green tea extract twice daily improved the elastic tissue composition histologically, other photoaging scores did not improve significantly compared to the control group (Chiu *et al* 2005). Similarly, long-term oral administration of green tea polyphenols failed to provide an adequate anti-photoaging effect as well. In a 2 year double-blind, placebo-controlled study, research, 56 women (25–75 years old) were prescribed 250 mg of green tea polyphenols twice daily. Although improvements of erythema and telangiectasia after sun exposure were observed, other photoaging characterization, including the histology of sun-exposed skin, did not show meaningful improvements (Janjua *et al* 2009). In addition, a 3 month double-blind, randomized, placebo-controlled trial, including 50 healthy patients (18–65 years old), was conducted to investigate the pharmacological effect of twice-daily administration of 540 mg green tea polyphenols on UV-induced inflammation. Again, oral treatment of green tea polyphenols failed to improve skin erythema or white blood cell infiltration in sun-exposed skin (Farrar *et al* 2015). According to these clinical trials, green tea polyphenols are not as effective in preventing photoaging as previously assumed.

Isoflavones, which are derived from soybeans and other plants, can scavenge free radicals, prevent lipid peroxidation, and decrease UV-induced DNA oxidative damage. According to an *in vitro* assay, soybean isoflavones reduced oxidative stress-driven cell death in keratinocytes by inhibiting the release of H_2O_2 caused by UVB (Huang *et al* 2010). It was also found that isoflavones diminished UVA-induced skin erythema and cutaneous inflammation in rodents (Wei *et al* 1996). Equol, a flavone metabolite, could minimize the area and depth of skin wrinkles of postmenopausal women (Oyama *et al* 2012).

Anthocyanins and procyanidins, derived from fruits, possess potent anti-inflammatory and antioxidant effects (Jiang 2010, Wei *et al* 1996). Petruk reported that cyanidin, a metabolite from acai fruit, protected fibroblasts from UVA-induced oxidative damage (Petruk *et al* 2017). Intriguingly, diets rich in procyanidins extracted from grape seeds attenuated the oxidative damage and carcinogenesis caused by UVB on a mouse model (Mittal *et al* 2003). Further, a cream containing anthocyanins extracted from pomegranate has been shown to be effective in reducing wrinkles and enhancing skin hydration (Abdellatif *et al* 2021).

In addition, silymarin and its derivatives from milk thistle also exhibit considerable photoprotective effects. Silymarin could inhibit the enzyme activities of hyaluronidase, collagenase, and elastase, which is associated with the breakdown of extracellular matrix (Vostálová *et al* 2019). Silymarin also worked as a scavenger of free radicals and prevented UVA damage to keratinocytes and fibroblasts (Rajnochová Svobodová *et al* 2019, 2018, Svobodová *et al* 2007). The sun protection factor (SPF) index of 10% silymarin cream is close to 9. When nano-scale silymarin was used, the SPF index was up to 14, indicating it might be an excellent component for sunscreen (Couteau *et al* 2012, Netto MPharm and Jose 2018).

3.2.3.2 Phenolic acids

Gallic acid and cinnamic acid are the most common phenolic acids. Gallic acid ameliorated UVB-induced dryness and wrinkles in hairless mice by lowering the expression of MMP-1 and elevating the accumulation of elastin and type I collagen in the dermis (Hwang *et al* 2014). In human skin fibroblasts, trans-cinnamic acid inhibited the activation of AP-1, which subsequently suppressed MMP1 and MMP-3 synthesis. Meanwhile, trans-cinnamic acid activated NRF2-mediated antioxidant enzymes to attenuate UVA-induced photoaging (Hseu *et al* 2018).

3.2.3.3 Astragalus

Resveratrol, the most well-known astragalus compound, is a small bioactive molecule rich in natural sources such as red grapes, red wine, and berries (Todaro *et al* 2008). Resveratrol serves as an potent antioxidant by activating the SIRT1 and NRF2 pathways, which subsequently inhibit the production of ROSs and slow down the process of photoaging (Baur *et al* 2006, Zhuang *et al* 2019). It has been demonstrated that resveratrol can block the expression of the AP-1 and NF-KB transcription factors, protecting skin cells from oxidative damage induced by free radicals in UV-irradiated area (Adhami *et al* 2003, Cui *et al* 2022). Moreover, resveratrol can suppress melanin synthesis and mitigate UVB-induced pigmentation on guinea pig skin (Lee *et al* 2016). Both resveratrol-containing lotions and dietary supplements have demonstrated anti-aging properties. Buonocore *et al* conducted placebo-controlled, double-blind research including 50 participants. Volunteers' skin appeared to have increased moisture retention and suppleness, as well as a reduction in roughness and wrinkle depth, after dietary intake of resveratrol and procyanidins for 60 days. The study clarified that resveratrol can enhance skin conditions and have anti-aging benefits (Buonocore *et al* 2012). In addition, creams containing resveratrol (1%), baicalein (0.5%), and vitamin E (1%) were verified to modify the skin condition in a 12 week clinical trial of 55 middle-aged women, with statistically significant improvements in skin elasticity, tiny wrinkles, and hyper-pigmentation (Farris *et al* 2014).

Grapes and pomegranates are ingredients rich in natural antioxidants. Their extracts have been used increasingly for photoprotection recently (Che *et al* 2017, Kim *et al* 2019, Liu *et al* 2022). Grapes contain various polyphenolic compounds, including β-carotene, resveratrol, catechin, epicatechin, gallic acid, procyanidins, anthocyanins, and so on. Creams with grape seed extracts demonstrate remarkable properties of skin whitening, hydration, and rejuvenation (Sharif *et al* 2015). Pomegranates are excellent sources of anthocyanins and ellagic tannins, which provide powerful photoprotection. These bioactive molecules can restore UV-induced cell apoptosis, suppress the activities of MMPs, and promote the synthesis of the extracellular matrix (collagen fiber and hyaluronic acid).

A growing number of basic studies have revealed that the application of antioxidants derived from natural sources exerts reliable safety and anti-aging effects. Antioxidants prevent photoaging by directly eliminating ROSs and free radicals or indirectly boosting endogenous antioxidant enzyme activity. In fact, antioxidants, particularly from plant and fruit extracts, are widely included in diets and cosmetics nowadays. However, long-term monitoring and rigorous *in vivo* studies and clinical trials are still required to evaluate whether topical or oral antioxidants effectively prevent photoaging and photodamaging. The precise mechanisms of antioxidants, such as the compatibility, stability, permeability and bioavailability in emulsions or sunscreens, remain to be clarified.

3.3 Osmolytes

Aging leads to the deterioration of tissue structure and function, and in skin, environmental factors such as ultraviolet radiation (UVR) can accelerate aging effects such as reduced barrier function and hydration loss. Water homeostasis is essential for cellular functions, with organic osmolyte transport playing a crucial role. Aging disrupts these physiological mechanisms, possibly due to decreased expression of organic osmolyte transporters.

Osmolytes are small, compatible organic molecules that regulate water and electrolyte balance. They are cytoprotective, aiding in protein folding, function, stability, and preventing aggregation. Osmolytes can be categorized into: (i) carbohydrates or polyols (e.g. myo-inositol, sorbitol), (ii) methylamines (e.g. betaine, sarcosine), and (iii) amino acids and their derivatives (e.g. taurine), as shown in table 3.1. They are common in anti-aging products for maintaining water homeostasis, improving epidermal barrier function, stabilizing proteins, and providing antioxidation (El-Chami *et al* 2014).

The cellular control of water homeostasis is critical; cell shrinkage can lead to cell death and if this goes unresolved, it ultimately leads to tissue dehydration and, potentially organismal death. Therefore, important cellular mechanisms exist to maintain tight control of cell volume. One of the mechanisms employed by cells is the expression of naturally occurring compounds known as organic osmolytes, such as betaine, myo-inositol, and sarcosine. Such osmolytes are transported into cells under water stress and can accumulate at high concentrations without adverse effect, thus preventing further water loss from cells. Conversely, under threat of excessive swelling, osmolytes are actively pumped out of cells. These mechanisms allow the movement of osmolytes and water molecules across the cell membrane via transporters and initiate cell volume recovery in response to osmotic fluctuation (Judy and Kishore 2016).

Homeostasis of most osmolytes is regulated by cells through biosynthesis, absorption from food and renal re-absorption. Under hyperosmotic stress, cells enable the transport of inorganic ions as well as of organic osmolytes, namely sorbitol, inositol, and betaine to maintain homeostasis. Organic osmolytes do not interfere with substrates and macromolecular–solvent interactions even at a high solute concentration (tens to hundreds of millions per litre), however, they are metabolically expensive and are accumulated in the cells at the expense of ATP.

Table 3.1. Major types of osmolytes.

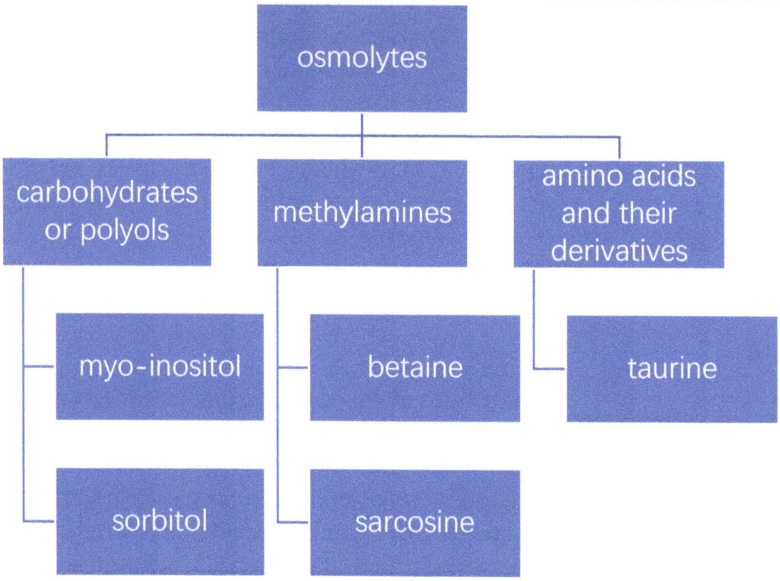

In skin, human epidermal keratinocytes do possess an osmolyte strategy that allows cell volume regulation to take place. Cell volume and osmolyte transporter expression are negatively impacted by age and both chronic and acute UVR exposure in human skin (Foster *et al* 2020). Organic osmolytes could play a key role in age-related alterations to skin structure and function.

Studies have shown that organic osmolytes act to maintain water homeostasis, not only by stabilizing cell volume, but also via protein stabilization and exerting antioxidant effects (Mueed *et al* 2020). Not all osmolytes can maintain protein metabolism. Amino acids which are branched, such as leucine, are a class of osmolytes with the ability to stimulate protein anabolism and reduce protein catabolism among others. Organic osmolytes also protect the development of tight junctions, maintain epidermal barrier functions, especially following exposure to ultraviolet (UV) (El-Chami *et al* 2020).

Understanding the functional mechanism of osmolytes in water homeostasis control is fundamental to the development of novel strategies for skin hydration management and to our understanding of the pathophysiology of skin aging. In this section, we briefly summarize the effects of various osmolytes on anti-aging.

3.3.1 Taurine

The beta-amino acid transporter, which acts as an organic osmolyte transporter in cells, is a substrate of taurine. Due to its humectant qualities, ability to initiate gene expression that results in the upregulation of tight junctions (TJ) proteins, and ability to improve barrier function even in skin that is in good health, taurine has been a longstanding addition to topical anti-aging skincare products (El-Chami *et al* 2021).

Only one transporter, taurine transporter (TAUT), has been identified and is expressed in the human epidermis' stratum granulosum and stratum spinosum. It has also been demonstrated that taurine, the substrate of TAUT, is expressed in these epidermal layers of rat and dog skin (Ito *et al* 2015). Research utilizing human skin samples from photoexposed and photoprotected areas in both young and old volunteers has demonstrated a downregulation of osmolyte transporters TAUT with both acute and chronic UVR exposure, as well as a reduction in keratinocyte cell size with age. Single-cell live imaging revealed that, after physiological stress, youthful keratinocytes have effective cell volume recovery mechanisms that are absent from elderly keratinocytes. Exogenous taurine supplementation, on the other hand, dramatically restored cell volume, a finding supported by a decrease in TAUT mRNA and protein in older keratinocytes relative to younger ones. All of these new findings show that osmolyte-mediated cell volume regulation mechanisms are present in human epidermal keratinocytes and may become weakened with age. As a result, taurine in particular, an organic osmolyte, may be important in cutaneous age-related xerosis and shed light on a basic process that led to the development of our knowledge of the pathophysiology of skin aging.

Furthermore, several TJ proteins are upregulated in the presence of taurine. In particular, taurine boosts the production of proteins crucial to TJs' ability to create barriers (El-Chami *et al* 2021). According to a quantitative reverse-transcriptase polymerase chain reaction assay, taurine increased the amounts of occludin, claudin-1, and claudin-4 at the mRNA level. Through enhanced mRNA synthesis and, most likely, the production of new protein, taurine drove the rise in particular TJ protein expression. By lowering transepidermal water loss (TEWL), taurine also enhanced epidermal barrier function in the rebuilt epidermis.

Cell shrinkage brought on by osmolyte loss, which is more common in later life stages, can lead to senescence and apoptosis. It has been documented that taurine's anti-senescent and anti-apoptotic properties control cell volume homeostasis.

Recent research has shown hitherto unknown effects of organic osmolytes that are frequently utilized in topical preparations. To be more precise, osmolytes have the ability to upregulate the expression of their own transporters and improve the structure and function of TJs. This would improve the function of the epidermal barrier, particularly after exposure to environmental stressors or in skin conditions such as atopic dermatitis that are linked to barrier dysfunction.

3.3.2 Betaine

Betaine is a naturally occurring compound which is widely distributed in plants, microorganisms, several types of food and medicinal herbs, including *Lycium chinense*, which has been demonstrated to have high levels of betaine.

Betaine is critical in maintaining skin barrier function, cellular transport of lipid molecules and modulating the optimal levels of, in particular, keratinocyte differentiation in the epidermis. It probably mediates its effects via stabilizing the existing tight junction's protein pool, and stimulates the expression of Filaggrin (FLG) and Filaggrin 2 (FLG2). Betaine may reduce transepidermal water loss (TEWL) due to an increase in skin barrier function along with enhancing lipid transport among the epidermal layers. Betaine increases the expression of the fatty acid synthesis enzyme ACACB and fatty acid binding proteins (FABPs). It is also well known for its ability to stabilize proteins, and might act via a reduction in protein turnover or increased mRNA lifetime. It also upregulates the expression of ABCA12, which supports the notion that betaine may improve keratinocyte differentiation. Another skin biomarker is UDP-glucose ceramide glucosyltransferase (UGCG), which can cause ichthyosis when levels are decreased, however, betaine treatment stimulates UGCG.

In addition, betaine has been reported to be beneficial for a number of conditions and diseases. It may be applied for photoprotection based on antioxidant properties (Punzo *et al* 2021). Betaine inhibits UVB-induced wrinkle formation, decreases the thickness of the epidermis, reduces collagen fiber damage, and inhibits the expression of matrix metalloprotein 9 (MMP-9) and the phosphorylation of MEK and ERK in UVB-irradiated hairless mice (Im *et al* 2016, Lee *et al* 2016). Betaine also stimulates oxidative stress target genes, such as heme oxygenase 1 (HMOX1) and NAD(P)H quinone dehydrogenase 1 (NQO1) in their expression to bolster the antioxidant system in the skin. Hydrogen peroxide (H_2O_2) was produced during the oxidative stress reaction and this process requires further detoxification by catalase or glutathione peroxidase 1 (GPX1), which is also stimulated by betaine. Notably, oral administration of betaine reduced the occurrence of characteristics associated with skin aging. In summary, betaine has protective effects on UV-induced skin damage and photoaging.

3.3.3 Sorbitol

Sorbitol is a type of carbohydrate known as a 'sugar alcohol' or 'polyol'. It is not only frequently used in skin care, but also commonly used as a sweetener in food products, since it has about 1/3 fewer calories compared to cane sugar. It is a naturally occurring compound that can be found in a wide variety of fruits and berries such as apples and blackberries. It can also be produced artificially from glucose in a laboratory setting, which is the most common type of sorbitol used for both food products and cosmetics.

Sorbitol is primarily used in skin care and cosmetics as a thickener, humectant, and for its sweetness (for lip products). It helps the skin absorb and retain moisture, has a sweet taste and exerts an anti-aging effect and it can help improve the texture of skincare products.

Sorbitol is a humectant, or hydrating ingredient. It has a similar composition to glycerin, another common humectant, and helps the skin absorb moisture from the air and retain it for a longer period of time. Human skin maintains an optimal permeability barrier function in a terrestrial environment that varies considerably in humidity. Cells cultured under hyperosmotic stress accumulate osmolytes including sorbitol (Zhang *et al* 2021). Epidermal keratinocytes experience similar high osmolality under dry environmental conditions because of increased TEWL and concomitant drying of the skin. Some studies showed that sorbitol protected epidermal keratinocytes from osmotic toxicity induced by sodium chloride. Clinical studies indicated that skin chronically exposed to a hot, dry environment appeared to exhibit a stronger skin barrier and a lower baseline TEWL. Sorbitol exhibited significant improvement in both skin barrier repair and moisturization, particularly in individuals subjected to arid environmental conditions. Topical application of sorbitol to skin exposed to a dry environment reduced TEWL. This means sorbitol is beneficial to dehydrated, cracked, and dry skin.

As the skin ages, it becomes less effective at retaining moisture in the skin and more vulnerable to oxidative stress. This leads to dry skin that is more prone to fine lines and wrinkles, while sorbitol helps the skin retain moisture for a plump, youthful complexion.

3.3.4 Inositol

Inositol, also called myo-inositol, plays a critical function in the body's cellular growth. Although it used to be referred to as vitamin B8, inositol is not actually a vitamin. It is a type of carbohydrate, which plays an important role in anti-aging through maintaining water homeostasis, especially under UV exposure.

The inositol requirement for cultured human keratinocytes is markedly higher than for any other normal or malignant cell type investigated to date. Inositol may subserve quantitatively or qualitatively different functions in the keratinocytes than in other cell types (Gordon *et al* 1988). Studies have shown that both normal human epidermal keratinocytes and the immortal human keratinocyte line (HaCaT) are osmo-sensitive and express the sodium-coupled myo-inositol transporter (SMIT). A small number of studies performed on cultured keratinocytes have revealed that normal human epidermal keratinocytes increase the gene expression of the sodium-coupled myo-inositol transporter (SMIT) in response to both hypertonicity and doses of UVA and UVB radiation (Namkoong *et al* 2018). This results in an increased uptake of radiolabeled myo-inositol in these cells. Another study performed in HaCaT cells has shown that these cells lose water and shrink after UVB exposure, which leads to an increase in SMIT mRNA expression, resulting in an intracellular accumulation of their respective osmolytes. One study examining improvement of moisture applied the skin lotions containing inositol to the arm skins of 45 Asian women aged between 20 and 40 for 7 weeks. As a result, wrinkle reduced and elasticity improved on average, respectively, further indicating that Inositol has an anti-aging effect.

3.3.5 Sarcosine

Sarcosine, a type of methylamine, is a naturally occurring osmolyte that functions as a skin conditioning and oil-control agent and may play a role in enhancing the penetration of other ingredients.

Research has shown that sarcosine can help minimize the impact UV light exposure has on the skin's surface, and thus it has been included in various types of sunscreens.

Sarcosine is also known to stabilize proteins and peptides at elevated temperatures. It has emerged as the most successful osmolyte, rendering the highest degree of protection against aggregation. In a rat model of aging, dietary restriction is a robust modulator of the metabolome in plasma and tissues, and effectively opposes a significant number of age-related changes in amino acids, glycerophospholipids, acylcarnitines, and sphingolipids in plasma and tissue. Circulating sarcosine as a metabolite that is similarly and uniquely modulated by aging and dietary restriction in rats and humans is dramatically elevated in the serum of long-lived mice (Walters *et al* 2018). Furthermore, supplementation studies in old animals identified sarcosine as a critical node linking amino acid and lipid metabolism, while mechanistic studies demonstrate sarcosine as a potent stimulator of basal macroautophagy, implicating an interesting link among aging, growth factors, metabolism, and proteostasis. Thus, these data identify sarcosine, which sits at the nexus of folate, methionine (Met), and glycine metabolism, as a potential functional biomarker of the aging and dietary restriction metabolic phenotype. The circulating sarcosine is decreased with aging and increased by dietary restriction, making it a potentially important biomarker of diet and aging in mammals.

3.4 DNA repair enzymes

3.4.1 DNA damage and repair upon UV irradiation

DNA damage induced by ultraviolet irradiation (UVR) is a major environmental factor for skin aging and cancer. UVR may lead to DNA damage in two different ways: direct formation of DNA photoproducts and indirect damage by UV-induced reactive oxygen species (ROSs). UVB radiation on cellular DNA results directly in structural changes by covalently binding two neighboring pyrimidines, thus forming two photoproducts: cyclobutene pyrimidine dimers (CPD, 75%) and pyrimidine-pyrimidone (6–4) photoproducts (6–4PPs, 25%) (Vechtomova *et al* 2021). The accumulation of CPD and 6–4PPs intervenes in normal DNA replication and transcription, leading to cell apoptosis, senescence, mutagenesis and even carcinogenesis. Meanwhile, UVA causes indirectly damage to DNA via oxidative stress and leads to the formation of 8-hydroxy-2′-deoxyguanosine (8-OHdG). The accumulation of DNA damage leads to DNA mutations and modifications, affecting gene expression of the antioxidant system, cell cycle control and extracellular matrix enzymes, thus contributing to the process of aging.

Normally, cells utilize DNA repair systems to rectify UVR-induced DNA damage spontaneously. For most organisms except mammals, DNA photolyases directly reverse CPD and 6–4PPs (a process called 'direct repair' or 'photoreactivation'), whereas in humans, DNA damage can only be repaired by excision repair system, where the damaged sites are removed first and then replaced by newly synthesized nucleotides. More specifically, nucleotide excision repair (NER) is for

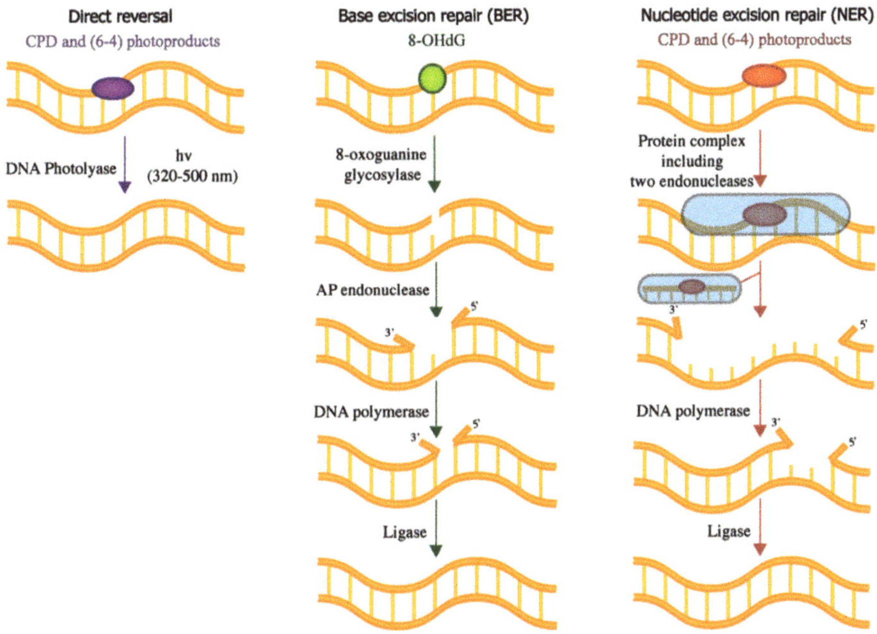

Figure 3.1. Excision repair system of humans.

removal of CPD and 6–4PPs, while base excision repair (BER) for 8-OHdG (Moriwaki and Takahashi 2008).

In general, the NER process can be triggered in two ways. The first is global genome repair (GGR) where the NER enzymes themselves find the damaged sites and start the repair process. The second way is transcription-coupled repair (TCR), which is initiated when an RNA polymerase encounters a damaged site during transcription. Next, a cascade of reactions is started, as follows. (i) A protein complex including endonucleases cuts out a DNA oligomer containing CPD. (ii) The DNA polymerase is attached and synthesizes a complementary intact DNA chain. (iii) The ligase completes the process by connecting the free ends of DNA chains (Vechtomova *et al* 2021). The BER system is similar to the NER but consists of fewer enzymes. The damaged base is removed by 8-oxoguanine glycosylase (OGG), and then the nucleotides including the damaged site are removed by AP endonuclease and synthesized again (figure 3.1).

In early 2000, researchers found there was a significant decrease of removal rate of CPD and 6–4PPs with aging detected in UV-irradiated dermal fibroblast from donors of different ages (Goukassian *et al* 2000, Yamada *et al* 2006). There was also an age-associated decrease in the protein levels of key players participating in the NER process including ERCC3, PCNA, RPA, XPA, and p53 (Goukassian *et al* 2000). Other studies revealed markedly decreased expression of DNA repair synthesis-related genes in the cells from elderly subjects as compared with those from young subjects, such as DNA polymerases, PCNA and RFC (Takahashi *et al* 2005). These results indicated that reduced post-UV DNA repair capacity during aging results in an accumulation of UV-induced DNA damage, which might contribute to the phenotypes of photoaged skin.

3.4.2 DNA repair enzymes in sunscreen and other skincare products

Technology aimed at DNA repair has been available in skincare and sunscreen products for several decades. Although conventional sunscreens can effectively prevent photoaging, they do not have a repairing effect on UVR-induced cell damage. The concept of 'active photoprotection' then arises, aiming at dual function of prevention and repair, promoting a compound sunscreen containing DNA repair enzymes and antioxidants in addition to the sun protection factor (SPF) (Yeager and Lim 2019). Recent studies have proved that topical DNA repair enzymes do enhance the removal of DNA damage and exert certain preventive and therapeutic effects on skin photoaging and carcinogenesis (Luze *et al* 2020). Most of the topical DNA repair enzymes are extracted from bacteria and plants, and then encapsulated in liposomes to assure their percutaneous absorption. Photolyases, T4 endonuclease V (T4N5) and OGG1 are major representatives of DNA repair enzymes which have been widely used and studied in skincare products (Yarosh *et al* 2019).

Photolyases, a class of flavoproteins, repair UVB-induced DNA damage through the absorption of blue light. Two different kinds of photolyases specifically repair CPDs and 6–4PPs and are usually classified as CPD photolyases or 6–4 photolyases, respectively. Photolyases are not innately present in humans who possess other DNA repair system such as the NER pathway. However, NER is more effective in repairing pyrimidine pyrimidines and relatively inefficient in repairing CPDs, in addition its efficiency decreases with advancing age (Leccia *et al* 2019). Therefore, recent studies have focused on CPD photolyases as a novel photoprotective strategy. Clinical studies revealed that the addition of photolyase to a traditional sunscreen decreased the number of UVR-induced CPDs by 40%–45% (Stege *et al* 2000), contributing significantly to the prevention of UVR-induced DNA damage and cell apoptosis when applied topically to human skin (Berardesca *et al* 2012). Emanuele *et al* conducted an experimental pilot study ($N = 12$) in healthy volunteers (Fitzpatrick skin types I and II) and found the combined use of a sunscreen plus photolyase 30 min before irradiation and an endonuclease preparation immediately after irradiation completely abrogated telomere shortening and c-FOS gene hyper-expression induced by the experimental irradiations (Emanuele *et al* 2013), indicating the potential of topical photolyases and endonuclease in prevention of skin aging and carcinogenesis.

Several commercially available sunscreens containing liposome-encapsulated photolyases have been used successfully in the prevention and treatment of actinic keratoses (AK) and other non-melanoma skin cancer (NMSC). For example, clinical studies found that Eryfotona® AK-NMSC (Isdin, SA) exerted beneficial effects on field cancerization in AK patients clinically and histologically, manifested as improvement in AK lesion count and extent of cancerization field (Puviani *et al* 2013, Navarrete-Dechent and Molgó 2017). Although studies on normal individuals and preventive effect focusing on photoaging are limited, the therapeutic effect on

AK could be transferred to photoaging as well, since DNA damage and ROS are similar initial events in both photoaging and photocarcinogenesis.

T4 endonuclease V (T4N5), initially isolated from *Escherichia coli* infected with T4 bacteriophage, is an enzyme that specifically recognizes UVR-induced CPD and initiates DNA repair by its two catalyzing activities: pyrimidine dimer-DNA glycosylase activity and the apurinic-apyrimidinic endonuclease activity. The efficacy of naturally occurring DNA repair can be enhanced approximately fourfold by T4N5. In addition to the removal of DNA damage, DNA repair enzymes also prevent the destruction of extracellular matrix components, and alleviates inflammation and immune response after UVR. An *in vitro* study found that both T4N5 and photolyases significantly inhibited UVR-induced MMP-1 release in human skin cells (Zattra *et al* 2009). *In vivo* study revealed that topical treatment with liposome containing T4N5 immediately and at 2, 4, and 5 h after UVR nearly completely prevented UVR-induced upregulation of interleukin-10 (IL-10) and tumor necrosis factor-alpha (TNF-α) at the tested sites (Wolf *et al* 2000).

The safety and efficacy of topical T4N5 have been studied extensively in clinical trials for xeroderma pigmentosum (XP), a genetic disease present with premature aging caused by defective DNA repair. Topical application of liposome-encapsulated T4N5 reduced the incidence of basal cell carcinomas (BCC) by 30% and of AK by >68% with minimal long-term adverse effects for 1 year daily application in patients with XP (Cafardi and Elmets 2008). Another pilot study ($N = 13$) evaluated the effect of intensified photoprotection by SPF 30 sunscreen and after-sun lotion both containing photolyases and T4N5 (Ateia® Kwizda Pharma,Vienna, Austria) for 1 year daily use in skin cancer risk patients. The results showed a statistical trend for fewer BCCs during the study period compared to the pre-study period. In addition, the patient's subjective ratings revealed a statistically significant improvement for smoothness, color spots and wrinkles, starting at the first 3 month visit with a maximum effect seen at 12 months (Wolf 2019).

A third class of DNA repair enzymes used to enhance DNA repair is 8-oxoguanine-DNA glycosylase 1 (OGG1), extracted from the mustard plant *Arabidopsis thaliana*, specifically repairing DNA oxidative damage 8-OHdG. Similar to photolyase and T4N5, most studies on OGG1 focus on its effect in the prevention of skin carcinogenesis. Topical application of liposome-encapsulated OGG1 significantly reduced skin tumor size and progression in a mouse model (Wulff *et al* 2008).

Recent clinical studies tend to use a mixed formula of different classes of DNA repair enzymes, such as a combination of photolyases, DNA endonuclease and OGG1. An observational study reported that the application of liposome-encapsulated UV endonuclease, photolyase and OGG1 twice daily for 4 weeks to 32 subjects with fine lines in the eye area resulted in a significant reduction in crow's feet lines, and 72% of subjects noted improvement (Yarosh *et al* 2019). Emanuele *et al* conducted a double-blind head-to-head clinical study ($N = 20$) of a novel topical product (TPF50) compared to traditional SPF50 sunscreen upon experimental irradiations. The TPF50 consists of three active components: traditional physical sunscreens (SPF50), liposome-encapsulated DNA repair enzyme complex (photolyase, endonuclease, and OGG1), and a potent antioxidant complex (carnosine, arazine, ergothionine). The study revealed that TPF50 showed better efficacy in reducing CPD, 8-OHdG and protein carbonylation (PC), a major form of UVR-induced oxidative protein damage, than sunscreen alone (Emanuele *et al* 2014). These results indicate a combination of traditional sunscreen with DNA repair enzyme and antioxidant could better promote the genomic and proteomic integrity of skin cells after UVR, thus reducing the risk of skin aging and NMSC.

In conclusion, topical DNA repair enzymes, mainly including liposome-encapsulated photolyases, endonuclease and OGG1, have been proved to effectively accelerate the removal of UVR-induced DNA damage in human skin, making it a promising supplementary treatment for skin aging and carcinogenesis. Clinical studies have demonstrated a notable preventive and therapeutic efficacy of topical DNA repair enzymes in skin NMSC especially in the treatment of AK. However, although experimentally topical DNA repair enzymes do inhibit UVB-induced aging-related markers such as p53, c-FOS and MMP-1 in human skin cells, there have been very few clinical studies focusing on the efficacy of topical DNA repair enzymes on skin aging. Future investigations with a double-blind design and long-term follow-up are of ultimate need to fill this gap and better confirm the evidence of topical DNA repair enzymes for skin aging.

3.5 Cytokines

Chronic UV exposure increases oxidative stress. This results in tissue damage, due to antioxidant depletion together with increased production of reactive oxygen species (ROSs, also known as free radicals). Multiple biochemical pathways that are triggered by ROS overload result in the suppression of transforming growth factor-β receptor II (TGF-β-R2) (Quan *et al* 2004), overexpression of matrix metalloproteinases (MMPs) which are collagenases (Schwartz *et al* 1993), and increased inflammation through the nuclear factor kappa β (NF-$k\beta$) pathway (Yamamoto and Gaynor 2001). During skin photoaging, TGF-β is upregulated and activated, inducing excessive MMPs and pro-inflammatory cytokines, and prolonged infiltration of neutrophils, leading to progressive collagen degradation and aberrant elastic fibers that contribute to extracellular matrix (ECM) destruction (Fisher *et al* 2002). UV radiation also causes direct damage to the structural proteins in the skin.

Growth factors and cytokines are a large group of regulatory proteins that attach to cell surface receptors and serve as chemical messengers. Through this interaction, they control cell growth, proliferation, and differentiation via a network of inter- and intracellular signaling pathways. The activity of growth factors and cytokines is confined to the vicinity of their production sites. In the skin, cytokines are synthesized from almost all nucleated cells, when stimulated or activated, including keratinocytes, melanocytes, Langerhans cells in the epidermis, and fibroblasts, endothelial cells, smooth muscle cells, mast cells, lymphocytes and other inflammatory cells in the dermis. Specific growth factors and cytokines regulate certain cellular activities, including mitogenesis, angiogenesis, chemotaxis, formation of the

ECM, and control of other growth factors and cytokines (Fabi and Sundaram 2014). Studies show that growth factors can directly affect collagen biosynthesis include platelet-derived growth factor (PDGF), vascular endothelial growth factor (VEGF), epidermal growth factor (EGF), granulocyte-colony stimulating factor (G-CSF), keratinocyte growth factor (KGF), and hepatocyte growth factor (HGF). Cytokines affecting collagen biosynthesis include TGF-β, interleukin (IL)–6, and IL-8 (Sproul and Argraves 2013, Uitto and Kouba 2000).

Growth factors and cytokines have been studied extensively in skin wound healing. When the skin is wounded, growth factors and cytokines accumulate at the site of injury and interact synergistically to initiate and coordinate the healing process. They can reverse the effects of collagenases, increase collagen levels, and decrease tissue inflammation. The healing of skin wounds is precisely regulated by complex interactions between growth factors and cytokines that result in signaling cascades. Successful wound healing requires a balance between development of inflammation and its resolution. This involves multiple growth factors and cytokines, including PDGF, vascular VEGF, TGF-β, EGF, G-CSF, KGF, HGF, as well as IL-6 and IL-8 (Eming *et al* 2007). Growth factors and cytokines relevant to wound healing induce dermal remodeling by stimulating synthesis of new collagen, elastin, and glycosaminoglycans, as well as by mediating angiogenesis (Martin 1997).

There are striking parallels between the pathways involved in skin wound healing and skin photoaging. Growth factors and cytokines levels in the body peak in youth and decline thereafter. It has been hypothesized that skin aging is analogous to a wound that is sufficiently extensive to overwhelm the skin's inherent repair mechanisms, which become attenuated with age (Sundaram *et al* 2009).

Thus, the aim of administering topical or injectable growth factors and cytokines is to replenish the skin's own depleted levels and to upregulate the activity of cells responsible for dermal remodeling, thereby slowing or even reversing the manifestations of skin aging. Cytokine signaling after topical application of exogenous or injection of growth factors and cytokines may mirror the interactions that occur during wound healing.

3.5.1 Topical application of growth factors and cytokines for skin rejuvenation

In recent years, topical and injectable growth factors and cytokines have emerged as intriguing therapeutic modalities, with increasing interest in their potential to serve as actives for skin rejuvenation. Growth factors and cytokines may be applied topically in cosmeceutical formulations, or injected in autologous platelet-rich plasma (PRP). Topical and injectable growth factors and cytokines have the potential to modulate complex cellular communication that ultimately results in upregulation of collagen synthesis and downregulation of collagen degradation (Fabi and Sundaram 2014). However, most of the growth factors and cytokines have large molecular size (generally > 15,000 Da). Studies have shown minimal penetration of intact stratum corneum by hydrophilic molecules that have a molecular weight greater than 500 Da. Thus the large molecular weight of growth factors limit their ability to penetrate the tightly packed stratum corneum (Mehta and Fitzpatrick 2007). One possible way for larger molecules such as growth factors and cytokines could reach epidermal keratinocyte receptors is via hair follicles and sweat glands. Another consideration is that the barrier function of skin is somewhat compromised, and this may permit better penetration, for example, after microneedling or laser treatment. In addition, it may be possible to improve the penetration of growth factors and cytokines by chemical modification with lipophilic molecules (Aldag *et al* 2016). Similarly, injection of PRP and its derivative, which contain growth factors and cytokines, can also potentially exert effects. Their direct intradermal or subdermal injection bypasses the epidermal barrier, and could accelerate and/or enhance the clinical effects (Cheng *et al* 2010).

Once growth factors and cytokines have penetrated the stratum corneum, they can bind to specific receptors on keratinocytes and can initiate a cytokine-signaling cascade. After binding, growth factors and cytokines secreted by the keratinocytes affects fibroblasts and other cells in the dermis. Fibroblast-derived growth factors and cytokines also stimulate keratinocyte proliferation, resulting in amplification of the initial signaling loop.

Clinical studies have demonstrated that topical application of human or animal-derived growth factors and cytokines or injection of autologous growth factors and cytokine may increase dermal collagen synthesis, and is associated with reduced signs of skin aging such as fine lines and wrinkles. Collagenesis and remodeling of the ECM has also been observed histologically, and can be correlated with clinical results (Fabi and Sundaram 2014).

Topical growth factors and cytokines are derived from a variety of sources including humans (epidermal cells, placental cells, foreskin, and colostrum), animals, plants, recombinant bacteria, and yeast (Bonin-Debs *et al* 2004). There are many types of growth factors to stimulate keratinocytes and fibroblasts to increase function. Using cosmetic products that contain growth factors can also achieve skin rejuvenation. Several cosmetic products aim for skin rejuvenation also contain growth factors and cytokines to promote collagen and elastin synthesis.

Nonrecombinant human growth factors and cytokines are commercially available in various topical cosmetic products. We briefly describe here products with clinical studies reported in peer-reviewed journal articles.

The Processed Skin Proteins (PSP) product line (Bio-restorative Skin Cream, Neocutis, Merz North America, Inc., Raleigh, NC, USA) contains processed skin proteins, which are a mixture of growth factors and cytokines obtained as the lysate of cultured human fibroblasts (De Buys Roessingh *et al* 2006). The anti-aging potential of PSPs was investigated in 20 women with facial wrinkling in the lateral canthal area. Participants in the study were randomized to twice-daily application for 2 months of Bio-restorative Skin Cream or its vehicle on respective face halves. The results showed that the cream significantly reduced periorbital and perioral wrinkles, as well as improving skin texture of the chin after 1 month of treatment. After 60 d of twice-daily application, 83% of the subjects showed an improved average wrinkle score in the eye area, while 50% showed an improved average wrinkle score in the mouth area (Gold *et al* 2007). Further details of the effects of

Bio-restorative Skin Cream were investigated in a study of 12 subjects who applied the cream twice daily to the entire face for 6 months. After the treatment, improved clinical appearance of periorbital and perioral wrinkles by 33% and 25% on average was demonstrated. Histologic evaluation indicated moderate changes in the epidermal thickness as well as an increased fibroblast density in the superficial dermis at the end of the 6 month treatment period. Ultrastructural changes consistent with new collagen formation were shown by electron microscopy (Hussain *et al* 2008).

The TNS (a combination of growth factors and cytokines, TNS recovery complex, SkinMedica, Carlsbad, CA, USA) products contain conditioned medium obtained from neonatal foreskin fibroblast culture.

In one clinical study, a growth factor serum containing TNS was applied to the facial skin of 14 patients twice daily for 60 d, with the aim of stimulating dermal remodeling. Eleven of 14 patients showed clinical improvement in at least one facial area. The periorbital region showed a statistically significant improvement and there was a significant change in objective measurements by optical profilometry, indicating a decrease in the depth and number of textural irregularities or fine lines. Biopsies revealed new collagen formation in the Grenz zone (37% increase in thickness) and thickening of the epidermis by 27%. Eight of 14 patients felt their wrinkles were improved, while 12 of 14 felt their skin texture was improved (Fitzpatrick and Rostan 2003). Another randomized, vehicle-controlled, double-blind study of the same growth factor mixture in 60 patients with a mean age of 55 years and facial photodamage (mild to moderate in 48 subjects and severe in 12 subjects) showed improvement in preauricular fine rhytids, skin tone and texture, and hyperpigmentation. Treatment with the active gel for 3 months produced greater reduction in fine lines and wrinkles. Optical profilometry of silicone skin surface impressions showed improvements in skin roughness (Mehta *et al* 2008).

ReGenica (Histogen Aesthetics, San Diego, CA, USA) is a third product line containing growth factors and cytokines from conditioned medium of fibroblasts (MRCx). MRCx contains VEGF, keratinocyte growth factor, and IL-8, but not TGF-β. In a proof-of-concept study including 49 participants who had been treated by laser resurfacing, a lotion containing MRCx produced a greater reduction in erythema and better re-epithelialization of the perioral and periocular areas than the placebo lotion (Kellar *et al* 2009). In a following split-face study with 42 subjects undergoing combination ablative and nonablative laser procedures, investigators assessed the effect of MRCx, formulated as a gel, produced a dose-dependent improvement in crusting and significantly improved barrier function as measured by transepidermal water loss. In addition, biopsies showed that skin areas treated with the active gel had less leukocyte infiltration as well as more normal stratum germinativum and more normal organization of epidermal layers than the biopsies taken from skin areas treated with the vehicle control (Zimber *et al* 2012).

3.5.2 Adverse effects and safety considerations

All topically applied products carry the risk of irritant or allergic contact dermatitis. Because some malignant cells have receptors for certain growth factors and cytokines, and some growth factors and cytokines may increase cellular proliferation, there has been concern as to whether growth factors and cytokines might have the potential for tumorigenesis or promotion of cellular atypia (Zemtsov and Montalvo-Lugo 2008). Others postulate that exogenous growth factors and cytokines have a normalizing effect on the growth and differentiation of target cells. Studies of the effects of individual growth factors and cytokines on animal skin and on tumor cells have shown conflicting results. For example, some study showed that exogenously added VEGF was able to stimulate up to 40% increased proliferation in melanoma cells (Liu *et al* 1995). In contrast, another study found exogenous VEGF had no significant effects on melanoma cell proliferation or on production of a transcriptional target for VEGF (Graeven *et al* 1999). The validity of extrapolating these results to topical or injectable application of growth factors and cytokines to human tissue remains to be clarified.

To date, there have been no investigations with sufficient evidence to indicate that topical or injectable growth factors and cytokines have either a stimulatory or inhibitory role in human carcinogenesis. Despite these potential threats, the study of topical and injectable growth factors and cytokines remains hopeful and fascinating. Ongoing research will ultimately advance our general understanding of dermal signaling mechanisms. This could provide deeper and more global insights into the potential roles of growth factors and cytokines.

The results of *in vitro* and clinical studies suggest that cosmetic products containing growth factors, cytokines, matrikines, or matrikine-like peptides can enhance the production of collagen and other ECM molecules and promote skin rejuvenation. However, data so far are limited. These products need to be further evaluated in more randomized trials to be able to evaluate their clinical effects and mechanisms of action.

References

Abdellatif A A H, Alawadh S H, Bouazzaoui A, Alhowail A H and Mohammed H A 2021 Anthocyanins rich pomegranate cream as a topical formulation with anti-aging activity *J. Dermatolog. Treat.* **32** 983–90

Adhami V M, Afaq F and Ahmad N 2003 Suppression of ultraviolet B exposure-mediated activation of NF-kappaB in normal human keratinocytes by resveratrol *Neoplasia* **5** 74–82

Aldag C, Nogueira Teixeira D and Leventhal P S 2016 Skin rejuvenation using cosmetic products containing growth factors, cytokines, and matrikines: a review of the literature *Clin. Cosmet. Investig. Dermatol.* **9** 411–9

Armstrong R B *et al* 1992 Clinical panel assessment of photodamaged skin treated with isotretinoin using photographs *Arch. Dermatol.* **128** 352–6

Bagatin E, Gonçalves H S, Sato M, Almeida L M C and Miot H A 2018 Comparable efficacy of adapalene 0.3% gel and tretinoin 0.05% cream as treatment for cutaneous photoaging *Eur. J. Dermatol.* **28** 343–50

Baumann L *et al* 2005 Open-label pilot study of alitretinoin gel 0.1% in the treatment of photoaging *Cutis* **76** 69–73

Baur J A *et al* 2006 Resveratrol improves health and survival of mice on a high-calorie diet *Nature* **444** 337–42

Berardesca E, Bertona M, Altabas K, Altabas V and Emanuele E 2012 Reduced ultraviolet-induced DNA damage and apoptosis in human skin with topical application of a photolyase-containing DNA repair enzyme cream: clues to skin cancer prevention *Mol. Med. Rep.* **5** 570–4

Berardesca E, Gabba P, Farinelli N, Borroni G and Rabbiosi G 1990 *In vivo* tretinoin-induced changes in skin mechanical properties *Br. J. Dermatol.* **122** 525–9

Bissett D L, Chatterjee R and Hannon D P 1990 Photoprotective effect of superoxide-scavenging antioxidants against ultraviolet radiation-induced chronic skin damage in the hairless mouse *Photodermatol. Photoimmunol. Photomed.* **7** 56–62

Bonin-Debs A L, Boche I, Gille H and Brinkmann U 2004 Development of secreted proteins as biotherapeutic agents *Expert Opin. Biol. Ther.* **4** 551–8

Buonocore D, Lazzeretti A, Tocabens P, Nobile V, Cestone E, Santin G, Bottone M G and Marzatico F 2012 Resveratrol-procyanidin blend: nutraceutical and antiaging efficacy evaluated in a placebocontrolled, double-blind study *Clin. Cosmet. Investig. Dermatol.* **36** 159–65.69214

Burke K E, Clive J, Combs G F, Commisso J, Keen C L and Nakamura C L 2000 Effects of topical and oral vitamin E on pigmentation and skin cancer induced by ultraviolet irradiation in Skh:2 hairless mice *Nutr. Cancer* **38** 87–97

Cafardi J A and Elmets C A 2008 T4 endonuclease V: review and application to dermatology *Expert Opin. Biol. Ther.* **8** 829–38

Che D N, Xie G H, Cho B O, Shin J Y, Kang H J and Jang S 2017 Il. Protective effects of grape stem extract against UVB-induced damage in C57BL mice skin *J. Photochem. Photobiol.* B **173** 551–9

Cheng M, Wang H, Yoshida R and Murray M M 2010 Platelets and plasma proteins are both required to stimulate collagen gene expression by anterior cruciate ligament cells in three-dimensional culture *Tissue Eng.* A **16** 1479–89

Chiu A E, Chan J L, Kern D G, Kohler S, Rehmus W E and Kimball A B 2005 Double-blinded, placebo-controlled trial of green tea extracts in the clinical and histologic appearance of photoaging skin *Dermatol. Surg.* **31** 855–60

Cho S, Lee D H, Won C H, Kim S M, Lee S, Lee M J and Chung J H 2010 Differential effects of low-dose and high-dose beta-carotene supplementation on the signs of photoaging and type I procollagen gene expression in human skin in vivo. *Dermatology.* **221**(2) 160–71

Couteau C, Cheignon C, Paparis E and Coiffard L J 2012 Silymarin, a molecule of interest for topical photoprotection *Nat. Prod. Res.* **26**(23) 2211–4

Cui B, Wang Y, Jin J, Yang Z, Guo R, Li X, Yang L and Li Z 2022 Resveratrol treats UVB-induced photoaging by anti-MMP expression, through anti-inflammatory, antioxidant, and antiapoptotic properties, and treats photoaging by upregulating VEGF-B expression *Oxid. Med. Cell. Longev.* **2022** 6037303

Cunningham W J 1990 Topical isotretinoin and photodamage *Dermatologica* **181** 350–1

Daly T J and Weston W L 1986 Retinoid effects on fibroblast proliferation and collagen synthesis *in vitro* and on fibrotic disease *in vivo J. Am. Acad. Dermatol.* **15** 900–2

Darlenski R, Surber C and Fluhr J W 2010 Topical retinoids in the management of photodamaged skin: from theory to evidence-based practical approach *Br. J. Dermatol* **163** 1157–65

de Jager T L, Cockrell A E and Du Plessis S S 2017 Ultraviolet light induced generation of reactive oxygen species *Adv. Exp. Med. Biol.* **996** 15–23

De Buys Roessingh A S, Hohlfeld J, Scaletta C, Hirt-Burri N, Gerber S, Hohlfeld P, Gebbers J O and Applegate L A 2006 Development, characterization, and use of a fetal skin cell bank for tissue engineering in wound healing *Cell Transplant.* **15** 823–34

Doldo E, Costanza G, Agostinelli S, Tarquini C, Ferlosio A, Arcuri G, Passeri D, Scioli M G and Orlandi A 2015 Vitamin A, cancer treatment and prevention: the new role of cellular retinol binding proteins *BioMed Res. Int.* **2015** 624627

Eberlein-Konig B, Placzek M and Przybilla B 1998 Protective effect against sunburn of combined systemic ascorbic acid (vitamin C) and d-alpha-tocopherol (vitamin E) *J. Am. Acad. Dermatol* **38** 45–8

El-Chami C *et al* 2014 Role of organic osmolytes in water homoeostasis in skin *Exp. Dermatol.* **23** 534–7

El-Chami C *et al* 2020 Author correction: organic osmolytes preserve the function of the developing tight junction in ultraviolet B-irradiated rat epidermal keratinocytes *Sci. Rep.* **10** 8639

El-Chami C *et al* 2021 Organic osmolytes increase expression of specific tight junction proteins in skin and alter barrier function in keratinocytes *Br. J. Dermatol* **184** 482–94

Emanuele E, Altabas V, Altabas K and Berardesca E 2013 Topical application of preparations containing DNA repair enzymes prevents ultraviolet-induced telomere shortening and c-FOS proto-oncogene hyperexpression in human skin: an experimental pilot study *J. Drugs Dermatol.* **12** 1017–21

Emanuele E, Spencer J M and Braun M 2014 An experimental double-blind irradiation study of a novel topical product (TPF 50) compared to other topical products with DNA repair enzymes, antioxidants, and growth factors with sunscreens: implications for preventing skin aging and cancer *J. Drugs Dermatol.* **13** 309–14

Eming S A, Krieg T and Davidson J M 2007 Inflammation in wound repair: molecular and cellular mechanisms *J. Invest. Dermatol.* **127** 514–25

Fabi S and Sundaram H 2014 The potential of topical and injectable growth factors and cytokines for skin rejuvenation *Facial Plast. Surg.* **30** 157–71

Farrar M D, Nicolaou A, Clarke K A, Mason S, Massey K A, Dew T P, Watson R E B, Williamson G and Rhodes L E 2015 A randomized controlled trial of green tea catechins in protection against ultraviolet radiation-induced cutaneous inflammation *Am. J. Clin. Nutr.* **102** 608–15

Farris P, Yatskayer M, Chen N, Krol Y and Oresajo C 2014 Evaluation of efficacy and tolerance of a nighttime topical antioxidant containing resveratrol, baicalin, and vitamin E for treatment of mild to moderately photodamaged skin *J. Drugs Dermatol.* **13** 1467–72

Fisher G J, Kang S, Varani J, Bata-Csorgo Z, Wan Y, Datta S and Voorhees J J 2002 Mechanisms of photoaging and chronological skin aging *Arch. Dermatol.* **138** 1462–70

Fitzpatrick R E and Rostan E F 2002 Double-blind, half-face study comparing topical vitamin C and vehicle for rejuvenation of photodamage *Dermatol. Surg.* **28** 231–6

Fitzpatrick R E and Rostan E F 2003 Reversal of photodamage with topical growth factors: a pilot study *J. Cosmet. Laser Ther.* **5** 25–34

Foster A R *et al* 2020 Osmolyte transporter expression is reduced in photoaged human skin: implications for skin hydration in aging *Aging Cell* **19** e13058

Godic A, Poljšak B, Adamic M and Dahmane R 2014 The role of antioxidants in skin cancer prevention and treatment *Oxid. Med. Cell. Longev.* **2014** 860479

Gold M H, Goldman M P and Biron J 2007 Efficacy of novel skin cream containing mixture of human growth factors and cytokines for skin rejuvenation *J. Drugs Dermatol.* **6** 197–201

Gordon P R, Mawhinney T P and Gilchrest B A 1988 Inositol is a required nutrient for keratinocyte growth *J. Cell. Physiol.* **135** 416–24

Goukassian D, Gad F, Yaar M, Eller M S, Nehal U S and Gilchrest B A 2000 Mechanisms and implications of the age-associated decrease in DNA repair capacity *FASEB J.* **14** 1325–34

Graeven U, Fiedler W, Karpinski S, Ergün S, Kilic N, Rodeck U, Schmiegel W and Hossfeld D K 1999 Melanoma-associated expression of vascular endothelial growth factor and its receptors FLT-1 and KDR *J. Cancer Res. Clin. Oncol.* **125** 621–9

Grether-Beck S, Marini A, Jaenicke T and Krutmann J 2015 Effective photoprotection of human skin against infrared A radiation by topically applied antioxidants: results from a vehicle controlled, double-blind, randomized study *Photochem. Photobiol.* **91** 248–50

Grether-Beck S, Marini A, Jaenicke T, Stahl W and Krutmann J 2017 Molecular evidence that oral supplementation with lycopene or lutein protects human skin against ultraviolet radiation: results from a double-blinded, placebo-controlled, crossover study *Br. J. Dermatol.* **176** 1231–40

Griffiths C E, Russman A N, Majmudar G, Singer R S, Hamilton T A and Voorhees J J 1993 Restoration of collagen formation in photodamaged human skin by tretinoin (retinoic acid) *N. Engl. J. Med.* **329** 530–5

Halpner A D, Handelman G J, Harris J M, Belmont C A and Blumberg J B 1998 Protection by vitamin C of loss of vitamin E in cultured rat hepatocytes *Arch. Biochem. Biophys.* **359** 305–9

Han A, Chien A L and Kang S 2014 Photoaging *Dermatol. Clin.* **32** 291–9

Hawkins S, Adamus J, Chiang C Y, Covell E, O'Leary J and Lee J M 2017 Retinyl propionate and climbazole combination demonstrates clinical improvement to the appearance of hyperpigmentation and deep wrinkling with minimal irritation *Int. J. Cosmet. Sci.* **39** 589–99

He H, Li A, Li S, Tang J, Li L and Xiong L 2021 Natural components in sunscreens: topical formulations with sun protection factor (SPF) *Biomed. Pharmacother.* **134** 111161

Hseu Y C, Korivi M, Lin F Y, Li M L, Lin R W, Wu J J and Yang H L 2018 Trans-cinnamic acid attenuates UVA-induced photoaging through inhibition of AP-1 activation and induction of Nrf2-mediated antioxidant genes in human skin fibroblasts *J. Dermatol. Sci.* **90** 123–34

Huang C C, Hsu B Y, Wu N L, Tsui W H, Lin T J, Su C C and Hung C F 2010 Anti-photoaging effects of soy isoflavone extract (aglycone and acetylglucoside form) from soybean cake *Int. J. Mol. Sci.* **11** 4782–95

Hughes M C B, Williams G M, Baker P and Green A C 2013 Sunscreen and prevention of skin aging: a randomized trial *Ann. Intern. Med.* **158** 781–90

Humbert P G, Haftek M, Creidi P, Lapière C, Nusgens B, Richard A, Schmitt D, Rougier A and Zahouani H 2003 Topical ascorbic acid on photoaged skin. Clinical, topographical and ultrastructural evaluation: double-blind study vs. placebo *Exp. Dermatol.* **12** 237–44

Hussain M, Phelps R and Goldberg D J 2008 Clinical, histologic, and ultrastructural changes after use of human growth factor and cytokine skin cream for the treatment of skin rejuvenation *J. Cosmet. Laser Ther.* **10** 104–9

Hwang E, Park S Y, Lee H J, Lee T Y, Sun Z W and Yi T H 2014 Gallic acid regulates skin photoaging in UVB-exposed fibroblast and hairless mice *Phytother. Res.* **28** 1778–88

Ichihashi M, Funasaka Y, Ohashi A, Chacraborty A, Ahmed N U, Ueda M and Osawa T 1999 The inhibitory effect of DL-alpha-tocopheryl ferulate in lecithin on melanogenesis *Anticancer Res.* **19** 3769–74

Im A R *et al* 2016 Orally administered betaine reduces photodamage caused by UVB irradiation through the regulation of matrix metalloproteinase-9 activity in hairless mice *Mol. Med. Rep.* **13** 823–8

Jagdeo J, Kurtti A, Hernandez S, Akers N and Peterson S 2021 Novel vitamin C and E and green tea polyphenols combination serum improves photoaged facial skin *J. Drugs Dermatol.* **20** 996–1003

Janjua R, Munoz C, Gorell E, Rehmus W, Egbert B, Kern D and Chang A L S 2009 A two-year, double-blind, randomized placebo-controlled trial of oral green tea polyphenols on the long-term clinical and histologic appearance of photoaging skin *Dermatol. Surg.* **35** 1057–65

Jiang Y 2010 Effects of anthocyanins derived from Xinjiang black mulberry fruit on delaying aging *Wei Sheng Yan Jiu* **39** 451–3 (in Chinese)

Jones D P, Kagan V E, Aust S D, Reed D J and Omaye S T 1995 Impact of nutrients on cellular lipid peroxidation and antioxidant defense system *Fundam. Appl. Toxicol.* **26** 1–7

Judy E and Kishore N 2016 Biological wonders of osmolytes: the need to know more *Biochem. Anal. Biochem.* **5** 304

Jurzak M, Latocha M, Gojniczek K, Kapral M, Garncarczyk A and Pierzchała E 2008 Influence of retinoids on skin fibroblasts metabolism *in vitro Acta Pol. Pharm.* **65** 85–91

Kammeyer A and Luiten R M 2015 Oxidation events and skin aging *Ageing Res. Rev.* **21** 16–29

Kang S, Leyden J J, Lowe N J *et al* 2001 Tazarotene cream for the treatment offacial photodamage *Arch. Dermatol.* **137** 1597–604

Katiyar S K, Perez A and Mukhtar H 2000 Green tea polyphenol treatment to human skin prevents formation of ultraviolet light B-induced pyrimidine dimers in DNA *Clin. Cancer Res.* **6** 3864–9

Katz B E, Lewis J, McHugh L, Pellegrino A and Popescu L 2015 The tolerability and efficacy of a three-product anti-aging treatment regimen in subjects with moderate-to-severe photodamage *J. Clin. Aesthet. Dermatol.* **8** 21–6

Kellar R S, Hubka M, Rheins L A, Fisher G and Naughton G K 2009 Hypoxic conditioned culture medium from fibroblasts grown under embryonic-like conditions supports healing following post-laser resurfacing *J. Cosmet. Dermatol.* **8** 190–6

Khalil S, Bardawil T, Stephan C, Darwiche N, Abbas O, Kibbi A G, Nemer G and Kurban M 2017 Retinoids: a journey from the molecular structures and mechanisms of action to clinical uses in dermatology and adverse effects *J. Dermatolog. Treat.* **28** 684–96

Khlebnikov A I, Schepetkin I A, Domina N G, Kirpotina L N and Quinn M T 2007 Improved quantitative structure-activity relationship models to predict antioxidant activity of flavonoids in chemical, enzymatic, and cellular systems *Bioorg. Med. Chem.* **15** 1749–70

Kim J, Oh J, Averilla J N, Kim H J, Kim J S and Kim J S 2019 Grape peel extract and resveratrol inhibit wrinkle formation in mice model through activation of Nrf2/HO-1 signaling pathway *J. Food Sci.* **84** 1600–8

Kim M S, Lee S, Rho H S, Kim D H, Chang I S and Chung J H 2005 The effects of a novel synthetic retinoid, seletinoid G, on the expression of extracellular matrix proteins in aged human skin *in vivo Clin. Chim. Acta* **362** 161–9

Kobayashi M and Yamamoto M 2006 Nrf2-Keap1 regulation of cellular defense mechanisms against electrophiles and reactive oxygen species *Adv. Enzyme Regul.* **46** 113–40

Kostyuk V, Potapovich A, Albuhaydar A R, Mayer W, De Luca C and Korkina L 2018 Natural substances for prevention of skin photoaging: screening systems in the development of sunscreen and rejuvenation cosmetics *Rejuvenation Res.* **21** 91–101

Kurashige M, Okimasu E, Inoue M and Utsumi K 1990 Inhibition of oxidative injury of biological membranes by astaxanthin *Physiol. Chem. Phys. Med. NMR* **22** 27–38

Leccia M T, Lebbe C, Claudel J P, Narda M and Basset-Seguin N 2019 New vision in photoprotection and photorepair *Dermatol. Ther.* **9** 103–15

Lee D D, Stojadinovic O, Krzyzanowska A, Vouthounis C, Blumenberg M and Tomic-Canic M 2009 Retinoid-responsive transcriptional changes in epidermal keratinocytes *J. Cell. Physiol.* **220** 427–39

Lee K S *et al* 1998 All-trans-retinoic acid down-regulates elastin promoter activity elevated by ultraviolet B irradiation in cultured skin fibroblasts *J. Dermatol. Sci.* **17** 182

Lee T H, Kang J H, Seo J O, Baek S H, Moh S H, Chae J K, Park Y U, Ko Y T and Kim S Y 2016 Anti-melanogenic potentials of nanoparticles from calli of resveratrol-enriched rice against UVB-induced hyperpigmentation in guinea pig skin *Biomol. Ther.* **24** 85–93

Li J, Li Q and Geng S 2019 All-trans retinoic acid alters the expression of the tight junction proteins Claudin-1 and -4 and epidermal barrier function-associated genes in the epidermis *Int. J. Mol. Med.* **43** 1789–805

Li Z, Niu X, Xiao S and Ma H 2017 Retinoic acid ameliorates photoaged skin through RAR-mediated pathway in mice *Mol. Med. Rep.* **16** 6240–7

Lin J Y, Selim M A, Shea C R, Grichnik J M, Omar M M, Monteiro-Riviere N A and Pinnell S 2003 UV photoprotection by combination topical antioxidants vitamin C and vitamin E *J. Am. Acad. Dermatol.* **48** 866–74

Lintner K, Gerstein F and Solish N 2020 A serum containing vitamins C & E and a matrix-repair tripeptide reduces facial signs of aging as evidenced by Primos® analysis and frequently repeated auto-perception *J. Cosmet. Dermatol.* **19** 3262–9

Liu B, Earl H M, Baban D, Shoaibi M, Fabra A, Kerr D J and Seymour L W 1995 Melanoma cell lines express VEGF receptor KDR and respond to exogenously added VEGF *Biochem. Biophys. Res. Commun.* **217** 721–7

Liu W *et al* 2022 Urolithin A protects human dermal fibroblasts from UVA-induced photoaging through NRF2 activation and mitophagy *J. Photochem. Photobiol.* B **232** 112462

Luze H, Nischwitz S P, Zalaudek I, Müllegger R and Kamolz L P 2020 DNA repair enzymes in sunscreens and their impact on photoageing—a systematic review *Photodermatol. Photoimmunol. Photomed.* **36** 424–32

Ma Q 2013 Role of Nrf2 in oxidative stress and toxicity *Annu. Rev. Pharmacol. Toxicol.* **53** 401–26

Martin P 1997 Wound healing-aiming for perfect skin regeneration *Science* **276** 75–81

Masaki H 2010 Role of antioxidants in the skin: anti-aging effects *J. Dermatol. Sci.* **58** 85–90

Mehta R C and Fitzpatrick R E 2007 Endogenous growth factors as cosmeceuticals *Dermatol. Ther.* **20** 350–9

Mehta R C, Smith S R, Grove G L, Ford R O, Canfield W, Donofrio L M, Flynn T C and Leyden J J 2008 Reduction in facial photodamage by a topical growth factor product *J. Drugs Dermatol.* **7** 864–71

Minami Y, Kawabata K, Kubo Y, Arase S, Hirasaka K, Nikawa T, Bando N, Kawai Y and Terao J 2009 Peroxidized cholesterol-induced matrix metalloproteinase-9 activation and its suppression by dietary beta-carotene in photoaging of hairless mouse skin *J. Nutr. Biochem.* **20** 389–98

Mittal A, Elmets C A and Katiyar S K 2003 Dietary feeding of proanthocyanidins from grape seeds prevents photocarcinogenesis in SKH-1 hairless mice: relationship to decreased fat and lipid peroxidation *Carcinogenesis* **24** 1379–88

Moriwaki S and Takahashi Y 2008 Photoaging and DNA repair *J. Dermatol. Sci.* **50** 169–76

Mozos I, Stoian D, Caraba A, Malainer C, Horbanczuk J O and Atanasov A G 2018 Lycopene and vascular health *Front Pharmacol.* **9** 521

Mueed Z *et al* 2020 Cross-interplay between osmolytes and mTOR in Alzheimer's disease pathogenesis *Curr. Pharm. Des.* **26** 4699–711

Namkoong J, Kern D and Knaggs H E 2018 Assessment of human skin gene expression by different blends of plant extracts with implications to periorbital skin aging *Int. J. Mol. Sci.* **19** 3349

Navarrete-Dechent C and Molgó M 2017 The use of a sunscreen containing DNA-photolyase in the treatment of patients with field cancerization and multiple actinic keratoses: a case-series *Dermatol. Online J.* **23** 18

Netto MPharm G and Jose J 2018 Development, characterization, and evaluation of sunscreen cream containing solid lipid nanoparticles of silymarin *J Cosmet Dermatol.* **17**(6) 1073–83

Orlandi A, Bianchi L, Costanzo A, Campione E, Spagnoli L G and Chimenti S 2004 Evidence of increased apoptosis and reduced proliferation in basal cell carcinomas treated with tazarotene *J. Invest. Dermatol.* **122** 1037–41

Oyama A, Ueno T, Uchiyama S, Aihara T, Miyake A, Kondo S and Matsunaga K 2012 The effects of natural S-equol supplementation on skin aging in postmenopausal women: a pilot randomized placebo-controlled trial *Menopause* **19** 202–10

Parshad R, Sanford K K, Price F M, Steele V E, Tarone R E, Kelloff G J and Boone C W 1998 Protective action of plant polyphenols on radiation-induced chromatid breaks in cultured human cells *Anticancer Res.* **18** 3263–6

Petruk G, Giudice R, Del Rigano M M and Monti D M 2018 Antioxidants from plants protect against skin photoaging *Oxid. Med. Cell. Longev.* **2018** 1454936

Petruk G, Illiano A, Del Giudice R, Raiola A, Amoresano A, Rigano M M, Piccoli R and Monti D M 2017 Malvidin and cyanidin derivatives from açai fruit (*Euterpe oleracea* Mart.) counteract UV-A-induced oxidative stress in immortalized fibroblasts *J. Photochem. Photobiol.* B **172** 42–51

Polcz M E and Barbul A 2019 The role of vitamin A in wound healing *Nutr. Clin. Pract.* **34** 695–700

Przybylska S 2020 Lycopene—a bioactive carotenoid offering multiple health benefits: a review *Int. J. Food Sci. Technol.* **55** 11–32

Pullar J M, Carr A C and Vissers M C M 2017 The roles of vitamin C in skin health *Nutrients* **9** 866

Punzo A *et al* 2021 Grape pomace for topical application: green NaDES sustainable extraction, skin permeation studies, antioxidant and anti-inflammatory activities characterization in 3D human keratinocytes *Biomolecules* **11** 1181

Puviani M, Barcella A and Milani M 2013 Efficacy of a photolyase-based device in the treatment of cancerization field in patients with actinic keratosis and non-melanoma skin cancer *G. Ital. Dermatol. Venereol.* **148** 693–8

Quan T, He T, Kang S, Voorhees J J and Fisher G J 2004 Solar ultraviolet irradiation reduces collagen in photoaged human skin by blocking transforming growth factor-beta type II receptor/Smad signaling *Am. J. Pathol.* **165** 741–51

Rajnochová Svobodová A, Gabrielová E, Michaelides L, Kosina P, Ryšavá A, Ulrichová J, Zálešák B and Vostálová J 2018 UVA-photoprotective potential of silymarin and silybin *Arch. Dermatol. Res.* **310** 413–24

Rajnochová Svobodová A, Gabrielová E, Ulrichová J, Zálešák B, Biedermann D and Vostálová J 2019 A pilot study of the UVA-photoprotective potential of dehydrosilybin, isosilybin, silychristin, and silydianin on human dermal fibroblasts *Arch. Dermatol. Res.* **311** 477–90

Rhie G E, Mi H S, Jin Y S, Won W C, Kwang H C, Kyu H K, Kyung C P, Hee C E and Jin H C 2001 Aging-and photoaging-dependent changes of enzymic and nonenzymic antioxidants in the epidermis and dermis of human skin *in vivo J. Invest. Dermatol.* **117** 1212–7

Riahi R R, Bush A E and Cohen P R 2016 Topical retinoids: therapeutic mechanisms in the treatment of photodamaged skin *Am. J. Clin. Dermatol.* **17** 265–76

Rice-Evans C 1999 Implications of the mechanisms of action of tea polyphenols as antioxidants *in vitro* for chemoprevention in humans *Proc. Soc. Exp. Biol. Med.* **220** 262–6

Roh E, Kim J E, Kwon J Y, Park J S, Bode A M, Dong Z and Lee K W 2017 Molecular mechanisms of green tea polyphenols with protective effects against skin photoaging *Crit. Rev. Food Sci. Nutr.* **57** 1631–7

Samuel M, Brooke R C, Hollis S and Griffiths C E 2005 Interventions for photodamaged skin *Cochrane Database Syst. Rev.* CD001782 (withdrawn from publication)

Schwartz E, Cruickshank F A, Christensen C C, Perlish J S and Lebwohl M 1993 Collagen alterations in chronically sun-damaged human skin *Photochem. Photobiol.* **58** 841–4

Seki T, Sueki H, Kohno H, Suganuma K and Yamashita E 2001 Effects of astaxanthin from *Haematococcus pluvialis* on human skin-patch test; skin repeated application test; effect on wrinkle reduction *Fragr. J.* **12** 98–103

Sendagorta E, Lesiewicz J and Armstrong R B 1992 Topical isotretinoin for photodamaged skin *J. Am. Acad. Dermatol.* **27** S15–8

Sharif A, Akhtar N, Khan M S, Menaa A, Menaa B, Khan B A and Menaa F 2015 Formulation and evaluation on human skin of a water-in-oil emulsion containing Muscat hamburg black grape seed extract *Int. J. Cosmet. Sci.* **37** 253–8

Sorg O, Antille C, Kaya G and Saurat J H 2006 Retinoids in cosmeceuticals *Dermatol. Ther.* **19** 289–96

Sproul E P and Argraves W S 2013 A cytokine axis regulates elastin formation and degradation *Matrix Biol.* **32** 86–94

Stege H *et al* 2000 Enzyme plus light therapy to repair DNA damage in ultraviolet-B-irradiated human skin *Proc. Natl Acad. Sci. USA* **97** 1790–5

Sumita J M, Leonardi G R and Bagatin E 2017 Tretinoin peel: a critical view *An. Bras. Dermatol.* **92** 363–6

Sumita J M *et al* 2018 Tretinoin (0.05% cream vs. 5% peel) for photoaging and field cancerization of the forearms: randomized, evaluator-blinded, clinical trial *J. Eur. Acad. Dermatol. Venereol.* **32** 1819–26

Sundaram H, Mehta R C, Norine J A, Kircik L, Cook-Bolden F E, Atkin D H, Werschler P W and Fitzpatrick R E 2009 Topically applied physiologically balanced growth factors: a new paradigm of skin rejuvenation *J. Drugs Dermatol.* **8** 4–13

Svobodová A, Zdařilová A, Walterová D and Vostálová J 2007 Flavonolignans from *Silybum marianum* moderate UVA-induced oxidative damage to HaCaT keratinocytes *J. Dermatol. Sci.* **48** 213–24

Szymański Ł, Skopek R, Palusińska M, Schenk T, Stengel S, Lewicki S, Kraj L, Kamiński P and Zelent A 2020 Retinoic acid and its derivatives in skin *Cells* **9** 2660

Takahashi Y *et al* 2005 Decreased gene expression responsible for post-ultraviolet DNA repair synthesis in aging: a possible mechanism of age-related reduction in DNA repair capacity *J. Invest. Dermatol.* **124** 435–42

Todaro A, Palmeri R, Barbagallo R N, Pifferi P G and Spagna G 2008 Increase of trans-resveratrol in typical Sicilian wine using β-glucosidase from various sources *Food Chem.* **107** 1570–5

Tominaga K, Hongo N, Fujishita M, Takahashi Y and Adachi Y 2017 Protective effects of astaxanthin on skin deterioration *J. Clin. Biochem. Nutr.* **61** 33–9

Tominaga K, Hongo N, Karato M and Yamashita E 2012 Cosmetic benefits of astaxanthin on humans subjects *Acta Biochim. Pol.* **59** 43–7

Uitto J and Kouba D 2000 Cytokine modulation of extracellular matrix gene expression: relevance to fibrotic skin diseases *J. Dermatol. Sci.* **24** S60–9 Dec

Valko M, Rhodes C J, Moncol J, Izakovic M and Mazur M 2006 Free radicals, metals and antioxidants in oxidative stress-induced cancer *Chem. Biol. Interact.* **160** 1–40

Varani J, Warner R L, Gharaee-Kermani M, Phan S H, Kang S, Chung J H, Wang Z Q, Datta S C, Fisher G J and Voorhees J J 2000 Vitamin A antagonizes decreased cell growth and elevated collagen-degrading matrix metalloproteinases and stimulates collagen accumulation in naturally aged human skin *J. Invest. Dermatol.* **114** 480–6

Vechtomova Y L, Telegina T A, Buglak A A and Kritsky M S 2021 UV radiation in DNA damage and repair involving DNA-photolyases and cryptochromes *Biomedicines* **9** 1564

Vostálová J, Tinková E, Biedermann D, Kosina P, Ulrichová J and Svobodová A R 2019 Skin protective activity of silymarin and its flavonolignans *Molecules* **24** 1022

Walters R O *et al* 2018 Sarcosine is uniquely modulated by aging and dietary restriction in rodents and humans *Cell Rep.* **25** 663–676.e6

Wei H, Cai Q and Rahn R O 1996 Inhibition of UV light- and Fenton reaction-induced oxidative DNA damage by the soybean isoflavone genistein *Carcinogenesis* **17** 73–7

Wertz K, Seifert N, Hunziker P B, Riss G, Wyss A, Lankin C and Goralczyk R 2004 Beta-carotene inhibits UVA-induced matrix metalloprotease 1 and 10 expression in keratinocytes by a singlet oxygen-dependent mechanism *Free Radic. Biol. Med.* **37** 654–70

Wolf P *et al* 2000 Topical treatment with liposomes containing T4 endonuclease V protects human skin *in vivo* from ultraviolet-induced upregulation of interleukin-10 and tumor necrosis factor-alpha *J. Invest. Dermatol.* **114** 149–56

Wolf P 2019 Use of an SPF30 sunscreen and an after-sun-lotion in skin cancer risk patients *Study Record of Clinical Trial* NCT00555633 National Library of Medicine, US Government https://clinicaltrials.gov/ct2/show/NCT00555633

Wulff B C, Schick J S, Thomas-Ahner J M, Kusewitt D F, Yarosh D B and Oberyszyn T M 2008 Topical treatment with OGG1 enzyme affects UVB-induced skin carcinogenesis *Photochem. Photobiol.* **84** 317–21

Yaar M and Gilchrest B A 2007 Photoageing: mechanism, prevention and therapy *Br. J. Dermatol.* **157** 874–87

Yamada M *et al* 2006 Aged human skin removes UVB-induced pyrimidine dimers from the epidermis more slowly than younger adult skin *in vivo Arch. Dermatol. Res.* **297** 294–302

Yamamoto Y and Gaynor R B 2001 Therapeutic potential of inhibition of the NF-kappaB pathway in the treatment of inflammation and cancer *J. Clin. Invest.* **107** 135–42

Yamashita E 2015 Let astaxanthin be thy medicine *PharmaNutrition* **3** 115–22

Yarosh D B, Rosenthal A and Moy R 2019 Six critical questions for DNA repair enzymes in skincare products: a review in dialog *Clin. Cosmet. Investig. Dermatol.* **12** 617–24

Yeager D G and Lim H W 2019 What's new in photoprotection: a review of new concepts and controversies *Dermatol. Clin.* **37** 149–57

Yoon H S, Cho H H, Cho S, Lee S R, Shin M H and Chung J H 2014 Supplementating with dietary astaxanthin combined with collagen hydrolysate improves facial elasticity and decreases matrix metalloproteinase-1 and -12 expression: a comparative study with placebo *J. Med. Food* **17** 810–6

Zattra E *et al* 2009 Polypodium leucotomos extract decreases UV-induced Cox-2 expression and inflammation, enhances DNA repair, and decreases mutagenesis in hairless mice *Am. J. Pathol.* **175** 1952–61

Zemtsov A and Montalvo-Lugo V 2008 Topically applied growth factors change skin cytoplasmic creatine kinase activity and distribution and produce abnormal keratinocyte differentiation in murine skin *Skin Res. Technol.* **14** 370–5

Zhang Y *et al* 2021 Sorbitol accumulation decreases oocyte quality in aged mice by altering the intracellular redox balance *Aging* **13** 25291–303

Zhuang Y, Wu H, Wang X, He J, He S and Yin Y 2019 Resveratrol attenuates oxidative stress-induced intestinal barrier injury through PI3K/Akt-mediated Nrf2 signaling pathway *Oxid. Med. Cell. Longev.* **2019** 7591840

Zimber M P *et al* 2012 Human cell-conditioned media produced under embryonic-like conditions result in improved healing time after laser resurfacing *Aesthetic Plast. Surg.* **36** 431–7

Chapter 4

Botulinum toxin for skin rejuvenation

Yan Wu, Daniel Meng-Yen Hsieh and Shaomin Zhong

Currently, botulinum toxin treatment is a common option for addressing facial photoaging. It effectively reduces wrinkles and fine lines by relaxing facial muscles. However, outcomes vary based on individual factors such as skin condition and treatment adherence. Ongoing advancements should improve precision and longevity, enhancing patient satisfaction.

4.1 Introduction

Botulinum toxin (BoNT) is an extracellular toxin produced by *Clostridium botulinum.* It is a neurotoxin which can cause neuromuscular paralysis. BoNT weakens or paralyzes skeletal muscle by inhibiting neurotransmission between peripheral nerve endings and muscle fibers. It is known as one of the most toxic biological toxins and can lead to extremely high mortality rates. The earliest account of the clinical symptoms of food-borne botulism was published between 1817 and 1822 by Dr Justinus Kerner, a German physician, who witnessed several outbreaks of food poisoning after the consumption of smoked sausage (Erbguth and Naumann 1999). Kerner went on to conduct experiments, including clinical experiments on himself by consuming small amounts of the sour sausage, and documented symptoms of botulism including vomiting and intestinal spasms, mydriasis, ptosis and strabismus, dysphagia, flaccid paralysis, and respiratory failure (Whitcup 2021). He hypothesized that the toxin paralyzed skeletal muscles and parasympathetic function and recognized its potential therapeutic use for treating conditions of muscle hyperactivity and autonomic dysfunction (Erbguth and Naumann 1999). In around 1895, another outbreak of food poisoning connected with ham in Belgium finally led to the discovery and isolation of *Clostridium botulinum* and its toxin by E Van Ermengem, a professor of microbiology at the University of Ghent (Devriese 1999). Subsequently, scholars have conducted a vast amount of pioneering research into the separation, purification, mechanism of action, and exploration of clinical applications of botulinum toxin, pushing this single drug to unprecedented heights.

doi:10.1088/978-0-7503-5112-6ch4

Currently, applications of BoNT have expanded into various fields including ophthalmology, neurology, rehabilitation, surgery, plastic surgery and dermatology for a variety of indications.

After extensive studies and clinical testing of BoNT, it was approved for clinical use in the United States in 1989 for the treatment of strabismus, blepharospasm, and other facial nerve disorders, including hemifacial spasm (Jankovic 2004). Other indications including cervical dystonia, laryngeal dystonia, migraine, palmar hyperhidrosis, axillary hyperhidrosis and bladder dysfunction were subsequently approved over the years (Said *et al* 2003). In 1992, Dr Carruthers and Dr Carruthers reported an unexpected effect of BoNT when they found by chance that botulinum toxin A (BoNTA) improved glabellar frown lines while treating blepharospasm (Carruthers and Carruthers 1992). In 2002, Botox, a commercially available botulinum toxin type A was approved by the US Food and Drug Administration for the treatment of glabellar wrinkles (FDA 2011). The use of BoNT has since grew exponentially in the field of aesthetic medicine and has become the leading method among all wrinkle treatments. Global annual sales of BoNT soared to $4.5 billion in 2018, with half of that spent on cosmetic treatments. With a growing understanding of the molecular mechanisms of BoNT, more than 50 indications have been approved globally, and new areas such as antidepressants and anti-inflammatory indications are being explored. With a deeper understanding of the indications, contraindications, local anatomical structure as well as operation skills, BoNT injection is becoming an extremely safe method for improving cosmetic defects. The clinical effects of BoNT are transient and reversible, usually lasting 3–6 months, as new synaptic endplates form and muscle function gradually recovers (Berry and Stanek 2012). Thus, repeated treatment is required for maintaining optimal efficacy.

In order to enhance clinicians and practitioners' understanding of BoNT, its structure, the mechanism of action/function of botulinum toxin as well as the clinical applications, complications and limitations of BoNT injection will be discussed in detail here.

4.2 Structure and function

BoNT is a neurotoxin protein produced by gram-positive, spore-forming, anaerobic *Clostridium botulinum* (Eisele *et al* 2011). It exists in seven different serotypes named A, B, C, D, E, F, and G. All of them are produced by different strains of *Clostridium botulinum* with different toxicity and antigenicity. They may differ in singular amino acids. Currently, botulinum neurotoxin serotype A is commonly used for the treatment of many disorders and it has been the most widely studied serotype for therapeutic purposes. The only one approved by the FDA for aesthetic use is serotype A1 (Peck *et al* 2017). More recently, serotype B has become commercially available, whereas BoNT types E, C, D and F have only been used experimentally in humans.

BoNT consists of a neurotoxin combined with various nontoxic proteins. The neurotoxin is produced as precursor named protoxin which is a single polypeptide.

The protoxin is inactive and it can be transformed into the active form through cleaving by a trypsin-like bacterial protease. The active neurotoxin is made up of two chains with a molecular mass of about 150 kD (Oguma *et al* 1995). The neurotoxin has three functional domains, a light chain, a heavy chain and the disulfide bond which joins them together. The light chain is the catalytic domain with a molecular mass of approximately 50 kD. It is a zinc-dependent metalloproteinase with endopeptidase activity, which can cleave target proteins in neurons. It also has a conserved sequence HEXXH, which is an important part of metalloproteinases and acts as a catalytic site. The molecular mass of the heavy chain is approximately 100 kD. The heavy chain can be divided into N-terminal and C-terminal. The N-terminal is transmembrane region which is about 50 kD and the C-terminal is receptor binding domain which is about 50 kD.

The botulinum neurotoxins described above along with the nontoxic components are designated as progenitor toxins. They are large compounds with a molecular mass varying from 300 to 900 kD. Different strains of *Clostridium botulinum* produce different sized progenitor toxins. Serotype A strain produces 900, 500 and 300 kD toxins. Serotype B, C, D strains produce 500 and 300 kD toxins. Serotype E, F and G strains produce 300 and 500 kD toxins. The progenitor toxins of 900 and 500 kD have hemagglutinin (HA) activity (Oguma *et al* 1995), whereas the 300 kD toxin is formed by association of a neurotoxin with a nontoxic component having no HA activity (Oguma *et al* 1995).

BoNT drugs consist of botulinum neurotoxin, complexing proteins and excipients. Complexing proteins are not necessary ingredients for BoNT drugs to function. Until now, their role has not been well studied. They may attract leucocytes into the injection area so they may enhance the BoNT drug's antigenicity (Dressler 2020). In order to stabilize BoNT drugs, excipients are added artificially. Human serum albumin is the most commonly used ingredient, however, other ingredients such as bovine gelatin, polysorbate sugars including maltose, lactose, sucrose and dextran also have been used as excipients. Currently, the most commonly used type A botulinum toxins including onabotulinumtoxinA (onaA; Botox®, Allergan Inc., Irvine, CA, USA), abobotulinumtoxinA (aboA; Dysport®, Ipsen Ltd, Slough, Berkshire,UK), and incobotulinumtoxinA (incoA; Xeomin, Merz Pharmaceuticals GmbH, Frankfurt am Main, Hessen, Germany). Among these toxins, incoA is the only one that needs to be noted for the lack of complexing proteins.

Botulinum toxin acts by blocking the release of the neurotransmitter acetylcholine at neuromuscular junctions in humans, leading to a range of symptoms and signs characterized by flaccid paralysis. There are three steps involved in neurotoxicity. The first step is the irreversible binding of BoNT to presynaptic cholinergic receptors via the heavy chain's 50 KD carboxy terminal. The C-terminal recognizes the receptor on the nerve endings. The second step involves internalization of the neurotoxin through a receptor-mediated endocytosis. The third step is neuromuscular blockade (Huang *et al* 2000). It indicates that BoNT-A, B, E, F, G could affect the human nervous system while the other two serotypes do not. The light chain could cleave substrate proteins in neurons including Syntaxin 1, SNAP-25 and VAMP 1,2,3, which is essential for the release of neurotransmitter, so the nerve transmission will be blocked. In fact, different serotypes of botulinum toxin act on different substrate proteins. For example, serotypes B, D, F and G could degrade synaptic vesicles proteins VAMP 1, 2, 3, serotypes A, C and E could degrade peripheral membrane proteins SNAP-25, and serotype C could degrade serous membrane proteins Syntaxin 1. All these three substrate proteins belong to the SNARE family and they form central complexes that mediate the fusion of the vesicle membrane and serosal membrane which is critical for neurotransmitter release (Südhof and Rothman 2009).

In the late 1980s, BoNT drugs began to be used in medical practice. Over the past few decades, they have been used worldwide for a range of indications such as strabismus, blepharospasm, hemifacial spasm, cervical dystonia, spasticity, cerebral palsy, hyperhidrosis, bladder dysfunction and chronic migraine. Their application in the field of aesthetics started from the early 1990s because of the improvement in facial wrinkles that was observed when BoNT drugs were used. From then on, BoNT-A has been shown to be a safe and effective treatment for hyperfunctional glabellar lines, crow's feet lines, and forehead lines (Huang *et al* 2000).

In addition to neuronal cells, in recent years, many studies have shown the effect of BoNT on non-neuronal cells including epidermal keratinocytes, mesenchymal stem cells from subcutaneous adipose, neutrophils and macrophages, dermal fibroblasts, mast cells, sebocytes and vascular endothelial cells (Guida *et al* 2018). Thus, more and more clinicians and dermatologists have begun to pay attention to the innovative applications of BoNT in the field of dermatology. To date, there have been many reports indicating that the BoNT can be applied and has shown varying degrees of improvements for palmar and axillary hyperhidrosis, hypertrophic scars and keloids, the Raynaud phenomenon, facial flushing, oily skin, psoriasis, Hailey–Hailey disease, cutaneous leiomyomas and periorbital syringomas (Guida *et al* 2018).

4.3 Clinical application

4.3.1 Wrinkles

Most of the facial expression muscles in the face and neck originate from the bone or fascia and insert into the skin. The contraction of the expression muscles pulls the skin to produce wrinkles perpendicular to the direction of muscle movement, known as dynamic lines. BoNT type A (BoNTA) can inhibit muscle contraction by blocking the release of acetylcholine from synaptic vesicles at the presynaptic membrane in the neuromuscular junction, which produces chemical denervation and temporary muscle paralysis. Therefore, BoNTA injection can effectively reduce dynamic wrinkles by relaxing local muscles in a way that is reversible, and alleviate the formation of static wrinkle lines (Camargo *et al* 2021, Zarringam *et al* 2020).

The clinical effects of BoNTA can be seen within 1–4 d after injection, followed by 1–4 weeks of maximum effect, and will resolve after 3–4 months because of the gradual degradation by proteases and the reformation of new neuromuscular endplate structures. So the treatment cycle of facial rhytids is usually recommended every 4–6 months (Giordano *et al* 2017, Sundaram *et al* 2016a).

BoNTA treatment requires accurate injection into the target muscle. As significant anatomical diversity presents between individuals, the operator should be familiar with the mechanism of wrinkle formation and the anatomical location of muscles so as to choose proper injecting points and depth of injection. In the meantime, understanding the varieties of different BoNTA drugs is needed for making an appropriate decision of the dose and dilution ratio. The following section will introduce the injection techniques of BoNTA for different indications such as glabellar lines, horizontal forehead rhytids, lateral canthal rhytids (crow's feet), transverse nasal rhytids (bunny lines), perioral rhytids, oral commissures ('marionette lines'), cobblestone chin, nefertiti neck lift, and treatment of platysmal bands.

4.3.1.1 Glabellar lines

The formation of glabellar lines arises from the activity of a group of brow-associated muscles known as the glabella complex, which includes the corrugator supercilii, depressor supercilii, procerus, orbicularis oculi, and frontalis fibers. The patterns of frown lines vary with different muscle participation.

Prior to injection, the patient is instructed to frown for the evaluation of wrinkle pattern and the assessment of muscle mass, muscle shape and participating muscles. A patient-tailored treatment strategy will be decided accordingly. A classic injection method is to inject one site into the procerus, which is a superficial muscle and should not be injected too deeply, and 1–2 sites into each side of the corrugator supercilii. The medial portion of the corrugator supercilli muscle lies relatively deep and therefore should be injected more deeply, with a vertical needle insertion and injected intramuscularly. The lateral portion of the corrugator muscle inserts superficially into the middle of the eyebrow, and therefore should be injected more superficially with the needle inclined to a 45° angle and injected into the subcutaneous layer towards the outer upper direction. An injection dose of 2–4 U (10 s.U) per site is recommended. Dosing requirements may be adjusted according to muscle activity and muscle mass. Note that injection near the supraorbital foramen should be avoided and should be more than 1 cm superior to the upper bony orbital ridge (Nestor *et al* 2021).

The effective duration of glabellar lines treatment is commonly 4–6 months. If adjustment is required, supplementary injection can be performed about 4 weeks after the last injection.

During injection, dosage and injection site placement should be treated with caution to avoid affecting non-target muscles. The involvement of the frontalis may lead to a unilateral or bilateral excessive eyebrows raise (i.e. hanging eyebrows); while the involvement of extraocular muscles may lead to an abnormal eye movement or diplopia. Equalizing the bilateral dosage is also crucially important to prevent obvious asymmetry of bilateral glabellar lines (Sundaram *et al* 2016a, Kim and Lee 2016).

4.3.1.2 Horizontal forehead rhytids

The contraction of the frontalis and the senescence of skin can cause horizontal wrinkles on the forehead. The quantity, morphology, depth of wrinkles and muscle strength should be observed before injection, and personalized injection schemes should be designed accordingly. Since the lower 1/3 part of the frontalis controls eyebrow lifting, the injection point is recommended to be placed more than 2 cm above the eyebrows, which can be divided into 1–2 rows, generally 6–8 points each row. An additional row can be placed 1–1.5 cm above if necessary, with injection points regularly spaced apart. Oblique and subcutaneous injection is recommended. The suggested dosage is 0.5–2 U per injection point and 10–20 U in total. Higher doses may be needed for people with stronger muscle strength (such as males).

Before injection, it is important to focus on the position of the subject's eyelids and eyebrows. Excessively low eyebrows, ptosis or upper eyelid fat hypertrophy will increase the potential risk of heavy eyelids and difficulty in lifting the eyebrows after injection. Note that the injection points should cover as much range of the entire frontalis as possible. If the lateral frontalis is not relaxed, its compensatory contraction will cause a rise of the eyebrow tail. Injection sites and dose per site should be distributed as symmetrical as possible. Adjustment may be made around 4 weeks after last treatment.The frontalis is usually injected once per 4–6 months (Sundaram *et al* 2016a, Beer *et al* 2016).

4.3.1.3 Lateral canthal rhytids (crow's feet)

Lateral canthal rhytids (crow's feet) are mostly produced by the orbicularis oculi contraction. The severity of crow's feet, skin laxity and eye bags should be comprehensively evaluated before injection. Injections points are usually 1.5 cm from the lateral canthus and spaced about 1 cm apart. Intradermal injection is preferred and 2–4 U per injection point is recommended.

For patients with suborbital fine lines, low-dose intradermal botulinum toxin injection can be applied, preferably combined with photoelectric therapy. If the injection site is too close to the medial side or the dose is too large, it will cause lower eyelid relaxation. Dispersal of BoNTA to the extraocular muscle may affect eye movements and should be avoided. If the injection site is too close to the inner eyelid, the toxin will disperse to the dacryocyst, weakening the control of the lacrimal gland, which will result in epiphora (Sundaram *et al* 2016a, Humphrey *et al* 2017).

4.3.1.4 Transverse nasal rhytids (bunny lines)

Transverse nasal rhytids are symmetrical radial wrinkles at the nasion and the lateral nasal alar, as known as 'Bunny Lines'. The transverse nasal rhytids are formed anatomically due to the contraction of the nasal muscle. During injection, choose a site on both sides of the nasal muscle, the injection point should be close to the midline, to avoid affecting the levator labii superioris alaeque nasi (LLSAN) or levator labii superioris, which will cause upper lip asymmetry or upper lip sagging. 2–4 U per point is recommended. It is suggested to evaluate and treat bunny lines simultaneously during the treatment of glabellar lines or crow's feet because bunny lines may be aggravated due to the compensatory contraction of the nasal muscle after the injection of the corrugator supercilii or orbicularis oculi (Sundaram *et al* 2016a, Sundaram *et al* 2016b).

4.3.1.5 Perioral rhytids

Peroral rhytids refer to the radial wrinkles in the vertical direction of lip line, also known as 'smoker lines'. In addition to skin aging and sun exposure, it is also closely related to the repeated contraction movement of orbicularis oris. The injection points and dosage should be distributed symmetrically near the vermilion border and injected in the shallow subcutaneous level. Four symmetrical points are generally injected around the upper lip and two points on the lower one, at least 1 cm away from the mouth corners. The recommended total dosage is 0.8–1.2 U. The perioral skin in Asian people is relatively thicker than that of Euramerican people, and therefore perioral wrinkles are less common. For Asian people, the dosage can be appropriately reduced and BoNTA can also be further diluted for a micro-droplet multi-point intradermal injection technique.

Attention should be paid to not inject into the philtrum to avoid a flat ridge. Meanwhile, injection sites should not be too close to the mouth corners to avoid incomplete lip closure, salivation, dysarthria, smile asymmetry or other conditions (Sundaram *et al* 2016a, Kane and Monheit 2017).

4.3.1.6 Oral commissures ('marionette lines')

Oral commissures are puppet-like wrinkles that extend from the mouth corners to lower edge of the mandibular margin, which is formed by the contraction of the depressor anguli oris (DAO). BoNTA can be used to relax the DAO to partially improve the oral commissure, but for more serious scenarios, a combination therapy is often needed, such as hyaluronic acid filler injection or surgical treatment (Sundaram *et al* 2016a, Giordano *et al* 2017).

During injection, patients can be instructed to bite their teeth and make an 'e' sound to observe the position and shape of the commissure's lines. The injection point is usually 1 cm above the intersection of the extension line of the nasolabial fold and the mandibular margin, and 2 U per site is recommended. Note that the injection site should be away from the orbicularis oris, otherwise it will lead to limited mouth function and speech impairment. The injection point should be close to the outer lower side of the mouth to avoid the mouth corners and the depressor labii inferioris.

4.3.1.7 Cobblestone chin

Cobblestone Chin is an uneven appearance of the chin, presenting an orange peel-like surface, which is formed by the excessive tension of the mentalis. The injection points are on the chin protrusion 5 mm to the midline on both sides, and 2–4 U per site is recommended. The needle should be perpendicular to the skin and injected as deep as possible.

The injection points should be as low as possible to avoid the orbicularis oris. The depressor labii inferioris should be avoided as well, which can be achieved by choosing a closer point to the midline.Overdose can cause the lower lip to be unable to cling to the lower dentition, which will give rise to salivation (Sundaram *et al* 2016a).

4.3.1.8 'Nefertiti neck lift' and treatment of platysmal bands

With aging, sagging of the skin in the lower face and neck blurs the mandibular margin, widens the jaw-neck angle, and vertical muscle bands may even appear. These problems are related to the continuous contraction of the platysma muscle which pulls the facial and neck skin downward. When performing neck lifts with BoNTA, 1–3 rows along the mandibular border can be injected, with 3 points on each row and 2 U per point. For the injection of the platysmal band, a vertical series of 3–4 injection points spaced 2 cm apart (2 U per point) can be administered per band. Intramuscular injections on the band are recommended.

When performing injections along the mandibular border, the depressor labii inferioris and orbicularis oris should be avoided to prevent smile asymmetry, asophia or other problems. In the treatment of platysmal bands, the midline of the neck should be avoided, which is rich in pharyngeal muscles and nerves, as well as the sternocleidomastoid muscle which may cause unwanted effects such as neck rotation disorder (Sundaram *et al* 2016a).

4.3.2 Muscle reduction

4.3.2.1 Masseter reduction for masseter muscle prominence (MMP)

From traditional cultural point of view, a square face is not viewed as aesthetically pleasing in Chinese females and a slimmer lower face with an overall oval shape is considered more favorable and attractive. BoNTA injection has been widely accepted and favored by Asian patients seeking aesthetic treatments. Square faces may be attributed to masseter hyperplasia or MMP, localized fat accumulation, parotid gland hyperplasia, excess bone or excessive sagging of the skin. Injection with BoNTA is only suitable for MMP (Wu *et al* 2023).

The masseter muscle is the largest facial muscle responsible for mandibular chewing motions. The masseter muscle has superficial, middle and deep parts. The superficial part originates from the maxillary process of zygomatic bone, the middle one originates from the zygomatic arch, and the deep part originates from the lower margin of the zygomatic arch and the deep temporal fascia. The masseter muscle inserts onto the lateral surface of ramus and angle of mandible. There is a deep fascia between the superficial part and the deep part of the masseter muscles in the lower portion (Sermswan *et al* 2021).

Several injection techniques for lower face contouring with BoNTA have been used in clinical practice. The borders of the safety zone are defined superiorly by the earlobe to mouth corner line and inferiorly by the lower border of the mandible. The anterior and posterior borders are the edges of the masseter muscle. The targeted injection area is 1 cm inward of each border of the safety zone. In general, for 3–5 injection points, the central point is injected on the masseteric prominence followed by two additional pairs of injection points. Two deposits are given 1 cm above the lower mandibular border spaced 1–2 cm apart and two other points above the masseteric prominence within the targeted area. We should carefully avoid injecting too close to the anterior border of the masseter or too superiorly, which can lead to a sunken cheek (Park *et al* 2018, Cheng *et al* 2019).

Both sides of the masseter muscle are commonly injected with the same dose. However, for individuals with different lower facial volumes on the left and right sides of their face, dose adjustments may be required to achieve symmetry. The side of the face with a larger volume or masseter thickness is usually injected with a higher dose than the side with less volume/thickness. The recommended total dosage for both sides of the masseter varied between 20 and 90 U, higher doses may be more appropriate for patients with moderate to severe MMP, and the contouring effect could last longer. The onset of action in masseter contouring ranged from 2 to 4 weeks, and its effects can last up to 6–9 months.

The thickness of the muscle was reduced as determined by ultrasound measurements. Multiple factors, including individual differences, injection techniques, and types of BoNTA should be considered for determining the optimal dosage. Repeating the treatments 2 to 3 times yr^{-1} is required to maintain optimal results (Almukhtar and Fabi 2019).

4.4 Contraindications

The contraindications for using botulinum toxin in aesthetic treatment include: pregnant or breastfeeding women; active skin or systemic infections, such as herpes simplex, acne, erysipelas, ect; active autoimmune disease, hemostatic or coagulation disorders, under anticoagulation medication, have hypersensitivity to botulinum toxin or excipient in preparation, including botulinum toxin type A, human blood albumin, gelatin, lactose, sodium chloride or sodium succinate; keloid scarring; some progressing inflammatory skin disease such as psoriasis or eczema; presence of movement weakness in injection area or the muscles in the treatment area are already unable to contract voluntarily; suffering from neuromuscular junction disorders (myasthenia gravis), amyotrophic lateralizing sclerosis myopathies, myopathies (Kattimani *et al* 2019); taking medications that inhibit neuromuscular conduction or impact the effects of toxins, such as aminoglycosides, penicillamine, quinine and calcium channel blockers; patients with severe systemic disease; and patients with psychological symptoms or with unrealistic treatment goals, or with body dysmorphic disorders. Extra caution should be warranted in treating people who make a living that depends highly on facial expression, such as actors and singers.

4.5 Conclusion

Botulinum toxin has become one of the most widely used drugs in modern medicine over the past 20 years. The procedure of botulinum toxin injection has the advantages of being minimally invasive, economical, simple to operate and considerably safe with few side effects. In addition, BoNT treatments provide nearly immediate results with a short recovery time, making it one of the most popular treatment procedures in clinics. This chapter provides a detailed description of its structure, function, clinical indications, and standardized injection techniques for each indication, while addressing common and uncommon adverse reactions and management measures. This chapter also includes contents on basic anatomy and current research to assist doctors on further understanding the location for injection of botulinum toxin and its pharmacological mechanisms, which are crucial for mastering BoNT injection techniques.

At present, more and more botulinum toxin products have been introduced to the market. These products are different in units, chemical properties and biological activity, which lead to difference in clinical usage and safety profiles. In this chapter, we give a broad review of all the products on the market and summarized their characteristics and precautions.

Botulinum toxin has been introduced to new areas in recent years and has blooming applications in these different areas. The wide application of botulinum toxin in clinical usage needs more large scale, randomized, controlled, and long-term follow-up studies to prove its safety and efficacy. More consensus and guidelines are required to standardize the injection methods and dosage for each remedial field and indication. At the same time, relevant basic research is urgently needed to provide a scientific basis for clinical safety and quantitative use of botulinum toxin.

Due to limitations of space we have not included all the indications and new applications in the aesthetic field. In terms of combination therapy, botulinum toxin injection is relatively safe and compatible, it can be used in conjunction with other aesthetic methods such as dermal fillers, ablative laser resurfacing, micro focused ultrasound, or facial cosmetic surgeries to improve clinical outcomes.

Acknowledgments

The authors would like to acknowledge Menglong Ran, Xiangxi Wang, Qianqian Bai, Shu Gong, Rui Wang, Yunying Wu, and Jinyu Xia for their contribution to the composing of this chapter.

References

Almukhtar R M and Fabi S G 2019 The masseter muscle and its role in facial contouring, aging, and quality of life: a literature review *Plast. Reconstr. Surg.* **143** 39e–48e

Beer J I, Sieber D A, Scheuer J F and Greco T M 2016 Three-dimensional facial anatomy: structure and function as it relates to injectable neuromodulators and soft tissue fillers *Plast. Reconstr. Surg. Glob. Open.* **4** e1175

Berry M G and Stanek J J 2012 Botulinum neurotoxin A: a review *J. Plast. Reconstr. Aesthet. Surg.* **65** 1283–91

Camargo C P *et al* 2021 Botulinum toxin type A for facial wrinkles *Cochrane Database Syst. Rev.* **7** CD11301

Carruthers J D and Carruthers J A 1992 Treatment of glabellar frown lines with C. botulinum-A exotoxin *J. Dermatol. Surg. Oncol* **18** 17–21

Cheng J, Hsu S H and Mcgee J S 2019 Botulinum toxin injections for masseter reduction in East Asians *Dermatol. Surg.* **45** 566–72

Chinese Medical Doctor Association Dermatology Physician Branch Injective Aesthetics Subspecialty Committee 2017 Expert consensus on the application of botulinum toxin injection in dermatological aesthetics *Chin. J. Aesth. Med.* **26** 3–8

Devriese P P 1999 On the discovery of *Clostridium botulinum J. Hist. Neurosci.* **8** 43–50

Dressler D 2020 Therapeutically relevant features of botulinum toxin drugs *Toxicon* **175** 64–8

Eisele K, Fink K, Vey M and Taylor H V 2011 Studies on the dissociation of botulinum neurotoxin type A complexes *Toxicon* **57** 555–65

Erbguth F J and Naumann M 1999 Historical aspects of botulinum toxin: Justinus Kerner (1786–1862) and the 'sausage poison' *Neurology* **53** 1850

FDA 2011 BOTOX Cosmetic (onabotulinumtoxinA) for injection, for intramuscular use *Botox Cosmetic US PI* https://www.accessdata.fda.gov/drugsatfda_docs/label/2018/103000s5306lbl.pdf

Giordano C N, Matarasso S L and Ozog D M 2017 Injectable and topical neurotoxins in dermatology: indications, adverse events, and controversies *J. Am. Acad. Dermatol.* **76** 1013–24

Giordano C N, Matarasso S L and Ozog D M 2017 Injectable and topical neurotoxins in dermatology: indications, adverse events, and controversies *J. Am. Acad. Dermatol.* **76** 1027–42

Guida S *et al* 2018 New trends in botulinum toxin use in dermatology *Dermatol. Pract. Concept.* **8** 277–82

Huang W, Foster J A and Rogachefsky A S 2000 Pharmacology of botulinum toxin *J. Am. Acad. Dermatol.* **43** 249–59

Humphrey S, Jacky B and Gallagher C J 2017 Preventive, cumulative effects of botulinum toxin type A in facial aesthetics *Dermatol. Surg.* **43** S244–51

Jankovic J 2004 Botulinum toxin in clinical practice *J. Neurol. Neurosurg. Psychiatry* **75** 951–7

Kane M A C and Monheit G 2017 The practical use of AbobotulinumtoxinA in aesthetics *Aesthet. Surg. J.* **37** S12–9

Kattimani V, Tiwari R C, Gufran K, Wasan B, Shilpa P H and Khader A 2019 Botulinum toxin application in facial esthetics and recent treatment indications (2013–2018) *J. Int. Soc. Prev. Community Dent.* **9** 99

Kim H J S K and Lee H K 2016 *Clinical Anatomy of the Face for Filler and Botulinum Toxin Injection* (Berlin: Springer)

Nestor M S, Han H, Gade A, Fischer D, Saban Y and Polselli R 2021 Botulinum toxin–induced blepharoptosis: anatomy, etiology, prevention, and therapeutic options *J. Cosmet. Dermatol.* **20** 3133–46

Oguma K, Fujinaga Y and Inoue K 1995 Structure and function of *Clostridium botulinum* toxins *Microbiol. Immunol.* **39** 161–8

Park G, Choi Y, Bae J and Kim S 2018 Does botulinum toxin injection into masseter muscles affect subcutaneous thickness? *Aesthet. Surg. J.* **38** 192–8

Peck M *et al* 2017 Historical perspectives and guidelines for botulinum neurotoxin subtype nomenclature *Toxins* **9** 38

Said S, Meshkinpour A, Carruthers A and Carruthers J 2003 Botulinum toxin A: its expanding role in dermatology and esthetics *Am. J. Clin. Dermatol.* **4** 609–16

Sermswan P, Tansatit T, Meevassana J and Panchaprateep R 2021 A cadaveric study of dye spreading: determining the ideal injection pattern for masseter hypertrophy *Dermatol. Surg.* **47** 1354–8

Südhof T C and Rothman J E 2009 Membrane fusion: grappling with SNARE and SM proteins *Science* **323** 474–7

Sundaram H *et al* 2016a Aesthetic applications of botulinum toxin A in Asians: an international, multidisciplinary, pan-Asian consensus *Plast. Reconstr. Surg. Glob. Open* **4** e872

Sundaram H *et al* 2016b Global aesthetics consensus: botulinum toxin type A—evidence-based review, emerging concepts, and consensus recommendations for aesthetic use, including updates on complications *Plast. Reconstr. Surg.* **137** 518e–29e

Whitcup S M 2021 *The History of Botulinum Toxins in Medicine: A Thousand Year Journey* (Cham: Springer International Publishing) pp 3–10
Wu Y, Zeng D and Wu S 2023 Botulinum Toxin Type A for the Treatment of Masseter Muscle Prominence in Asian Populations *Aesthet. Surg. J. Open Forum* **5** ojad005
Zarringam D, Decates T, Slijper H P and Velthuis P 2020 Increased usage of botulinum toxin and hyaluronic acid fillers in young adults *J. Eur. Acad. Dermatol. Venereol* **34** e602–4

Chapter 5

Intradermal fillers for skin rejuvenation

Zhuanli Bai

Dermal fillers are injectable substances used to smooth wrinkles and augment tissue volume, serving as a surgical adjunct in cosmetic procedures. They are classified into absorbable and non-absorbable types, with various options such as hyaluronic acid, collagen, and calcium hydroxylapatite. Autologous options include fat, fibroblasts, and platelet-rich plasma. Surgeons must consider patient factors and cosmetic goals for optimal outcomes. Detailed preoperative communication and assessment are crucial for safety and satisfaction.

5.1 Introduction

With the improvement of people's economic level, their recognition and pursuit of beauty have also reached a higher level. They increasingly hope to achieve a natural-looking aesthetic through safer and more convenient means. Under the joint action of various factors such as genetics, age, hormone levels, UV exposure, smoking, and environmental pollution, various physiological changes occur in the skin and subcutaneous collagen content, fat atrophy or displacement, and facial bone atrophy or depression. These changes are mainly manifested clinically as skin laxity, sagging, increased wrinkles, and uneven facial contours (Kimball *et al* 2018, Mokos *et al* 2018), all of which are the 'aging' appearance that people want to reverse. In recent years, injectable filler products have become one of the minimally invasive treatment options for facial volume tissue loss, contour changes, and static wrinkles due to their minimally invasive and easy-to-operate characteristics. Injectable filler products are divided into temporary fillers and permanent fillers according to the absorbability of the material, and into inert fillers and collagen-stimulating fillers according to the degree of interaction with local tissues (Katsambas *et al* 2015). They can also be classified as autologous tissue injections and non-autologous tissue injections according to their different sources. Currently, autologous tissue fillers such as fat grafts and platelet-rich plasma are mainly used, while non-autologous tissue injectables are more widely used in clinical practice, including hyaluronic acid

doi:10.1088/978-0-7503-5112-6ch5

(HA), polycaprolactone (PCL), poly-L-lactic acid (PLLA), and others. When using injectable fillers, we need to fully consider their safety and efficacy.

5.1.1 Safety

For injectable fillers, safety is the first aspect that needs to be considered. The safety of medical products is a relative concept. A safe injectable filler product refers to the product that, based on conclusive scientific evidence, brings potential benefits to medical practitioners that outweigh potential risks for confirmed intended uses and conditions of use. Furthermore, due to its special properties as a cosmetic medical product, the population using it is different from the common patient population. The benefits for them rise from simple relief of illness to changes in appearance and psychological satisfaction. In order to maintain an acceptable risk/benefit ratio, the requirements for risk control for products are higher than for medical devices used for disease and injury diagnosis and treatment. In addition to pre-clinical product performance, biocompatibility and biosafety evaluations, the evaluation of its safety mainly relies on scientific evidence related to safety in clinical evaluations, including clinical trials.

As the number of injectable filler cases increases, the reported adverse reactions also significantly increase. The more common local adverse reactions include pain, swelling, bruising, hematoma, nodules, itching, and so on. In rare cases, complications such as infection, allergy, nodules, granulomas, skin pigmentation, and other complications may occur. Most of these adverse reactions gradually disappear within two weeks, and some inflammatory reactions occur by complex mechanisms, which may be mixed with type I immune reactions, type IV immune reactions, innate immune reactions, biofilm, or even specific infections, and may recur. For delayed adverse reactions, Bjarnsholt *et al* recommend the use of antibiotics for treatment (Bjarnsholt *et al* 2009). The most worrying adverse reaction is the acute complication caused by the mis-injection of fillers into blood vessels. Although rare, its consequences are very serious, and it may cause local skin necrosis, and even fainting, blindness, stroke or death. Because of the unpredictability of the position of some individual blood vessels, this risk cannot be completely avoided. The occurrence of adverse reactions is related to the chemical properties of the filler product itself, such as material type, degradation properties, impurities, and also to its physical properties such as shape, surface condition, and mechanical stress caused to tissues after implantation (Snozzi and van Loghem 2018). In reported cases, treatment with fillers in non-regular medical institutions with unknown qualifications of the doctor is a major factor in adverse reactions after filling, as well as a lack of understanding of the properties of filling materials, anatomical characteristics of the filling site, injection techniques and even aseptic concepts. Therefore, choosing a regular medical institution and an experienced doctor, avoiding injection for patients with immune or infectious diseases, choosing the appropriate filling material according to the filling site, and strictly performing aseptic operations can to some extent avoid adverse reactions.

5.1.2 Effectiveness

Injectable filling products are typically classified as medical devices, and their primary working principle for achieving effectiveness is physical filling. These products are generally in the form of gels or particle suspensions, and their supportive performance in the body depends on factors such as the mechanical properties of the materials (such as compressive strength and elasticity), rheological properties, and degradation characteristics. Additionally, the biological effects that go beyond the physical filling action and are induced by the mechanical stress generated by filling products in the body are also a working principle for their effectiveness. The effectiveness of injectable filling products is based on their filling effect, and therefore the gradual degradation of the material will limit the duration of its effectiveness. However, this also means that filling products made from absorbable materials have the advantage of being able to correct and adjust the filling effect. Therefore, in clinical use, in order to achieve satisfactory filling for the patient, the choice of product based on its physicochemical properties, the doctor's injection level, technique, injection frequency, and single-dose injection, as well as the individual characteristics of the patient, all need to be fully considered based on the filling site. The effectiveness of filling products is a comprehensive concept, as the filling effect does not necessarily represent the final cosmetic effect or psychological benefits for the patient. However, comprehensive effectiveness is subjective and uncertain, and cannot be used as a criterion for evaluation. Therefore, when evaluating the effectiveness of injectable filling products, more objective and quantitative filling effect indicators, such as the wrinkle severity rating scale (WSRS) (Day *et al* 2004), are often chosen as the main efficacy indicators, while global cosmetic effect and satisfaction are considered secondary efficacy indicators.

There are many factors that can affect the safety and effectiveness of filling products. For products of the same material but different brands, their safety and effectiveness may be completely different. Even for products from the same company, if their performance parameters (such as particle size range) or application scope are different, their safety and effectiveness cannot be generalized and require definitive clinical and non-clinical scientific evidence to support them.

After having introducing the concept of filling products, the following sections will now provide a summary and outlook on the classification and selection of filling materials.

5.2 Hyaluronic acid

Hyaluronic acid (HA) has good biocompatibility and extremely low immunogenicity, so it is mostly used as a raw material in currently commercialized filler products. There are many types of HA fillers, and their *in vivo* degradation time is usually around 1 year. From a safety perspective, HA fillers have significant advantages over other filling materials because in the early stage of mistaken injection into blood vessels, some mild-to-moderate subjects can dissolve them in time by injecting hyaluronidase to slow down the occurrence and development of complications. The core component of HA fillers is hyaluronic acid. In order to prolong its *in vivo* degradation time, hyaluronic acid is usually chemically crosslinked. Hyaluronic acid is a naturally occurring linear macromolecular viscous polysaccharide. The molecule is composed of disaccharide structural units connected by β-1,4-glycosidic bonds, and each structural unit is connected by a β-1,3-glycosidic bond to one D-glucuronic acid and one N-acetyl-D-glucosamine molecule. The relative molecular weight of each disaccharide structural unit of HA is about 400, and the number of structural units can reach more than 25 000 pairs, with an average relative molecular weight of 1×10^5–1×10^7. It is a biopolymer and mainly exists in the extracellular matrix, vitreous humor, and cartilage. The total amount of HA found in a 70 kg person is about 15 g, with an average replacement rate of 5 grams per day. About 50% of the total amount of HA in the human body is concentrated in the skin, with a half-life of 24–48 h. Natural HA dissolved in water is a viscous liquid, so it is necessary to use a crosslinking agent to cross-link many hyaluronic acids that are originally single-chain saccharides (this type of hyaluronic acid can be naturally absorbed and metabolized in the body within a few days) to form a three-dimensional honeycomb-like structure connected to each other, which must take a lot of time to metabolize in the human body. The crosslinked HA is in the form of a gel and is insoluble in water, which constitutes the effective HA products we discuss about. 1,4-butanediol diglycidyl ether (BDDE) is the crosslinking agent used for most HA dermal fillers on the market today. Its crosslinking ability is attributed to the reactivity of the epoxy groups at both ends of the molecule. Under basic ($pH > 7$) conditions, these epoxy groups preferentially react with the primary alcohols that are most easily accessible in the HA backbone to form ether bonds. The superior stability of ether bonds (compared to ester or amide bonds) is one of the reasons why BDDE crosslinked HA fillers can last for or exceed 1 year clinically. In addition, the toxicity of BDDE is significantly lower than that of other ether-bond crosslinking chemical preparations (such as diethylene glycol dimethyl ether), it is biodegradable, and has been well researched. All of these factors have made BDDE the industry standard crosslinking agent. However, even though BDDE is currently the safest crosslinking agent, as an allogeneic component, it still carries the potential risk of adverse reactions. Therefore, the degree of crosslinking of HA products, that is, the concentration of added BDDE, needs to be precisely controlled. The higher the degree of crosslinking of the product, the better the shaping effect and the longer the maintenance time.

All HA fillers have their own unique physicochemical characteristics, and this differentiation is the basis for selecting appropriate HA fillers according to different locations and aesthetic needs in clinical practice. For example, the HA performance required to correct fine lines by placing the filler in the shallow layer of the skin is different from that required to supplement volume in the deep layer (Pierre Liew and Bernardin 2015). This is related to the different forces acting on the filler after injection into different facial regions. Currently, it is believed that after HA injection into the face, it is mainly subjected to the combined effect of shear or torsional stress parallel to the surface (such as muscle movement) and compressive/tensile stress in the vertical direction (such as sleep, fat weight, skin tension) (Heitmiller *et al* 2021). Horizontal shear/torsional stress changes the shape of HA and preserves volume, while vertical compressive/tensile stress changes the volume and preserves shape. The performance of HA against these two forces is mainly related to HA hardness (viscoelasticity) and cohesion; factors affecting the final filling effect and duration of HA also include its support force, tissue integration, water absorption, etc (Heitmiller *et al* 2021). As mentioned above, the biological effects of HA products go beyond volumizing effects.

5.2.1 Degree of crosslinking

Tezel and Fredrickson proposed the concept of degree of crosslinking (Tezel and Fredrickson 2008), which is the percentage of disaccharide units in HA that are crosslinked with crosslinking agents. If the degree of crosslinking of an HA filler is 4%, it means that on average, every 100 disaccharide units of HA contain 4 crosslinking agent molecules. Kablik *et al* introduced the concept of degree of modification (Kablik *et al* 2009), which is the ratio of the total number of crosslinking agent molecules connected to HA to the number of disaccharide units in HA. Because crosslinking agents can bind to HA molecules at both ends or only at one end and remain free (un-crosslinked) at the other end during crosslinking, the degree of modification is the sum of the degree of crosslinking and the degree of un-crosslinked modification. Edsman *et al* further proposed the concept of crosslinking ratio (Edsman *et al* 2012), which is the ratio of the number of crosslinking agent molecules that form crosslinked structures with HA to the total number of cross-linking agent molecules.

The degree of crosslinking is closely related to the properties of the filler. When other factors are the same, a higher degree of crosslinking results in a harder HA gel, which means that it is less likely to deform under the tension of the skin and can better achieve a shaping effect, leading to longer-lasting effectiveness. If the degree of un-crosslinked modification is high, the degree of crosslinking of the filler will be lower for the same degree of modification, and the resulting HA gel will also have a lower hardness. HA fillers often consist of two parts, crosslinked and un-crosslinked. The un-crosslinked HA mainly acts as a lubricating filler during injection, enhancing its flowability and making it easy to be discharged from the syringe, while the crosslinked part mainly maintains the filling effect of the filler. When other parameters are kept constant, a higher degree of crosslinking leads to a longer duration of action for the HA filler. However, this does not mean that the higher the degree of crosslinking, the more suitable the filler is for injection, because excessively high degree of crosslinking can reduce the hydrophilicity of HA, thereby weakening its ability to lift and fill. Edsman *et al* pointed out that when the degree of crosslinking is too high, the biocompatibility of HA will also be reduced due to the crosslinking agent as an alien component (Edsman *et al* 2012). When injected into the human body, it will be recognized as a foreign substance and cause an immune response. The human body cannot degrade this type of filler biologically, and instead, it is encapsulated and can form sterile nodules or even lumps. Therefore, when producing HA fillers, it is necessary to strictly control the degree of crosslinking to be within the clinically required range, which can provide sufficient duration of action and ideal filling effect while minimizing the occurrence of adverse reactions due to reduced biocompatibility.

5.2.2 Complex modulus, elastic modulus, and viscous modulus

Most HA fillers are colloids with viscoelastic properties, and their rheological properties can be represented by the complex modulus (G^*). G^* represents the ability of the HA colloid as a whole to resist deformation and is the sum of the elastic modulus and the viscous modulus. The elastic modulus (G') is a measure of the hardness of the HA colloid and is also known as the storage modulus because it can be used to describe the energy stored during the force application to the filler. G' can also be used to describe the interaction between external forces and the colloid's own elasticity. The viscous modulus (G') is a measure of the colloid's viscosity and is also known as the loss modulus. It can be used to describe the energy lost during the colloid's viscous dissipation process. Tanδ is the ratio of G' to G' and indicates whether the filler is more elastic ($\tan\delta < 1$) or more viscous ($\tan\delta > 1$) (Pierre *et al* 2015).

Tezel and Fredrickson proposed that the concentration of HA filler, the degree of crosslinking, the amount of non-crosslinked HA, and the crosslinking technique used can all affect G' and thus the hardness of the colloid (Tezel and Fredrickson 2008). In low crosslinked HA colloids, the HA molecular chains are loosely connected, requiring less force for deformation, resulting in a lower G' and lower hardness. When the HA concentration is kept constant, increasing the crosslinking degree leads to a tighter connection between HA molecules and increased hardness. When the mass concentration of HA is greater than 20 mg ml^{-1}, colloid with a higher G' has lower viscosity, and colloid with a lower G' has relatively higher viscosity (Edsman *et al* 2015). When the mass concentration is less than 20 mg ml^{-1}, the viscosity of HA colloids with different G' is lower (Edsman and Ohrlund 2018).

Borrell *et al* pointed out that the viscous modulus can also be used to measure the cohesion of the colloid, which reflects the ability of the colloid to maintain its integrity under external forces (Borrell *et al* 2011). It is generally believed that high cohesion helps the filler resist the perpendicular stress applied to the soft tissue. Cohesion balances the vertical projection and relative plasticity to ensure the versatility of the filler. Low cohesion helps the plasticity and extensibility of the filler (Heitmiller *et al* 2021). Bentkover believed that the elastic modulus can also affect the initial injection force of the filler, that is, the higher the G', the greater the force required for initial injection (Bentkover 2009). Before exceeding the yield point, the injection force and filler displacement show a linear relationship, while after exceeding the yield point, the injection force required is mainly determined by G'.

Kablik *et al* pointed out that in the production of HA fillers, crosslinked HA needs to be cut into colloidal particles for easy injection through a fine syringe needle, but the colloid with high G' is difficult to inject even if its particle volume is small (Kablik *et al* 2009). Therefore, some manufacturers use non-crosslinked HA as a lubricant in the production of HA fillers to reduce injection resistance. The higher the G' of the HA colloid, the less deformation it undergoes under pressure, and the more energy it stores. Therefore, Sundaram *et al* believed that HA fillers with a higher G' coefficient can provide better facial support and are suitable for lifting and filling in the mid and lower face (Sundaram *et al* 2010). However, fillers with high hardness are more likely to cause local tissue damage after injection, resulting in a higher incidence of inflammation, edema, and pain. Gutowski believed that although low hardness HA fillers are not as good as high hardness fillers in providing facial support and resistance to deformation, they provide more natural and smooth results after injection and are suitable for filling low dynamic wrinkles such as tear troughs and soft tissues such as periorbital and perilabial areas (Gutowski 2016).

5.2.3 Concentration

The filler is composed of two parts: insoluble HA gel and soluble HA liquid. The total amount of HA in the filler is referred to as the total concentration, commonly measured in $mg\,ml^{-1}$ (Beasley Weiss and Weiss 2009). Concentration is an important indicator that affects the duration of filler effect. Soluble HA mainly refers to free HA, which is HA that has not undergone chemical modification or has not been fully crosslinked during the modification process. It is often a by-product of the chemical modification process and serves as a soluble liquid component that can assist in filler injection and reduce injection resistance. This portion of HA is easily degraded and is usually completely metabolized within a few days after injection; therefore, it has no significant impact on the duration of filler effect and filling results. Insoluble HA refers to crosslinked HA gel, which can resist enzymatic degradation and free radical degradation to extend the duration of effect, so this part of HA plays a major role in fillers. Falcone and Berg conducted a 6 month follow-up on clinical users and compared the severity of wrinkles before and after injection using the Wrinkle Severity Grading Scale (WSRS) (Falcone and Berg 2008). The results showed that increasing the concentration of HA can effectively extend the duration of filler effect. Bentkover also believed that fillers with high concentrations of HA usually do not reach a state of hydration equilibrium before injection, meaning that they still have strong binding ability to water after injection, resulting in a strong swelling effect (Bentkover 2009). The concentration of crosslinked HA determines the duration of filler effect. However, Falcone *et al* suggested that the HA mass concentration of fillers should not exceed 25 $mg\,ml^{-1}$, because fillers with concentrations higher than this have excessive viscosity and cannot be injected with a 27 or 30 G injection needle (Falcone and Berg 2007).

5.2.4 Particle size

In order to minimize local bleeding and inflammatory reactions during the injection of HA fillers, finer needles such as 27 or 30 G are often used (Kablik *et al* 2009), which requires that the particle size of the HA gel must be small enough to pass through these finer needles. HA gels made using crosslinking technology are usually in the form of a block and need to be sieved or homogenized before they can be used for injection. Compared to homogenization, sieving makes the HA gel have uniform particle sizes and smoother and easier to inject characteristics (Borrell *et al* 2011). Beasley *et al* proposed that particle gels provide stronger support and longer duration by relying on particle size, with larger particle sizes suitable for deep wrinkles and smaller particle sizes suitable for filling light to moderate wrinkles, lips, eyes, and other areas with thin skin (Beasley *et al* 2009). However, a randomized double-blind clinical trial was conducted on fillers with different particle sizes, and the results showed that fillers with larger particle sizes did not effectively prolong their duration (Carruthers *et al* 2005). Falcone and Berg confirmed through experiments that particle size has no significant impact on the viscoelasticity of HA fillers (Falcone and Berg 2008).

5.2.5 Biological effects

In recent years, it has been found that HA fillers have functions in stimulating the synthesis of collagen, elastin, and aquaporin-3 (AQP3), which positively affect skin elasticity and hydration in addition to their physical volumizing effect. This biological effect of HA goes beyond the volumizing effect and is often reported by patients to result in longer improvement of skin quality than the expected average half-life of the filler (La Gatta *et al* 2019, Paliwal *et al* 2014). Although the mechanism is not yet fully understood, the most convincing hypothesis is that HA fillers stimulate fibroblasts to synthesize collagen through mechanical tension, affecting skin metabolism. Furthermore, according to existing reports, the metabolic response of the skin to mechanical stress decreases with age (Turlier *et al* 2013).

5.2.6 Facial fillers in different areas of the face

Under the effects of natural aging and photoaging, the face undergoes sagging and aging due to loss of volume and uneven distribution of fat. Moreover, the occurrence and development of these changes have their own characteristics in different areas of the face. Some experts believe that the early onset of facial aging is the loss of volume around the eyes, followed by the deepening of tear troughs, nasolabial folds, and volume loss in the forehead and temple areas at around the age of 40. After the age of 50, the soft tissues of the face begin to show significant sagging and drooping, manifested by further deepening of the tear troughs and nasolabial folds, drooping of the outer corners of the eyes, and downward and backward displacement of the skin and soft tissues around the mouth. HA fillers can be used to reduce facial wrinkles and sagging, replenish lost volume, reshape facial contours, and improve skin quality. However, the different stress characteristics, aging processes, and aesthetic demands in different areas of the face require personalized characteristics of HA fillers. The facial aesthetics zoning in clinical practice is rather complicated, and a simple way of dividing it is into the upper, middle, and lower parts, separated by the line from the zygomatic arch to the upper edge of the ear, the line from the mouth corner to the earlobe.

5.2.6.1 Upper facial filling

Compared to the middle and lower facial regions, the application of HA in the upper facial area is limited. Due to the dynamic nature of wrinkles and volume loss in the forehead and glabellar regions, more often than not, clinicians will choose to treat with neurotoxin agents such as botulinum toxin (Ogilvie *et al* 2020b). When combining HA with other cosmetic treatments such as neurotoxin, the complexity of the anatomical structures in this area should be taken into consideration. The upper facial area is a relatively risky filling area, and serious acute adverse reactions such as skin necrosis or blindness caused by vascular occlusion are more likely to occur in this area.

The brow ridge is a bony structure that protrudes in an arch shape and is located at the junction of the upper face and upper eyelid area, above the superior orbital rim. It plays an important role in providing support and contour to the upper face, and the morphology of the brow ridge is often used as an important aesthetic standard to evaluate the upper face. In East Asian populations, many people exhibit facial features such as low brow ridge, underdeveloped orbital bones, prominent eyeballs, and sagging upper eyelid skin (Liew *et al* 2020, Wu *et al* 2020). In pursuit of a three-dimensional aesthetic, many patients choose to use fillers to create a high and arched brow ridge to achieve an overall improvement in perceived beauty. In the past, injection and filling methods have led to adverse reactions such as filler displacement, swelling (Kim *et al* 2014a), and poor brow ridge morphology due to inadequate preoperative evaluation, unfamiliarity with the anatomical structure of the brow ridge area, and improper selection of injection materials. With advances in existing research, filling the brow ridge involves improving the bony contour, and the brow area involves complex facial expressions in the upper face. Therefore, considering the physicochemical properties of the filler, fillers with a higher elastic modulus (G') are preferred to provide better support and resist the shear force generated by gravity, pressure from overlying soft tissues, and facial muscle activity. Fillers with higher cohesive force are also preferred to resist filler diffusion and displacement that may occur after injection, reducing the chances of poor injection area morphology caused by multiple factors (Jones and Murphy 2013).

5.2.6.2 Mid-facial filling

The mid-face is one of the earliest areas to show signs of facial aging, and it is an important area for the application of hyaluronic acid (HA) fillers. A thorough understanding of the local anatomy by the operator is essential for ensuring the clinical effectiveness of mid-facial filling and reducing the incidence of complications. The mid-face includes the nose, cheeks, and the lower eyelid area around the orbit. Apart from the area around the lower eyelid, the anatomy of the mid-face can be divided into five layers: the skin, subcutaneous adipose tissue, the superficial musculoaponeurotic system (SMAS), deep adipose tissue, and deep fascia. The inferior orbital hollow is divided into two parts based on the distribution of the facial artery (medial palpebral artery) and vein 4–6 mm inside the pupil center: the outer and inner parts (Cotofana *et al* 2015). The outer part of the inferior orbital hollow contains seven layers of structures, including the skin, subcutaneous fat, orbicularis oculi muscle, infraorbital fat, deep fascia, subperiosteal fat layer, and periosteum. The inner part of the inferior orbital hollow (medial to the facial vein) consists of two layers, the skin, and the orbicularis oculi muscle, which is firmly attached to the subperiosteal area of the inferior orbital region (De Maio *et al* 2017). Therefore, injections into the inner part of the inferior orbital hollow are typically submuscular injections. The distribution of blood vessels in the cheek area is relatively rich and connected to several major blood vessels, such as the medial palpebral artery, facial artery, and infraorbital artery and its branches (Kim *et al* 2014b). The medial palpebral artery connects with the ophthalmic artery, and injecting into the medial palpebral artery may cause retinal artery occlusion and even blindness (Park *et al* 2012). The nasal area also has a relatively rich distribution of blood vessels, and its blood supply comes from the facial artery. The lateral nasal artery and dorsal nasal artery are its important branches. The dorsal nasal artery anastomoses with the angular artery, and injecting in this area has a high risk of severe complications, such as blindness (De Maio *et al* 2017). The blood vessels in the subperiosteum or cartilage of the nasal dorsum are less abundant and more suitable for injection (Saban *et al* 2012). The HA filler selected for this area not only needs to withstand shear and compression forces caused by the gravity and tension of the overlying soft tissues but also needs to withstand dynamic traction caused by lip and cheek movements and external forces affecting the face. To maintain the shape and projection of the filler, the selected HA needs to have a medium to high G' to withstand shear forces and a medium to high cohesive force to resist compression, which is critical for reducing filler diffusion and displacement caused by repeated muscle and soft tissue compression in the mid-face. In addition, for the special area of the mid-face, the nose, it mainly needs to resist the compression force produced by skin tension and muscle tension on prominent bone structures and is not subject to strong shear forces. Therefore, by selecting HA fillers with high cohesive force and high G', their advantages of small lateral diffusion, long vertical projection time while maintaining plasticity can be used to maintain good nose augmentation for 12 months after injection (Heitmiller *et al* 2021). Additionally, by using HA fillers with relatively lower G', although their structural enhancement for volume is relatively weak, they can provide a softer and less uneven appearance while increasing volume. This is suitable for more subtle volume enhancement, such as in the perioral area. Existing studies also advocate for combining different HA products in layers, using HA fillers with relatively larger G' for deep injections to enhance volume, and using HA fillers with relatively smaller G' for superficial injections to provide a softer, more natural appearance, in order to achieve a more effective and aesthetically pleasing enhancement effect.

5.2.6.3 Lower face filling

Lower face filling has two characteristics. First, this area has a rich range of facial expressions, and the texture and wrinkles are relatively complex, including marionette lines, nasolabial folds, wrinkles, and submental creases, which represent various signs of aging (Goodman and Remington 2015). Second, the jawline is an important component of facial contour, and with age and changes in hormone levels, it often forms signs of an inadequate jawline contour, such as posterior displacement, defects or imbalances (Ogilvie *et al* 2020a). For the filling and repair of various wrinkles and textures in the lower face, a highly pliable HA filler that can integrate well with the lower face's movement and achieve a natural and dynamic effect without being palpable should be selected. Considering that the lower face is mainly subject to shear force rather than compressive force, it is believed that products with medium G' and low to moderate cohesion are more suitable for filling and repairing wrinkles in this area. For some patients with a short and wide face, according to their individual aesthetics, filling the chin can be selected to stretch the proportion of the lower face and achieve a more balanced distribution of facial features. To achieve this sculpting effect, a highly cohesive and high G' HA filler is needed.

5.2.6.4 Fine line filling

As facial skin becomes loose, the entire facial area may be involved in fine line filling. Unlike fillers used for volume enhancement, fine line fillers aim to treat shallow skin indentations and improve skin quality, so their focus is on the pliability and extensibility of the filler. HA fillers with low cohesive forces and low to moderate G^* and G' will be the ideal products for fine line filling (Fagien *et al* 2019, Ogilvie *et al* 2020b).

5.3 Collagen

The dermal matrix in human skin is mainly composed of 80%–85% type I collagen and 10%–15% type III collagen, as well as small amounts of glycosaminoglycans and elastic fibers. Collagen is synthesized and secreted by dermal fibroblasts, and as age increases, collagen secretion decreases (Fenske and Lober 1986). Additionally, external factors such as UV radiation, smoking, and environmental pollution can increase levels of collagenase in the body (Bauman 2004), leading to a decrease in collagen volume and resulting in aging symptoms such as skin atrophy, reduced thickness, and wrinkle formation (Lavker 1979). Therefore, as a natural protein and the main component of skin tissue, collagen is an ideal material for correcting soft tissue defects in the face. As a facial soft tissue filler, collagen has the following advantages. (i) Good biocompatibility: Implanted collagen in tissue is non-toxic and non-irritating to the body. (ii) Biological effects: It can interact with cells. As the main component of the extracellular matrix, collagen provides attachment and scaffold support for cell growth. After collagen is injected into a correction site, it can induce the proliferation, differentiation, and transplantation of epithelial cells and other cells, promoting the synthesis of new collagen and the production of new tissue similar to the host, and working together with normal surrounding skin to achieve a corrective effect and aesthetic result. (iii) Good mechanical properties: The right-handed helix formed by the three α-chains of collagen molecules and the crosslinking of intra- and inter-molecular groups all give collagen good strength and elasticity. It can fully exert the volumizing effect required for filling, achieving a good shaping effect. (iv) Moisturizing: Collagen molecules contain a large number of hydrophilic groups such as carboxyl and hydroxyl, which makes it very easy for collagen molecules to form hydrogen bonds with water, and thus collagen has good water retention and moisturizing properties. (v) Low immunogenicity: Collagen is a type of protein macromolecule with low immunogenicity, which has a repeating structure of glycine (Gly)-*X*–Y (proline often occupies the *X* position and hydroxyproline occupies the *Y* position). The main antigenic determinant cluster exists in the non-helical region at the ends of the collagen molecule. In the process of collagen extraction, the immunogenic terminal peptides can be selectively cleaved with pepsin to further reduce the already low immunogenicity. Based on this characteristic, collagen fillers that meet clinical standards on the market currently include bovine collagen, human collagen, and porcine collagen. The biggest disadvantage of bovine collagen is that it has a very short retention time and potential hypersensitivity reactions, with about 3% of individuals experiencing such reactions (Baumann 2006, Zeide 1986). Compared with the hypersensitivity reaction problem of bovine collagen, human collagen has better biocompatibility. Although human collagen avoids the disadvantage of requiring hypersensitivity reaction testing for bovine collagen, it still has the problem of short retention time and difficult material sourcing. Porcine collagen has many advantages, such as a widely available source and good biocompatibility, and existing clinical trials have shown that porcine collagen has better performance in retention time than human collagen and bovine collagen (Lorenc Nir and Azachi 2010). At the same time, the complications of porcine collagen are very mild, limited to injection reactions such as swelling, redness, bleeding, and papules (Narins *et al* 2008). (vi) Biodegradability: Collagen peptide bonds can be broken down by the hydrolytic action of collagenase, and the helical structure of collagen is subsequently destroyed. The broken collagen peptides can be hydrolyzed by most proteases. Although collagen protein as a biological material has many advantages as a filler product, there are still certain risks from production to clinical application. First, the animal-derived raw materials used to prepare collagen products have the potential risk of carrying infectious diseases. Second, although collagen has lower immunogenicity compared to other filling materials, it may still cause immune rejection reactions in the human body due to its origin from complex biological tissues, cell membrane epitopes, DNA in allogeneic or xenogeneic materials, and certain proteins. In order to meet clinical needs, some additives (such as crosslinking agents that control the degradation time and increase strength, currently mostly using glutaraldehyde, and lidocaine drugs that can have a local anesthetic effect) may be introduced in the preparation process of fillers. Although these exogenous substances are present in trace amounts in the product, they may cause adverse reactions if used in large quantities, used improperly, or in the presence of allergies.

5.4 Calcium hydroxylapatite (CaHA)

CaHA is the main inorganic component of human bones and teeth, with a chemical formula of $Ca_{10}(PO_4)_6(OH)_2$. Due to its good biocompatibility, it has been widely used as a degradable biomaterial in the field of orthopedics and maxillofacial surgery for nearly 20 years. In 2006, the US FDA approved CaHA for the treatment of facial wrinkles such as nasolabial folds and facial lipoatrophy in patients with human immunodeficiency virus (Graivier *et al* 2007). Its mechanism of action is similar to other dermal stimulants, where the hydrogel carrier is gradually absorbed after 3–6 months, and the CaHA microspheres form a scaffold where fibroblasts produce collagen fibers on the scaffold, maintaining the effect for 1–2 years, after which CaHA is finally degraded into calcium ions and phosphate absorbed by the body (Coleman *et al* 2008). CaHA is mainly used for filling wrinkles on the face and body, and there is expert consensus on dilution levels and injection techniques depending on the filling site, serving as a guideline (de Almeida *et al* 2019, Van Loghem 2018). Moreover, considering the safety of CaHA as a filling material, its complications are very mild and mainly caused by injection factors.

5.5 Other fillers

Currently, non-autologous injectables are the main materials used for soft tissue fillers in the face. In addition to those mentioned above, there are also polycaprolactone (PCL) and poly-L-lactic acid (PLLA). In addition, there are some fillers derived from autologous tissue, such as autologous fat, autologous fibroblasts, and platelet-rich plasma.

5.5.1 Polycaprolactone (PCL)

In 2009, a new type of bio-stimulating soft tissue filler based on polycaprolactone (PCL)-Ellansé® (a product of Sinclair Pharmaceuticals) appeared on the European cosmetic market, and has since been promoted and used in more than 60 countries (Christen and Vercesi 2020, De Melo *et al* 2017, Galadari *et al* 2015, Moers-Carpi *et al* 2021). Ellansé® is composed of PCL microspheres and carboxymethylcellulose (CMC) gel. It is divided into four types, and the duration of action of each product depends on the initial polymer (polymer material) chain length and the time it takes for the product to be completely absorbed (Kim *et al* 2020). PCL undergoes biodegradation and absorption through ester bond hydrolysis, and the end products of metabolism are carbon dioxide (CO_2) and water (H_2O), both of which can be completely eliminated from the body. The CMC gel carrier is gradually engulfed and absorbed by cells within 6–8 weeks. According to existing animal and clinical experiments, PCL microspheres have the ability to stimulate new collagen synthesis, and type I collagen gradually exceeds type III collagen, so it can achieve better and faster results than other absorbable long-acting products (Chouzouri and Xanthos 2007, Kim and Van Abel 2015). The product also has the advantages of high tissue compatibility, high elasticity modulus, and high viscosity.

5.5.2 Poly-L-lactic acid (PLLA)

PLLA is an artificially synthesized polymer from the α-hydroxy acid family. As early as 1999, PLLA was used in the European market under the brand name New-Fill. In 2004, it was approved by the FDA under the brand name Sculptra for the treatment of facial fat loss in HIV-infected patients. In 2009, Sculptra was once again approved by the FDA for cosmetic purposes on the face (Ezzat and Keller 2011). PLLA has good biocompatibility and biodegradability. When injected subcutaneously or in the deep dermis, it stimulates fibroblast proliferation and collagen production, achieving a cosmetic effect and gradually being degraded by the body (Engelhard and Mest 2005, Gogolewski *et al* 1993). Unlike other fillers in gel form, Sculptra is stored in sterile lyophilized powder form in ampoules. Also, unlike other skin fillers, the therapeutic effect of PLLA is not immediate and often requires multiple injections to achieve the desired therapeutic effect. Once newly synthesized collagen is deposited in the skin, it can last for a long time, with most therapeutic effects lasting up to 2 years, and some studies reporting up to 3 years (Salles *et al* 2008). Therefore, Sculptra is not suitable for local contour filling, but is more suitable for facial rejuvenation to increase skin thickness and elasticity.

5.5.3 Autologous fat

Autologous fat refers to the excess subcutaneous fat cells taken from certain parts of the body and then processed and purified into composite fat particles by injection of medication. The intact particle fat cells are then transplanted back into the parts of the body where fat filling is needed. Although the history of autologous fat filling dates back nearly a century, its further application has been limited by its high absorption rate and complications. It was not until the 1980s, with the improvement of fat suction technology, that fat transplantation gradually became widely used in the treatment of various soft tissue defects (Illouz 1986). 'Survival rate' has always been the core issue faced by fat transplantation, and current research focuses mainly on three areas: fat acquisition, fat injection, and fat processing, in order to improve survival rate and reduce complications (Khater *et al* 2009).

In terms of fat acquisition, existing studies have shown that the use of low negative pressure suction and fine-caliber suction tubes can effectively reduce damage to fat cells and improve cell viability (Gonzalez *et al* 2007). In terms of fat injection, the use of multi-level, multi-tunnel, and multi-point injection methods can increase the chances of contact between fat cells and surrounding tissue, and reduce the problem of fat particle necrosis due to inadequate blood supply (Locke and de Chalain 2008). According to studies comparing the survival rates of fat injected into three different injection sites, subcutaneous, muscle surface, and muscle below, the results showed that the survival rate of fat transplanted on the muscle surface was the highest, so the selection of different injection sites is important for improving the effectiveness of fat transplantation. In terms of fat processing, it is necessary to purify the fat particles, and washing and settling methods and centrifugation methods are currently the two commonly used purification methods. The washing and settling method causes the least damage to cells, but is less effective at removing impurities than the centrifugation method; the centrifugation method can remove impurities well, but may damage cells (Kurita *et al* 2008, Toledo and Mauad 2006). At the same time, many researchers are enthusiastic about adding biologically active substances to fat particles to promote the vascularization of transplanted fat, thereby increasing fat survival rates, such as basic fibroblast growth factor (bFGF), vascular endothelial growth factor (VEGF), angiopoietin-1 (Ang-1), platelet-derived growth factor (PDGF), insulin-like growth factor (IGF), etc (Pallua *et al* 2009). In 2001, Zuk *et al* isolated a type of fibroblast-like cell group from adipose tissue, called adipose-derived stem cells (ASCs), and discovered their multidirectional differentiation potential, including differentiation into adipocytes, osteocytes, chondrocytes, endothelial cells, skeletal muscle cells, cardiomyocytes, and neural cells (Zuk *et al* 2001). Subsequently, some scholars proposed the concept of cell-assisted lipotransfer (CAL), which involves using half of the obtained fat particles to isolate stromal vascular fraction (SVF) containing adipose stem cells and adding it to the other half of the fat particles to increase the content of stem cells in the transplanted fat (Yoshimura *et al* 2020, Yoshimura Suga and Eto 2009). Experimental results have shown that CAL technique has an average survival rate of transplanted fat that is 35% higher than that of non-CAL technique, and significant microvascular generation can be observed in the transplanted fat using CAL technique (Matsumoto *et al* 2006). Meanwhile, a study of 147 human immunodeficiency virus-induced lipoatrophy patients who were treated with PLLA, CaHA, PMMA, and autologous fat injections showed that although there was no statistically significant difference between the number and amount of treatments of autologous fat filling and synthetic fillers, the cost of autologous fat filling was significantly lower than that of all synthetic fillers, suggesting that autologous fat cells may become the most cost-effective filling material (Vallejo *et al* 2018).

5.5.4 Autologous fibroblasts

Human skin fibroblasts have the function of secreting collagen. In 1995, researchers first used autologous fibroblasts to fill facial small indentations (Boss *et al* 2000). In 2007, Robert *et al* conducted a phase III clinical trial on the use of autologous fibroblasts for facial contour defects, and the results showed that autologous fibroblasts can safely and effectively improve facial wrinkles and acne scars, and have long-term efficacy, with a duration of at least 12 months (Weiss *et al* 2007). However, to achieve the ideal therapeutic effect through autologous fibroblast filling, multiple injections are often required and there is no immediate effect.

In order to address this limitation, Yoon *et al* first combined human skin fibroblasts with hyaluronic acid gel (Restylane) for soft tissue filling research (Yoon *et al* 2003). The results showed that fibroblasts can survive and secrete collagen, and can also enhance the therapeutic effect of hyaluronic acid. Subsequent studies by other researchers have also shown the potential significance of adding fibroblasts for skin filling, which helps to extend the filling effect of the material (Hoben *et al* 2011).

However, there are also many shortcomings in autologous fibroblast transplantation. There is no fixed process for the industrialization of autologous fibroblasts: the uncertainty of *in vitro* cultivation and the lack of clear guidance on cultivation time, as well as the need for multiple injections and the inability to achieve immediate therapeutic effects. In addition, further research is needed on issues such as cell concentration, interval between transplants, and post-transplant cell fate.

5.5.5 Platelet-rich plasma (PRP)

Platelet-rich plasma (PRP) is plasma rich in platelets, which are separated from whole blood by centrifugation. It contains a variety of human growth factors, including platelet-derived growth factor (PDGF), transforming growth factor (TGF), vascular endothelial growth factor (VEGF), insulin-like growth factor (IGF), among others. These factors can influence cellular migration, adhesion, proliferation, and differentiation through specific cell-surface receptors. PRP has been widely used in various clinical treatments, including wound healing and oral and maxillofacial surgery. Recently, PRP has gained significant attention in the field of skin beauty. After platelets in PRP are activated, they can promote the proliferation of skin fibroblasts and the secretion of type I collagen by releasing abundant growth factors, thereby achieving the goal of skin beauty (Kim *et al* 2011).

There are currently various methods for preparing PRP, and there is no unified standard, which leads to differences in the proportions of components in the final product. According to the concentration of leukocytes and fibrin, PRP can be classified into four types: pure PRP, leukocyte-rich PRP, pure platelet-rich fibrin (P-PRF), and leukocyte-rich PRF (Frautschi *et al* 2017). However, when people refer to PRP, they are actually referring to a suspension of liquid platelets containing leukocytes. Currently, leukocyte-rich PRP has been widely used in wound repair, skin beauty, and other fields, and it is often used in combination with adipocytes to improve the success rate of transplantation. However, a series of issues related to the growth factor content measurement, injection concentration, and working mechanism of current PRP products are still unclear, and their effectiveness and safety need to be further verified.

5.6 Conclusion

With the development of social and economic progress and the increasing pursuit of beauty by people, there is a growing expectation for the emergence of more safe, efficient, convenient, minimally invasive, and low-cost soft tissue fillers. However, each existing filler has its own advantages and disadvantages. Although synthetic materials have advantages such as stable physicochemical properties, good plasticity, and long-term retention, their development is severely constrained by issues of biocompatibility and safety. The improvement of preparation technology for bio-sourced filler materials has further increased their biocompatibility, but their short retention time is their biggest drawback. Therefore, how to prolong the degradation time of biofillers in the body is an urgent problem to be solved. In view of the advantages and disadvantages of these two types of fillers, the combination of biological and synthetic materials to complement each other's shortcomings and prepare fillers with more perfect indicators is a very worthwhile direction to explore. At the same time, in order to further improve the safety and effectiveness of existing fillers, it is necessary for production companies, regulatory authorities, medical personnel, and patients to exchange information and achieve a consensus on safety and effectiveness. Production companies need to control the entire life cycle of their products, including pre-market design-validation-confirmation, full non-clinical and clinical evaluation, and good operation of the production quality system. Regulatory authorities need to regulate the standardized sales, storage, transportation, and medical personnel training of products after they are marketed and provide support. Medical personnel need to have a deep understanding of the characteristics of the different products used and master the anatomical basis of the filling site. They should strictly follow the expert guidance for filling and collect and analyze adverse events, and receive training on corrective and preventive measures. Patients need to understand their own physical condition, choose legitimate and compliant medical institutions, fully and completely explain their own situation to medical personnel, accept the advice and guidance of medical personnel, and reasonably and safely choose filling projects.

References

Bauman L 2004 CosmoDerm/CosmoPlast (human bioengineered collagen) for the aging face *Facial Plast. Surg.* **20** 125–8

Baumann L 2006 Collagen-containing fillers: alone and in combination *Clin. Plast. Surg.* **33** 587–96

Beasley K L, Weiss M A and Weiss R A 2009 Hyaluronic acid fillers: a comprehensive review *Facial Plast. Surg.* **25** 86–94

Bentkover S H 2009 The biology of facial fillers *Facial Plast. Surg.* **25** 73–85

Bjarnsholt T, Tolker-Nielsen T, Givskov M, Janssen M and Christensen L H 2009 Detection of bacteria by fluorescence *in situ* hybridization in culture-negative soft tissue filler lesions *Dermatol. Surg.* **35** 1620–4

Borrell M, Leslie D and Tezel A 2011 Response of authors: 'Lift capabilities of hyaluronic acid fillers *J. Cosmet. Laser Ther.* **13** 200–1

Boss W K Jr *et al* 2000 Autologous cultured fibroblasts as cellular therapy in plastic surgery *Clin. Plast. Surg.* **27** 613–26

Carruthers A *et al* 2005 Randomized, double-blind comparison of the efficacy of two hyaluronic acid derivatives, restylane perlane and hylaform, in the treatment of nasolabial folds *Dermatol. Surg.* **31** 1591–8 discussion 8

Chouzouri G and Xanthos M 2007 *In vitro* bioactivity and degradation of polycaprolactone composites containing silicate fillers *Acta Biomater.* **3** 745–56

Christen M O and Vercesi F 2020 Polycaprolactone: how a well-known and futuristic polymer has become an innovative collagen-stimulator in esthetics *Clin. Cosmet. Investig. Dermatol.* **13** 31–48

Coleman K M, Voigts R, Devore D P, Termin P and Coleman W P 2008 Neocollagenesis after injection of calcium hydroxylapatite composition in a canine model *Dermatol. Surg.* **34** S53–5

Cotofana S *et al* 2015 Midface: clinical anatomy and regional approaches with injectable fillers *Plast. Reconstr. Surg.* **136** 219S–34S

Day D J, Littler C M, Swift R W and Gottlieb S 2004 The wrinkle severity rating scale: a validation study *Am. J. Clin. Dermatol.* **5** 49–52 de Almeida A T *et al* 2019 Consensus recommendations for the use of hyperdiluted calcium hydroxyapatite (Radiesse) as a face and body biostimulatory agent *Plast. Reconstr. Surg. Glob. Open* **7** e2160

de Maio M, Deboulle K, Braz A and Rohrich R J 2017 Facial assessment and injection guide for botulinum toxin and injectable hyaluronic acid fillers: focus on the midface *Plast. Reconstr. Surg* **140** 540e–50e

de Melo F *et al* 2017 Recommendations for volume augmentation and rejuvenation of the face and hands with the new generation polycaprolactone-based collagen stimulator (Ellanse(R)) *Clin. Cosmet. Investig. Dermatol.* **10** 431–40

Edsman K, Nord L I, Ohrlund A, Larkner H and Kenne A H 2012 Gel properties of hyaluronic acid dermal fillers *Dermatol. Surg.* **38** 1170–9

Edsman K L, Wiebensjo A M, Risberg A M and Ohrlund J A 2015 Is there a method that can measure cohesivity? Cohesion by sensory evaluation compared with other test methods *Dermatol. Surg.* **41** S365–72

Edsman K L M and Ohrlund A 2018 Cohesion of hyaluronic acid fillers: correlation between cohesion and other physicochemical properties *Dermatol. Surg.* **44** 557–62

Engelhard P, Humble G and Mest D 2005 Safety of Sculptra: a review of clinical trial data *J. Cosmet. Laser Ther.* **7** 201–5

Ezzat W H and Keller G S 2011 The use of poly-L-lactic acid filler in facial aesthetics *Facial Plast. Surg.* **27** 503–9

Fagien S, Bertucci V, Von Grote E and Mashburn J H 2019 Rheologic and physicochemical properties used to differentiate injectable hyaluronic acid filler products *Plast. Reconstr. Surg.* **143** 707e–20e

Falcone S J and Berg R A 2008 Crosslinked hyaluronic acid dermal fillers: a comparison of rheological properties *J. Biomed. Mater. Res.* A **87** 264–71

Falcone S J, Doerfler A M and Berg R A 2007 Novel synthetic dermal fillers based on sodium carboxymethylcellulose: comparison with crosslinked hyaluronic acid-based dermal fillers *Dermatol. Surg.* **33** S136–43 discussion S43

Fenske N A and Lober C W 1986 Structural and functional changes of normal aging skin *J. Am. Acad. Dermatol.* **15** 571–85

Frautschi R S, Hashem A M, Halasa B, Cakmakoglu C and Zins J E 2017 Current evidence for clinical efficacy of platelet rich plasma in aesthetic surgery: a systematic review *Aesthet. Surg. J.* **37** 353–62

Galadari H, Van Abel D, Al Nuami K, Al Faresi F and Galadari I 2015 A randomized, prospective, blinded, split-face, single-center study comparing polycaprolactone to hyaluronic acid for treatment of nasolabial folds *J. Cosmet. Dermatol.* **14** 27–32

Gogolewski S, Jovanovic M, Perren S M, Dillon J G and Hughes M K 1993 Tissue response and *in vivo* degradation of selected polyhydroxyacids: polylactides (PLA), poly(3-hydroxybutyrate) (PHB), and poly(3-hydroxybutyrate-co-3-hydroxyvalerate) (PHB/VA) *J. Biomed. Mater. Res.* **27** 1135–48

Gonzalez A M, Lobocki C, Kelly C P and Jackson I T 2007 An alternative method for harvest and processing fat grafts: an *in vitro* study of cell viability and survival *Plast. Reconstr. Surg.* **120** 285–94

Goodman G J, Swift A and Remington B K 2015 Current concepts in the use of Voluma, Volift, and Volbella *Plast. Reconstr. Surg.* **136** 139S–48S

Graivier M H *et al* 2007 Calcium hydroxylapatite (Radiesse) for correction of the mid-and lower face: consensus recommendations *Plast. Reconstr. Surg.* **120** 55S–66S

Gutowski K A 2016 Hyaluronic acid fillers: science and clinical uses *Clin. Plast. Surg.* **43** 489–96

Heitmiller K, Ring C and Saedi N 2021 Rheologic properties of soft tissue fillers and implications for clinical use *J. Cosmet. Dermatol.* **20** 28–34

Hoben G, Schmidt V J, Bannasch H and Horch R E 2011 Tissue augmentation with fibrin sealant and cultured fibroblasts: a preliminary study *Aesthetic Plast. Surg.* **35** 1009–15

Illouz Y G 1986 The fat cell 'graft': a new technique to fill depressions *Plast. Reconstr. Surg.* **78** 122–3

Jones D and Murphy D K 2013 Volumizing hyaluronic acid filler for midface volume deficit: 2-year results from a pivotal single-blind randomized controlled study *Dermatol. Surg.* **39** 1602–12

Kablik J, Monheit G D, Yu L, Chang G and Gershkovich J 2009 Comparative physical properties of hyaluronic acid dermal fillers *Dermatol. Surg.* **35** 302–12

Katsambas A D, Lotti T M, Dessinioti C and Angelo Massimiliano D (ed) 2015 *European Handbook of Dermatological Treatments* (Berlin: Springer)

Khater R, Atanassova P, Anastassov Y, Pellerin P and Martinot-Duquennoy V 2009 Clinical and experimental study of autologous fat grafting after processing by centrifugation and serum lavage *Aesthetic Plast. Surg.* **33** 37–43

Kim D H *et al* 2011 Can platelet-rich plasma be used for skin rejuvenation? Evaluation of effects of platelet-rich plasma on human dermal fibroblast *Ann. Dermatol.* **23** 424–31

Kim J A and Van Abel D 2015 Neocollagenesis in human tissue injected with a polycaprolactone-based dermal filler *J. Cosmet. Laser Ther.* **17** 99–101

Kim J H, Ahn D K, Jeong H S and Suh I S 2014a Treatment algorithm of complications after filler injection: based on wound healing process *J. Korean Med. Sci.* **29** S176–82

Kim J S, In C H, Park N J, Kim B J and Yoon H S 2020 Comparative study of rheological properties and preclinical data of porous polycaprolactone microsphere dermal fillers *J. Cosmet. Dermatol.* **19** 596–604

Kim Y S *et al* 2014b The anatomical origin and course of the angular artery regarding its clinical implications *Dermatol. Surg.* **40** 1070–6

Kimball A B *et al* 2018 Age-induced and photoinduced changes in gene expression profiles in facial skin of Caucasian females across 6 decades of age *J. Am. Acad. Dermatol.* **78** 29–39.e7

Kurita M *et al* 2008 Influences of centrifugation on cells and tissues in liposuction aspirates: optimized centrifugation for lipotransfer and cell isolation *Plast. Reconstr. Surg.* **121** 1033–41

La Gatta A *et al* 2019 Hyaluronan-based hydrogels as dermal fillers: the biophysical properties that translate into a 'volumetric' effect *PLoS One* **14** e0218287

Lavker R M 1979 Structural alterations in exposed and unexposed aged skin *J. Invest. Dermatol.* **73** 59–66

Liew S *et al* 2020 Consensus on changing trends, attitudes, and concepts of Asian beauty *Aesthetic Plast. Surg.* **44** 1186–94

Locke M B and De Chalain T M 2008 Current practice in autologous fat transplantation: suggested clinical guidelines based on a review of recent literature *Ann. Plast. Surg.* **60** 98–102

Lorenc Z P, Nir E and Azachi M 2010 Characterization of physical properties and histologic evaluation of injectable Dermicol-p35 porcine-collagen dermal filler *Plast. Reconstr. Surg.* **125** 1805–13

Matsumoto D *et al* 2006 Cell-assisted lipotransfer: supportive use of human adipose-derived cells for soft tissue augmentation with lipoinjection *Tissue Eng.* **12** 3375–82

Moers-Carpi M *et al* 2021 European multicenter prospective study evaluating long-term safety and efficacy of the polycaprolactone-based dermal filler in nasolabial fold correction *Dermatol. Surg.* **47** 960–5

Mokos Z B, Curkovic D, Kostovic K and Ceovic R 2018 Facial changes in the mature patient *Clin. Dermatol.* **36** 152–8

Narins R S *et al* 2008 Twelve-month persistency of a novel ribose-cross-linked collagen dermal filler *Dermatol. Surg.* **34** S31–9

Ogilvie P *et al* 2020a VYC-25L hyaluronic acid injectable gel is safe and effective for long-term restoration and creation of volume of the lower face *Aesthet. Surg. J.* **40** NP499–510

Ogilvie P *et al* 2020b Expert consensus on injection technique and area-specific recommendations for the hyaluronic acid dermal filler VYC-12L to treat fine cutaneous lines *Clin. Cosmet. Investig. Dermatol.* **13** 267–74

Paliwal S *et al* 2014 Skin extracellular matrix stimulation following injection of a hyaluronic acid-based dermal filler in a rat model *Plast. Reconstr. Surg.* **134** 1224–33

Pallua N, Pulsfort A K, Suschek C and Wolter T P 2009 Content of the growth factors bFGF, IGF-1, VEGF, and PDGF-BB in freshly harvested lipoaspirate after centrifugation and incubation *Plast. Reconstr. Surg.* **123** 826–33

Park S W *et al* 2012 Iatrogenic retinal artery occlusion caused by cosmetic facial filler injections *Am. J. Ophthalmol.* **154** 653–62.e1

Pierre S, Liew S and Bernardin A 2015 Basics of dermal filler rheology *Dermatol. Surg.* **41** S120–6

Saban Y, Amodeo C A, Bouaziz D and Polselli R 2012 Nasal arterial vasculature: medical and surgical applications *Arch. Facial Plast. Surg.* **14** 429–36

Salles A G, Lotierzo P H, Gimenez R, Camargo C P and Ferreira M C 2008 Evaluation of the poly-L-lactic acid implant for treatment of the nasolabial fold: 3-year follow-up evaluation *Aesthetic Plast. Surg.* **32** 753–6

Snozzi P and Van Loghem J A J 2018 Complication management following rejuvenation procedures with hyaluronic acid fillers-an algorithm-based approach *Plast. Reconstr. Surg. Glob. Open* **6** e2061

Sundaram H, Voigts B, Beer K and Meland M 2010 Comparison of the rheological properties of viscosity and elasticity in two categories of soft tissue fillers: calcium hydroxylapatite and hyaluronic acid *Dermatol. Surg.* **36** 1859–65

Tezel A and Fredrickson G H 2008 The science of hyaluronic acid dermal fillers *J. Cosmet. Laser Ther.* **10** 35–42

Toledo L S and Mauad R 2006 Fat injection: a 20-year revision *Clin. Plast. Surg.* **33** 47–53 vi

Turlier V *et al* 2013 Association between collagen production and mechanical stretching in dermal extracellular matrix: *in vivo* effect of cross-linked hyaluronic acid filler. A randomised, placebo-controlled study *J. Dermatol. Sci.* **69** 187–94

Vallejo A *et al* 2018 Comparing efficacy and costs of four facial fillers in human immunodeficiency virus-associated lipodystrophy: a clinical trial *Plast. Reconstr. Surg.* **141** 613–23

Van Loghem J A J 2018 Use of calcium hydroxylapatite in the upper third of the face: retrospective analysis of techniques, dilutions and adverse events *J. Cosmet. Dermatol.* **17** 1025–30

Weiss R A, Weiss M A, Beasley K L and Munavalli G 2007 Autologous cultured fibroblast injection for facial contour deformities: a prospective, placebo-controlled, phase III clinical trial *Dermatol. Surg.* **33** 263–8

Wu W T L *et al* 2020 Consensus on current injectable treatment strategies in the Asian face *Aesthetic Plast. Surg.* **44** 1195–207

Yoon E S, Han S K and Kim W K 2003 Advantages of the presence of living dermal fibroblasts within restylane for soft tissue augmentation *Ann. Plast. Surg.* **51** 587–92

Yoshimura K *et al* 2020 Cell-assisted lipotransfer for cosmetic breast augmentation: supportive use of adipose-derived stem/stromal cells *Aesthetic Plast. Surg.* **44** 1258–65

Yoshimura K, Suga H and Eto H 2009 Adipose-derived stem/progenitor cells: roles in adipose tissue remodeling and potential use for soft tissue augmentation *Regen. Med.* **4** 265–73

Zeide D A 1986 Adverse reactions to collagen implants *Clin. Dermatol.* **4** 176–82

Zuk P A *et al* 2001 Multilineage cells from human adipose tissue: implications for cell-based therapies *Tissue Eng.* **7** 211–28

IOP Publishing

Skin Photoaging (Second Edition)

Rui Yin, Yang Xu and Chengfeng Zhang

Chapter 6

Chemical peels for skin rejuvenation

Xian Jiang and Xiaoxue Li

Chemical peels are a therapeutic method used to improve skin appearance and texture by applying chemical substances to the skin surface to remove the epidermal layer. Common chemical peels include alpha-hydroxy acids, fruit acids, salicylic acid, and trichloroacetic acid, each achieving therapeutic effects through varying degrees of exfoliation. The application of chemical peels in skin photoaging helps improve skin texture, fade pigmentation, and reduce wrinkles, while also stimulating the synthesis of collagen and elastin, promoting skin regeneration and repair. However, potential complications such as irritant reactions, pigmentation changes, and excessive peeling may occur during chemical peel treatments. Therefore, thorough assessment and guidance by a healthcare professional are essential to ensure the safety and efficacy of the treatment.

6.1 Introduction

Chemical peel is one kind of skin reconstruction technique to improve skin quality, texture and appearance. Different peeling agents induce controlled skin damage at different levels, thus initiating the wound-healing process and thereby promoting epidermal and dermal reconstruction.

Chemical peels are one of the oldest skin rejuvenation forms. The history of chemical peels traces back to 1550 BC when ancient Egyptians applied mixtures of animal oils, salt, and alabaster to the skin for rejuvenation (Brody *et al* 2000). Records of chemical peels in modern medicine began in the mid-nineteenth century, when European dermatologists discovered that phenol had the effect of peeling off the skin and used it to treat freckles and melasma. Despite the current application of various light-based devices, chemical peels, as a safe, efficacious, cost-effective and flexible procedure, are still widely used in clinical practice today for treating photoaging, acne and varied pigmentation dermatosis.

doi:10.1088/978-0-7503-5112-6ch6

6.2 Classification

6.2.1 Depth of penetration

The penetration depth of a chemical peel determines its efficacy and safety. Traditionally, based on the depth of injury, chemical peels are divided into very superficial peels, superficial peels, medium-depth peels and deep peels. Table 6.1 lists common peeling agents used for different depth of penetration. However, one should always remember that the penetration depth of a chemical peel is closely related to the properties of the peeling agents as well as the conditions of treated skin areas and peeling technique. For example, glycolic acid peel is mostly time dependent. While 20%–50% glycolic acid applied for 1–4 min can be used for very superficial peels, 20%–70% glycolic acid applied for 2–10 min causes superficial peels. In contrast, trichloroacetic acid (TCA) peel is mostly dosage dependent. TCA applied in different coats can result in different depths of peels. The pH of a solution represents the fraction of free acid, which is also a key factor considering the depth of peel. In addition, the formulation of peeling agents, the application pressure, skin type and condition, anatomic location, pre-peel and post-peel care all contribute to the penetration depth of chemical peels. Therefore, it is always important to observe what is happening during the procedure.

6.2.2 Very superficial peel

Very superficial peels induce exfoliation of stratum corneum and results in improvement of skin tone, texture and epidermal pigmentation. Alpha-hydroxy acid (AHA) and salicylic acid are commonly used for this depth of peel. Recommended

Table 6.1. Classification of chemical peels based on penetration depth.

Classification	Depth of injury	Example of peeling agents[a]
Very superficial peels	Stratum corneum	20%–35% glycolic acid 20%–25% mandelic acid 20%–30% salicylic acid
Superficial peels	Above basal cell layers	20%–70% glycolic acid 20%–30% salicylic acid 10%–30% TCA Jessner's solution
Medium-depth peels	Epidermis to upper dermis	70% glycolic acid 35%–50% TCA
Deep peels	Epidermis to upper reticular dermis	$\geqslant$ 50% TCA Phenol-croton

[a]Concentration is one of the most important factors that decides the depth of chemical peel. However, many other factors can also influence the depth injury.

treatment interval for very superficial peel is 2 weeks. Slight erythema can be visible during the procedure. Very superficial peel is a very safe depth, since no desquamation is seen after the procedure. Theoretically, the risk of post-inflammatory hyperpigmentation (PIH) does not exist when performing very superficial peel. However, moisturization and sun-avoidance are still important.

6.2.3 Superficial peel

Superficial peel penetrates the epidermis above the basal layer, i.e. it induces exfoliation of the stratum corneum as well as acanthosis. Superficial peels accelerate epidermal turnover and increase epidermal thickness. Therefore it is more effective than very superficial peel for the treatment of epidermal pigmentosa such as actinic keratoses, solar lentigines and epidermal melasma. Superficial peel is also a good choice for the treatment of mild to moderate acne as it can unclog the openings of folliculosebaceous units. The recommended treatment interval for superficial peel is 2–4 weeks. Erythema is visible during superficial peel. White pinpoint (frosting) can be noticed if the basement layer is injured, which requires neutralization immediately. After superficial peel, scaling and slight erythema can be seen within the first few days. Superficial peel is not dangerous but cautions should be taken in patients with dark or sensitive skin because they are at risk of PIH. AHA, salicylic acid and 10%–30% TCA are commonly used for this depth of peel.

6.2.4 Medium-depth peel

Medium-depth peels penetrate the entire epidermis and dermal papillary layer. Thus, it has the effect of superficial peel as well as stimulation on the upper dermis. The commonly used peeling agents for this depth include high concentration glycolic acid, 35%–50% TCA and combined chemical peels. During chemical peel, erythema and frosting develop aggressively, representing coagulation of dermal protein. Medium-depth peel is useful for epidermal and dermal pigmentation, fine wrinkles and laxity. However, the risks increase as the efficacy improves. Medium-depth peel takes a longer time to heal (about 7 d), so the treatment interval is longer (at least 4 weeks). Because the basal layer is injured during medium-depth peel, the risk of PIH is higher. Therefore, medium-depth peel should be conducted in appropriate patients. For those with Fitzpatrick skin type IV–VI, or prone to PIH, or unable to take strict sun protection, medium-depth peel should not be performed. Infection and scaring are not common at this depth.

6.2.5 Deep peel

Deep peels induce injury to the level of the dermal reticular layer, which is comparable to ablative CO_2 laser resurfacing. Deep peels can improve moderate wrinkling and acne scars. However, there are great risks of post-inflammatory pigmentation abnormalities (including hyperpigmentation, hypopigmentation and depigmentation), scarring and infection. Phenol-croton solution and TCA higher than 50% concentration are used for deep peel. Due to the above adverse reaction and systemic toxicity, phenol peel is not common these days. Chemical reconstruction of skin scars (CROSS), which is a local deep peel by high concentration TCA, is very useful for treating acne scars, especially the icepick type.

6.2.6 Peeling agents

6.2.6.1 AHAs

6.2.6.1.1 Chemistry and properties

AHAs, also known as fruit acids, represent a group of carboxylic acids extracted from various fruits and dairy products. Based on their chemical structure, AHAs can be divided into hydrophilic aliphatic (glycolic, lactic, citric and tartaric acid) and lipophilic aromatic acids (mandelic acid).

Glycolic acid is the most used AHAs. Glycolic has the simplest molecular structure and smallest molecular weight, so it penetrates the skin most easily. The pKa of glycolic acid is 3.83. Four concentrations glycolic acid for chemical peel are available, 20%, 35%, 50%, and 70%, with pHs ranging from 0.6 to 1.6. Glycolic acid of low concentration is also popular in daily cosmetics.

Mandelic acid contains the benzene ring structure, giving it a slower penetration, so the action is gentle. In addition, mandelic acid has a prominent antimicrobial effect compared to other AHAs. For chemical peel, the concentration ranges from 20% to 50%. Mandelic acid peel causes little irritation, making it a great choice for sensitive skin, melasma and post-inflammatory pigmentation.

Lactic acid can be extracted from sour milk, and only L-lactic acid is biologically active. Concentrations for peeling should be 10%–20% or stronger. At a concentration of 50%–70%, lactic acid produces the same amount of exfoliation as glycolic acid. In addition, it had been shown that lactic acid improved skin hydration.

Malic acid, tartaric acid and citric acid have two or more carboxyl groups in their structure and can therefore also be classified as β-hydroxy acids. Citric acid chelates metal ions, has antioxidant effects, has three carboxyl groups in its structure, and has a pH range of 1.0–2.0, so it can be used as a pH regulator.

6.2.6.2 Mechanism of action

AHAs promote epidermal cell turnover and dermal extracellular matrix reconstruction.

6.2.6.2.1 Epidermis

At low concentrations (<30%), AHAs disrupts intercellular connections by interference with the formation of ionic bonds, and that this action is mediated by interference with the functions of enzymes that form O–S and O–P linkages of sulfate and phosphate bonds, respectively (Van Scott and Yu 1984). To determine the effect of AHAs (containing lactic acid 28%–32%, glycolic acid 12%–17%, citric acid 2%–6%, malic acid 1% maximum, tartaric acid 1% maximum) on the rate of cell renewal, twenty female volunteers were patched with 5% dansyl chloride in petrolatum base for 24 h on the vulnar forearm prior to treatment and examined by UV lamps to evaluate stain uptake by the stratum corneum. The results showed that the rate of cell turnover increased by 34% compared with the control group. Also, the epidermal renewal rate correlated with AHA concentration (Scholz Brooks *et al* 1994).

By reducing corneocyte cohesion, AHAs (<30%) reduce the thickness of the hyperkeratotic stratum corneum for around 2 weeks, after which the superficial layers of the epidermis will heal. The action becomes more aggressive as the concentrations increases. So, it is always important to know which depth of peeling we want to perform. However, a long-term study showed significant increase in epidermal thickness, reversal of basal cell atypia and a return to a more normal rete pattern (Ditre *et al* 1996).

6.2.6.2.2 Dermis

For superficial peel, a histological study after AHAs peels (concentration 25%, pH 3.5, lactic acid, glycolic acid and citric acid) revealed increased thickness, increased acid mucopolysaccharides, improved quality of elastic fibers, and increased density of collagen in papillary dermis (Ditre *et al* 1996). Although superficial peeling only acts on epidermal cells, it is likely that crosstalk between keratinocytes and fibroblasts via various cytokines contribute to the improvements in the dermis. At high concentration or lower pH, AHAs exert a caustic effect, which has a more significant effect on the dermis and can stimulate dermal mucopolysaccharide and collagen synthesis.

6.2.6.2.3 Melanin

As chemical peel accelerates epidermal turnover, melanin pigmentation in the keratinocytes is dispersed. In addition, AHAs suppress melanin formation by directly inhibiting tyrosinase activity (Usuki *et al* 2003).

6.2.6.2.4 Anti-bacterial

The anti-bacterial effect of AHAs is due to its acidic nature, as the vast majority of bacteria cannot grow normally in an acidic environment.

6.2.6.3 Indications and contraindication

AHAs are usually used for superficial peel. Indications of AHA peel include:

- Mild skin photoaging.
- Coarse skin.
- Mild and moderate acne vulgaris.
- Dyschromia of the skin including solar lentigines, flat seborrheic keratoses, melasma and PIH.
- Keratosis pilaris.
- Other conditions including ichthyosis and skin amyloidosis.

Skin photoaging is the most common indication of AHA peels, especially glycolic acid peels. Glycolic acid peel is a relatively more suitable and cost-effective choice for mild photoaging. It can improve fine lines, uneven skin tone, pigmentation of the epidermis and rough skin texture. For moderate and deep skin photoaging, it is better to combine glycolic acid peel with other therapies such as intense pulsed light, lasers and radiofrequency technique. Acne vulgaris is another common indication of AHAs. Its main mechanism of action is to reduce abnormal keratosis in the funnel of hair follicles. In addition, glycolic acid has inhibitory and bactericidal effects on *Propionibacterium acnes.* Melasma, post-inflammatory pigmentation and rosacea are diseases accompanied by skin barrier damage. Choosing when to perform chemical peel is important in these cases, as the chemical peel itself causes skin injury first. If the skin condition of a patient is not stable, performing a chemical peel may sometimes be disastrous.

Contraindications of superficial chemical peel (including AHA and other peels) include:

- Poor general condition, psychosis or emotional instability, immunodeficiency, severe diabetes or cardiopulmonary disease, and coagulation dysfunction.
- Allergic to peeling agents or their components.
- Gestational period.
- Acute inflammation (such as acute eczema), or active bacterial, fungal or herpes virus infection, or unhealed wounds, or malignant tumors at the treatment site.
- Unable to follow the doctor's orders for skin care.
- Unrealistic expectations.

Superficial chemical peel by AHA should be performed with caution during lactation. Retinoids was formerly suggested to be avoided before and during chemical peel, as combination therapy can lead to excessive peeling and increase the risk of adverse reactions such as delay of healing and scarring, which is now argued to be a relative contradiction when performing medium-depth and deep peel. For superficial chemical peel, combined with oral isotretinoin does not increase the risk of postoperative scarring or delayed healing.

6.2.6.4 Peeling technique

AHAs are tolerable for most patients. AHAs cannot neutralize themselves. They must be neutralized using a weak buffer or their consistent action would burn the skin. Glycolic acid is the most used AHA. Because glycolic acid penetrates fast and unevenly, a higher pH value is recommended (Obagi 2020). However, lower pH means weaker acidifying power on skin. During the procedure, careful observation of skin reaction is mandatory. Usually, it takes 3–5 min for glycolic acid to achieve a superficial peel. Because glycolic acid peel is time dependent, it must be neutralized after a certain time. However, if epidermolysis occurs within the certain time, which appears as a grayish white discoloration of the skin (though this sign is also called 'frosting', it is different from that during the salicylic acid and TCA peel), it must be neutralized immediately regardless of the time.

6.2.6.5 Salicylic acid

6.2.6.5.1 Chemistry and properties

Salicylic acid, chemically known as o-hydroxybenzoic acid, is a hydrolyzed product of acetylsalicylic acid that can be extracted from willow bark, sweet birch and holly leaves or synthesized artificially. Salicylic acid has often been described as β-hydroxy acid, but in fact, β-hydroxy acid refers to the carboxylic acid that is connected to the aliphatic or aliphatic β carbon atoms, and the hydroxyl group of salicylic acid is connected to the carbon atom on the benzene ring. The hydroxyl group of β-hydroxy acid is neutral, while the hydroxyl group of salicylic acid is acidic. Therefore, it may be more appropriate to classify salicylic acid as an aromatic acid.

Salicylic acid is an organic weak acid with a pKa of 2.98 and pH of 2.4 in saturated solution. It is lipid soluble. There are various preparations of salicylic acid, including ethyl alcohol solution containing 10%–50% salicylic acid, 30% salicylic acid in polyethylene glycol vehicle and 30% salicylic acid in poloxamer 407 vehicle. Topical salicylic acid has been used to treat dermatosis as early as the first century when people used willow bark to treat calluses. But it was not until the 1760s that scientists determined the ingredient in willow bark that softened and peeled off the stratum corneum was salicylic acid. German dermatologist P G Unna first elaborated on the properties of salicylic acid in 1882 (Brody *et al* 2000). Salicylic acid began to be used as a peeling agent in the mid-nineteenth century, when Joseph Eller described the Jessner's solution consisting of salicylic acid, lactic acid, and resorcinol. By the 1990s, AHAs became popular, and shortly thereafter, salicylic acid was described as a β-hydroxy acids to compete with AHAs (Brody *et al* 2000). Currently, low concentrations of salicylic acid are still used in cosmetics designed for acne, psoriasis, dandruff and keratosis pilaris.

Salicylic acid and AHAs are the most used peeling agents in clinical practice, but there are differences between them. Salicylic acid penetrates the skin very slowly and within the epidermis. It is absorbed well by the sebaceous follicle. In addition, salicylic acid does not need neutralization.

The common concentrations of salicylic acid used for chemical peel are 20% and 30%. The damage level is limited to the epidermis, which belongs to very superficial and superficial peel. Salicylic acid can be used in combined with other peeling agents to enhance penetration.

Although salicylic acid has limited effect on the dermis, the penetration is slower and more uniform due to the large molecular weight. So, it is gentle and safe for sensitive skin and dark skin. Moreover, salicylic acid is lipophilic and therefore good for acne. Its anti-inflammatory property also makes it suitable for rosacea.

6.2.6.6 Mechanism of action

Salicylic acid has been shown to have multiple effects on the skin, as described in the following sections.

6.2.6.6.1 Epidermis

Salicylic acid works as both a keratolytic and comedolytic agent, which means it extracts desmosomal proteins as well as comedos. It does not affect keratinocytes with mitosis. The effects of salicylic acid on the skin can be observed at the histological level through animal experiments. After applying varied concentrations of salicylic acid on the skin of hairless mice, it was found that 7.5% and 15% salicylic acid caused only a very slight change in histology. After applying 30% salicylic acid peeling, the thickness of the epidermis underwent transient thickening, then thinning and finally returning to normal roughly 48 h after peeling. At the same time, the hair follicle opening expands after 30% salicylic acid peel and falls off within 48 h. After salicylate decortication, epidermal cells are more neatly arranged and basal cell proliferative activity increases (Imayama *et al* 2000).

6.2.6.6.2 Dermis

As a superficial peeling agent, salicylic acid does not act directly on the dermis, but it can still activate dermal fibroblasts and induce collagen synthesis to a certain extent by cell cross-talking. By observing the changes in the dermis layer of the skin of hairless mice before and after salicylic acid peel, it was found that there may be temporary mild edema of the papillary layer of the dermis, and sparse lymphocyte infiltration can be seen around the hair follicle. Different formulations cause different reactions in the dermis. In the ethanol solution group, edema and inflammation occur earlier and subside faster, while in the polyethylene glycol solution group, the situation is the opposite. An increase in the proliferative activity of partially spindle cells in the dermis layer was seen 48 h postoperatively, suggesting that salicylic acid was able to activate dermal fibroblast activity without directing damaging the dermis and causing inflammation (Imayama *et al* 2000). In addition, a clinical study of ten patients looked at histological changes in the skin before and after 30% salicylic acid peel (once a week for a total of six times) and found that the content of dermal collagen fibers and elastin fibers increased after treatment (Abdel-Motaleb *et al* 2017).

6.2.6.6.3 Sebaceous glands

Salicylic acid is lipid soluble and can be miscible with sebum and intercellular lipids. In addition, salicylic acid has been shown to inhibit proliferative activity of sebaceous gland cells, inhibit sebaceous gland synthesis through the AMPK/SEEBP-1 pathway, and inhibit the inflammatory response of acne lesions through the NF-κB pathway (Lu *et al* 2019). Thus, some scholars believe that salicylic acid has a superior sebum inhibition effect compared with other peeling agents. However, the results of clinical trials vary. Sebum secretion by 30% salicylic acid has been reported to inhibit sebum secretion stronger than 50% pyruvate, but no significant reduction in sebum content in patients after salicylate decortication has also been reported (Marczyk *et al* 2014). In fact, the effect on sebum secretion correlates with the acidity of the peeling solution. In addition, clinical studies measured sebum content at a certain point after chemical peels and did not reflect the dynamic change process well.

6.2.6.6.4 Anti-inflammatory effect

Acetylsalicylic acid is a classic nonsteroidal anti-inflammatory drug that exerts an anti-inflammatory effect by inhibiting prostaglandin production. As a hydrolyzed product of acetylsalicylic acid, topical salicylic acid may have a mild anti-inflammatory effect. In clinical studies, salicylic acid peel has shown a good effect on inflammatory acne lesions. In an early study, the anti-inflammatory effect of topical salicylic acid and other drugs were tested in animal experiments. By investigating their inhibitory effect on the development of erythema mimic UV dermatitis, the tested drugs were ranked in the following ascending order of activity (percent of maximum possible score): bufexamac = 36%, salicylic acid = 37%, hydrocortisone = 44%, acetylsalicylic acid = 48%, flumethasone pivalate = 51%, fluocinolone acetonide = 51%, phenylbutazone = 56%, and indomethacin = 58% (Weirich *et al* 1976). However, the measurement of anti-inflammatory effect by subjective observation was not persuasive enough.

6.2.6.6.5 Photoprotective effect

A low concentration of salicylic acid has a photoprotective effect on both UVA and UVB, which can weaken the acute erythema reaction and chronic carcinogenic effect induced by ultraviolet rays. The National Toxicology Program funded a 52-week animal study to explore the effect of topical AHA (0%, 4%, or 10% glycolic acid, pH 3.5) and BHA (0%, 2%, or 4% salicylic acid, pH 4.0) on the photocarcinogenesis of simulated solar radiation (SSL). The results showed that glycolic acid had no effect on the time required to induce tumors by SSL; however, topical salicylic acid at 4% increased the time required for SSL to induce skin tumors (National Toxicology Program Technical Report 2007). It should be pointed out that high concentrations of salicylic acid have obvious keratolytic effects, i.e. salicylic acid at peel removes the stratum corneum. So strict sun protection is still required after salicylic acid peel. In UVB-irradiated hairless mice, chemical peel with salicylic in polyethylene glycol (SA-PEG) was found to inhibit the expression of p53 protein and to improve the expression of filament globin and lychein as well as the maturation of the cornified envelope (CE), suggesting a restorative effect of salicylic acid on photo-damaged skin (Dainichi *et al* 2006).

6.2.6.6.6 Melanin

Salicylic acid can have some whitening effect by accelerating melanin in the epidermis to shed as well as mild inhibitory effect on melanin synthesis.

6.2.6.7 Indications and contraindications

Indications and contraindication of salicylic acid peel are almost the same as those for AHAs peel. Salicylic acid peel should be avoided during pregnancy and lactation.

6.2.6.8 Peeling technique

Salicylic acid penetration is slow and superficial. Different formulations of salicylic acid induce different skin reactions. Ethanol preparation of salicylic acid will form so-called ‘pseudofrost’ when peeling, which is a sign of crystallization of ethanol vapors. Salicylic acid gels or creams do not induce obvious pseudofrost. Some gel scan even be left on the skin for hours.

6.2.6.9 TCA

6.2.6.9.1 Chemistry and properties

TCA is a strong acid with a pKa of 0.54. It is the strongest acid currently used for chemical peels. TCA is a white or colorless crystalline solid and extremely soluble in water. Because of its hydrophilicity, TCA is difficult to store. TCA solutions, however, do not hydrate by themselves and are therefore safer to keep.

The toxicity of TCA is a concern as industry accidents have been reported in which accidental ingestion or inhalation of concentrated vapors caused systemic toxicity. TCA is harmful for both organisms and the environment. The lethal dose or rats is 5000 mg kg^{-1}. Short exposure to TCA can cause serious temporary or long-lasting injuries. However, a toxicity event has never been reported when performing a TCA peel. As in a chemical peel, the amount of TCA applied to the skin is coagulated by skin protein and therefore not absorbed into blood circulation (Obagi 2020).

6.2.6.10 Mechanism of action

The histological effect of TCA peel is due to its destruction of the living cells of the epidermis and dermis depending on the depth of injury. A histological study compared the effects of four different types of peels on mice, i.e. 50% glycolic acid, 30% TCA, 50% TCA and phenol peel (Baker Gordon’s formula). Dermal thickness examination on day 60 after peeling revealed that dermal thickness was significantly increased after 50% TCA and phenol peels. However, no statistical difference between glycolic acid or 30% TCA and the control were found. Collagen increase was visible by day 28 and elastic fibers lost their thickened appearance and assumed a more normal horizontal arrangement in the papillary dermis after glycolic acid and 30% TCA peel and in both papillary and upper reticular dermis after 50% TCA and phenol group. Biochemical analysis showed collagen increased in the initial 3 d, decreased on day 7 and peaked on day 28 and lasted for 60 d. Glycosaminoglycan increased significantly after 50% TCA and phenol peel and was maintained for 60 d (Butler *et al* 2001).

The effects of TCA on photoaged skin are long lasting as TCA penetrates to the papillary dermis at least. However, deeper penetration comes with risks as dermal fibroplasia causes hypertrophic scarring or even keloids, especially in dark skins and areas with high skin tone such as the jaw. There is also risk of skin atrophy when all the appendages of a treated area undergo necrosis.

6.2.6.11 Indications and contraindications

TCA can be used for all depths of chemical peels. For a superficial peel, 10%~30% TCA is used and indications include mild photoaging, epidermal pigmentation, acne and coarse skin. For a medium-depth peel, 35%~50% TCA is used and indications include mild to moderate photoaging, wrinkles, seborrheic keratosis, coarse pores, rough skin, mild acne scar, warts and xanthelasma. For a deep peel, TCA above 50% is used for local treatment of acne scars (so-called TCA CROSS).

Contraindications of TCA peel include:

- Poor general condition, psychosis or emotional instability, immunodeficiency, severe diabetes or cardiopulmonary disease, and coagulation dysfunction.
- Allergy to TCA.
- Gestational and lactation period.
- Acute inflammation (such as acute eczema), or active bacterial, fungal or herpesvirus infection, or unhealed wounds, or malignant tumors at the treatment site.
- Keloids at the treatment site.
- Inability to follow the doctor's orders for skin care.
- Unrealistic expectations.
- For medium-depth and deep peels, patients with keloids, Fitzpatrick skin type IV–VI, and a history of isotretinoin use within half a year should be carefully evaluated.

6.2.6.12 Peeling technique

TCA is usually used for medium-depth and deep peeling. As we mentioned above, the volume of TCA is critical in determining the depth of peeling. Once applied to the skin, TCA coagulates with protein rapidly, generating frost. TCA cannot be neutralized, so to avoid over-peeling, it is always important to assess the skin signs of different depths of peeling. Generally, there are three levels of TCA peeling. Level 1 refers to peeling above the basal layer, which shows scattered pinpoint frost on the ground of the erythema. Level 2 refers to peeling of the entire epidermis, which shows cloudy-white frost on the ground of the erythema. Level 3 refers to peeling at the depth of the dermis, which shows pure white frost. Only when the desired endpoint is not achieved, should more coats of TCA be applied.

6.3 Priming and post-peel care

6.3.1 Priming

Priming is important because it ensures efficacy and safety. There are two steps of priming: medium-term and long-term preparation and immediate preparation. Medium-term and long-term preparation refer to pretreatment of the skin before chemical peel, usually 1 month before. Immediate preparation refers to preparation of the skin just before the application of the peeling agent. The benefits of priming include enhancing penetration depth, even penetration, preventing PIH, and accelerating skin healing. However, priming can sometimes be harmful since most priming agents produce irritation and enhanced penetration depth is accompanied by higher risk.

Medium-term and long-term preparation of the skin starts 2–4 weeks before the chemical peel, including strict sun protection, skin moisturization and using a priming agent. A different priming agent is chosen for different purposes. To accelerate healing and achieve deeper and more even penetration, retinoids and AHAs are commonly used. An animal study showed tretinoin pretreatment before TCA peel sustained the effects of TCA longer and showed synergistic effects of TCA and induced enhanced wound healing (Kim *et al* 1996). To create synergy with chemical peel in treating hyperpigmentation such as melasma, or to reduce the risk of PIH, tyrosinase inhibitor such as hydroquinone can be used. These priming agents should be stopped 1–3 d before the peel to prevent excessive peeling.

Immediate preparation before a chemical peel refers to using alcohol or acetone to degrease the skin, which is believed to promote even penetration.

6.3.2 Post-peel care

Good post-peel care maximizes the efficacy and minimize the side effects. Immediately after chemical peel, a cold compress is needed to sooth the skin. After chemical peel, there are different degrees of erythema, desquamation and even scabs, depending on the depth of chemical peels. For superficial peel such as glycolic acid peel or salicylic acid peel, re-epithelization usually takes 3–5 d. For medium-depth peel such as TCA peel or combined peel, it will take 6–7 d. Before complete healing of the skin, patients are told to avoid makeup, sunscreens and scratching the skin. A gentle cleanser is advised and patients are instructed to apply moisturizer at least three to four times a day. Patients are told to avoid out-door activities in the first week and if they must go outside, a hat, sunglasses and mask for sun protection are advised. Daily sunscreen can be restarted afterwards. If there is risk of bacterial infection, especially in medium-depth peeling, mupirocin ointment can be applied in the first few days. For those having a history of herpes simplex and who are about to have medium-depth or deep peeling, prophylactic antiviral therapy starting 2–3 d before the procedure and ending 14 d after it is advised (Costa *et al* 2017).

6.4 Application in photoaging

Skin photoaging is a series of clinical and histological changes caused by long-term and repeated exposure of the skin to sunlight, also known as premature aging of the skin. It is different from intrinsic aging, which is a physiological phenomenon that progresses synchronously with the aging of the body, along with gradual histological, functional and clinical changes and declines. Photoaging is not only a cosmetic concern, but it may also have an etiological link to the development of skin cancer.

Clinical manifestations and histological changes of photoaging are manifested as loose skin at the exposed area, thick and deep wrinkles, nodules, a leather-like appearance, increased pigmentation, telangiectasia, and often a grayish-yellow skin color. On occasions, these clinical manifestations may be accompanied by a benign, pre-cancerous or malignant tumors. Intrinsic aging is characterized by skin atrophy, roughness, prominent blood vessels, decreased or lost elasticity, dryness, desquamation, relaxation, and a gray complexion—there is a difference between the two.

Histologically, the main manifestations of photoaging skin are: an obviously uneven epidermal thickness due to hyperplasia or severe atrophy; increased fibroblasts and abundant and partially degranulated mast cells, and inflammatory cell infiltration is more common; collagen basophilic degeneration of mature collagen fibers; significantly increased elastin content, thickened, disordered or aggregated elastic fibers; increased amino dextran content; and dilated and twisted, and eventually sparse skin micro vessels (Xiao and Li 2003).

Chemical peels have a certain effect in the treatment of skin photoaging, often used alone or in combination with other techniques such as laser or dermabrasion, to improve the appearance of the skin. Chemical peel can reshape the surface of the skin, which presents a more uniform appearance than previously, which is beyond the reach of general surgical procedures. After a chemical peel, the fine wrinkles and pigmentation spots of the skin are usually reduced, and the skin elasticity is restored. After a chemical peel, however, tissues surrounding the treated area may have obvious inflammatory cell infiltration. The regeneration of the epidermis is generally completed within days after the procedure, while that for the dermis might be slower. After the wound surface has healed, the arrangement of epidermal cells tends to be in better order, with uniform staining and consistent polarity. The dermis will thicken, and new collagen will be synthesized in the papillary layer of the dermis, collagen fibers are rearranged, changing the original wavy disordered form into a parallel and dense one arranged in the horizontal direction. A large number of new elastic fibers are synthesized, and the pigment cells in the basal layer are reduced and evenly distributed.

A general chemical peel procedure may include some of the following steps: before exfoliation, clean the skin with liquid soap and water, and then use ether to remove surface oil; apply the exfoliation liquid carefully to the surface of the skin with moderate and even force, and let it stay on the surface of the skin for an appropriate time (the length of time depends on the type of exfoliating agent); use dry gauze to absorb the residual liquid on the surface, wash with cold water to neutralize the exfoliating agent and relieve the burning sensation; then apply Vaseline cream, and apply waterproof adhesive tape on the outside. Attention should be paid after the procedure to avoiding light and preventing infection, sunscreen might be considered (He *et al* 2020).

6.5 Complications

Some complications may occur after chemical peels, such as pigmentation, persistent depigmentation, erythema, colloid milia and scars, etc. Although certain complications can be avoided or resolved with standardized operations, close post-procedure observation and follow-up, and timely symptomatic treatment, some complications, such as post-inflammatory hyperpigmentation and persistent depigmentation, may persist. In general, as the the exfoliation layer goes deeper, the effect of skin reconstruction is more obvious, but also the risk of adverse reactions post-procedure is higher. The occurrence of adverse reactions of chemical peels is related to factors such as the patients' skin condition, the concentration of peeling agent, the residence time of the peeling agent, and postoperative care (Vemula *et al* 2018). Therefore, it is necessary to carefully evaluate the patients' skin condition before treatment and choose the appropriate type and concentration of peeling agent to reduce the occurrence of adverse reactions.

6.5.1 Erythema

Temporary erythema of the skin in the treated area is usually due to the local irritation of the chemical peel. The severity and duration of erythema are closely related to the depth and location of chemical peels. Usually, erythema lasts for 1–5 d after superficial chemical peels, 15–30 d after medium-depth chemical peels, and 60–90 d after deep chemical peels. The duration of erythema may be prolonged when chemical peels remain on the skin for too long or in areas of thin skin thickness. If the erythema persists after the expected time, it is called persistent erythema. Persistent erythema is caused by angiogenic factors stimulating vasodilation, suggesting that chemical peels stimulate the skin for too long and increase skin fibrosis, which may lead to skin thickening and scarring. Risk factors for persistent erythema include use of vitamin A acid drugs, alcohol, contact dermatitis, contact with sensitizing substances, and certain skin diseases (such as rosacea, atopic dermatitis, lupus erythematosus, etc).

To avoid persistent erythema reaction, it is necessary to evaluate the patient's skin condition in detail before the procedure, closely observe the reaction of the treatment area during the procedure, neutralize it in time when there is a significant erythema reaction, and apply cold compress or cold spray immediately after the operation. Mild erythema usually recovers gradually in approximately 3 d. At the same time, patients need to pay attention to moisturization and sun protection after surgery. If the erythema lasts longer than the expected number of days or is accompanied by symptoms such as edema and exudation after treatment, it may increase the risk of post-inflammatory hyperpigmentation, hypopigmentation and scarring. Once diagnosed as persistent erythema, symptomatic treatment such as cold spray, cold compress, and application of cosmeceuticals should be given in time, and postoperative moisturizing and sun protection should be strengthened. Some scholars suggest local application of potent glucocorticoids for 1–2 weeks, and oral or intramuscular injections of glucocorticoids if necessary. In addition, persistent erythema also responds well to intense pulsed light or pulsed dye lasers (Maity *et al* 2011).

6.5.2 Pain and burning sensation

Pain and burning are common treatment reactions to chemical peels and are more pronounced in medium-depth and deep chemical peels. Pain levels and pain scores vary from person to person. In intermediate-depth chemical peels, pain lasts only a few minutes after the peel is applied. Deep chemical peels are usually more painful and the pain can increase in the hours after treatment, lasting up to 8–12 h.

Since the pain and burning sensation are related to the concentration and residence time of the peeling agent, the principle of gradually increasing the concentration should be followed during chemical peels, and the treatment plan should be adjusted according to the skin reaction and recovery time of the previous treatment. In patients with compromised skin barrier function, pain or burning may often occur during the chemical peel. Prolonged sun exposure after surgery, or immediate topical tretinoin or glycolic acid can also worsen pain and burning. Therefore, exposure to adverse stimuli, such as sun exposure, external use of irritating drugs, hot springs, and saunas, should be avoided for 7 d after surgery.

Potent pain relievers may be needed for deep chemical peels. Patients can apply cold compresses when they feel skin burning, tingling, or itching. Topical corticosteroids can be applied, if necessary, which can reduce inflammation and relieve pain, but routine use is not recommended.

6.5.3 Frosting

Frosting is caused by the aggregation of keratin in the stratum corneum caused by peeling agents. Frosting indicates that the skin is overreacting, and the occurrence of frosting should be avoided during superficial peels. To avoid the occurrence of frosting as much as possible, it is usually necessary to use petroleum jelly to protect the thin skin area before the operation. When using glycolic acid and other peeling agents that need to be neutralized, the patient's skin reaction should be closely observed during treatment, and neutralizing solution should be given in time for neutralization; for peeling agents that do not need to be neutralized, such as salicylic acid and TCA, the dosage should be strictly controlled to avoid applying too much peeling agent on the treatment area. When the terminal reaction or white frost is observed, the excessive peeling agent should be quickly wiped off with clean and soft gauze. Scabs may appear on the part where hoarfrost occurs 3–7 d after the operation. Patients should be instructed to avoid scratching until the scabs fall off naturally. Symptomatic treatment is required in the case of blisters and oozing.

6.5.4 Edema

Chemical peels may cause edema. Deep chemical peel is especially more likely to cause skin edema. Edema usually appears within 24–72 h after the procedure and may take several days to disappear. Edema is usually caused by excessive concentration of chemical peeling agents or excessive sensitivity of the skin, so the concentration of chemical peeling agents should be strictly controlled. In most cases, the edema is mild, but there may be obvious edema in special parts such as the eyelids, and the healthcare professionals should inform the patient in advance. Oral NSAIDs (such as ibuprofen), antihistamines (such as loratadine, cetirizine), and proper skin care can help reduce discomfort and avoid severe edema. For patients with severe edema, systemic corticosteroids such as prednisone or methylprednisolone can be used, but prophylactic use is not recommended to avoid poor skin healing.

6.5.5 Desquamation and crusting

After chemical peels, due to peeling of the epidermis and impaired skin barrier function, patients may experience varying degrees of skin tightness and local skin desquamation, which are usually mild and can be relieved after a few days or about a week. Patients should be informed to pay attention to strengthening moisturization postoperatively. Since the essence of a chemical peel is to peel off the skin at different levels and form controllable damage within a certain range, different degrees of scabs often appear after surgery. The degree of scabbing depends on the depth of the chemical peel: only a thin scab appears when the peel is superficial; if the chemical peel penetrates too deeply, it can cause a significant scab, at which time the risk of pigmentation abnormalities and scarring increases. Care should be taken when scabs appear, the patient should wait for the scabs to fall off by themselves and not uncover the scabs by themselves.

6.5.6 Hyperpigmentation

PIH is a common adverse reaction after chemical peels, which is caused by inflammatory mediators activating melanocytes in the basal layer and producing excessive melanin. PIH usually occurs 4 d to 2 months after surgery. Factors that increase the risk of PIH include not using adequate sun protection, skin types III and VI, sun-exposed skin types I and II, the use of photo-sensitizers, the use of estrogen-containing medications such as oral contraceptives and hormone replacement therapy. Improper postoperative care (especially failure to comply with required sun protection) is the most common cause of PIH. The occurrence of PIH is related to the patient's skin type and individual skin differences. The duration of pigmentation in light-skinned patients is shorter, and pigmentation may persist in dark-skinned patients; patients with other skin pigmentation history are at higher risk of developing PIH, so a spot test is recommended before proceeding with a full-face treatment. The occurrence of PIH is related to the depth of peeling, and the risk of hyperpigmentation and hypopigmentation after medium and deep chemical peels is higher. The higher the concentration of peeling agent, the deeper the penetration depth, the greater the probability of pigmentation. PIH is also the most common complication after TCA chemical peels, and if not handled properly, pigmentation can persist. In addition, photo-sensitizers can promote the occurrence of PIH. Hormonal changes in the body can also cause hyperpigmentation. Patients who become pregnant within 6 months of chemical peels are at higher risk of developing hyperpigmentation, even with sun protection. Pregnant women with skin types III and VI need sun protection for up to 1 year after surgery (Costa *et al* 2017).

Patients should be strictly sun-protected before and after chemical peels. Try to suspend taking contraceptives during treatment. If hyperpigmentation occurs after surgery, the patient should be informed to strengthen sun protection, and externally use drugs with a whitening effect, such as 4%–6% hydroquinone, 0.025% retinoic acid, 10%–20% azelaic acid, etc, or use vitamin A, licorice and other skin care products with whitening ingredients. The use of epidermal exfoliates (20%–35% glycolic acid) can speed up the removal of melanin.

6.5.7 Hypopigmentation

After chemical peels, the epidermis is exfoliated, and the removal of melanin in the keratinocytes is accelerated, making the overall skin color whiter. However, if the peeling is excessive and the basal melanocytes are damaged, permanent hypopigmentation or loss may occur, especially in dark skin. The risk of hypopigmentation also increases when the penetration of chemical peeling is not uniform and there is an obvious difference in distribution. For example, severe white precipitate reactions may occur in some areas when high concentration glycolic acid is peeled. In addition, improper postoperative care leading to infection and scarring can also cause hypopigmentation, which is very evident in patients with type III and VI skin. Treatment of hypopigmentation is difficult and should therefore be carefully evaluated prior to treatment, especially when chemical peels are used in patients with skin types above III. Some scholars have suggested that micro-needling treatment can induce re-coloration of some hypopigmented spots, but in most cases, hypopigmentation or depigmentation will persist.

6.5.8 Reactive acne

A chemical peel is an effective treatment for acne. Abnormal keratinization of the pilosebaceous duct is an important factor in the formation of acne. In the lower part of the hair follicle funnel, the lamellar granules in the keratinocytes are reduced and replaced by a large number of tension filaments, desmosomes and lipid inclusion bodies. The peeling agent can promote the dissolution and peeling of the horn plug at the opening of the pilosebaceous gland, so that the excessively accumulated sebum can be excreted through the dredged duct. However, due to the certain irritation of chemical peeling agents, in susceptible patients, it may lead to reactive inflammation in the skin of some acne patients, and temporary increase in skin lesions or aggravation of inflammation within 1 week after treatment (Yang and Jiang 2019). And because most topical acne treatments are irritating to the skin, reactive acne can be relatively difficult to treat. Therefore, before treatment, patients should be informed that there may be transient aggravation of skin lesions after surgery. For patients with reactive acne, oral tetracyclines can be given as appropriate.

6.5.9 Scarring

Scarring is one of the most serious complications of chemical peels. Although the incidence of scarring is low, doctors must inform patients before surgery. The scars that appear after chemical peels are mainly atrophica and hypertrophic scars, which usually appear within 2–3 months after the peel. Scarring is caused by abnormal wound healing. There are two forms of wound healing, one is complete repair (superficial damage to the skin), and the other is cicatricial healing with epithelialization. When the skin tissue is damaged to a certain depth, an inflammatory reaction occurs locally, and a variety of cytokines are released. Skin fibroblasts proliferate in large numbers and synthesize a large amount of collagen and matrix, resulting in abnormal collagen metabolism and arrangement. At the same time, it is affected by macrocirculation and free oxygen radicals, which may lead to the formation of scars. The first signs of scarring are persistent erythema, pruritus, and delayed healing (epithelialization over 2 weeks). If the patient has a history of poor wound healing, keloid, or hypertrophic scar formation, the risk of scar formation after chemical peels increased. The pros and cons of chemical peels should be carefully evaluated, and the patient should be fully informed of the high risk of scar formation after surgery. Medium and deep chemical peels are not recommended for such patients.

The choice of peeling agent should also be undertaken with caution. If the concentration of peeling agent is too high, the residence time is too long, there is forced scab removal, improper care or infection after the operation, this could lead to a scar forming. When performing a chemical peel, the residence time should be reasonably controlled. If scabbing or desquamation occurs after the operation, the patient should be educated to avoid scratching and to remove the scab naturally. Scar formation is also related to the type of exfoliating agent and the use of drugs. Trichloroacetic acid (TCA) is more corrosive than phenol and can penetrate deep into the deep dermis, so scars are more likely to occur, especially in thinner skin. Isotretinoin is associated with delayed wound healing and increased incidence of scarring. Patients who use tretinoin peels should wait at least 6 months before undergoing medium and deep chemical peels again.

Other causes of scarring include smoking, dermabrasion or laser treatments within 6 months, multiple applications of TCA, and chemical peels in areas prone to scarring, such as the mandible, neck, and chest. Some scholars have inferred that scar formed after intermediate-to-deep exfoliation in patients who have recently undergone laser hair removal may be difficult to recover due to re-epithelialization of adnexal structures.

Strictly standardized procedures and enhanced postoperative care can reduce the risk of scar formation. The scars caused by chemical peels are usually produced in the middle and lower parts of the face (such as the upper lip, perioral area, cheeks and other areas of excessive facial movement). Hyper-motion areas such as the inner corners of the eyelids, and other body areas such as the neck, chest, and backs of the hands are at higher risk for scarring after chemical peels compared to the face. Hypertrophic scars are prone to form in the neck, suprasternal and submental areas, and usually these areas should be limited to a superficial chemical peel, and intermediate- and deep-layer peeling should be avoided. Smaller scars usually vanish on their own, which can be helped by applying pressure bandages, massage, or topical scar softening creams. If the skin shows signs of delayed healing after a chemical peel, aggressive intervention with biological dressings and antibiotics is required. If scarring occurs, the most effective treatment is intralesional injection of glucocorticoids, and patients should be informed of the risks of skin atrophy and telangiectasia with long-term use of glucocorticoids. Intense pulsed light and pulsed dye lasers can help relieve red scars. Hypertrophic scars and contractures that affect facial function and movement require multiple treatments including surgical intervention to maximize the return of function and movement, but surgical correction is at least 6 months after initial treatment.

6.5.10 Infection

Chemical peels may cause tissue damage such as a compromised skin barrier, which can lead to infection of the skin. Infections after chemical peels are less common because the acidic chemical peels are bactericidal. Predisposing factors for the development of localized skin infections include being immunocompromised, systemic disease, localized skin breakdown, and poor wound care, which can promote the growth of micro-organisms such as streptococci, staphylococci, or pseudomonas, so preoperative-ly patients should be carefully evaluated for general condition and skin condition. Clinical features of infection are delayed wound healing, folliculitis, and ulceration.

Chemical peels may also cause herpes recurrence, so a history of herpes simplex must be screened before procedure. Particular attention should be paid when medium-depth or deep chemical peel is planned. If the patient has a history of herpes-related diseases, antiviral treatment should be considered first. The duration of herpes attack after chemical peel is about 5–12 d. Since the epidermis has not fully recovered after exfoliation, the lesions of herpes are not vesicular, but mostly appear in the form of ulcers, with a size of 2–3 mm, round, isolated or with extensive confluent erythema at the base. Treatment is oral antiviral drugs such as acyclovir and valacyclovir. For patients with a history of facial herpes simplex infection in the past, preventive antiviral therapy can be given before medium-depth to deep chemical peel, and the regimen is oral acyclovir (200–400 mg/d) or valacyclovir (500 mg/day). The treatment time should start from 2–3 d before the operation to 14 d after the end of treatment. Although herpes infections usually resolve without scarring, early treatment is recommended to prevent scarring.

Although chemical peels rarely result in bacterial or fungal infections, infectious diseases remain a relative contraindication to chemical peels. Common infection bacteria are *Staphylococcus*, *Streptococcus*, and *Pseudomonas*, and fungi are mainly *Candida*. Predisposing factors include recent oral antibiotics, being immuno-compromise, diabetes mellitus, and prolonged use of topical corticosteroids. To reduce the risk of infection, the overall health status and skin condition of the patient should be carefully evaluated before surgery, and active treatment should be taken after infection occurs.

If the patient develops fever, syncope, hypotension, vomiting, diarrhea, or scarlet fever-like or erythroderma-like rash 2–3 d after peeling, toxic shock syndrome should be screened for and treated in a timely manner. Other symptoms of toxic shock syndrome include myalgia, mucosal congestion, and hepatic, renal, hematological, or central nervous system involvement.

6.5.11 Milia

Milia are benign tumors or retention cysts originating in the epidermis or adnexal epithelium. Colloid milia, which are cysts formed by the closure of tiny hair follicles, may appear shortly after chemical peels. In addition, improper care after deep chemical peels, topical cream or foundation blocking hair follicles and sebaceous ducts can also lead to secondary milia. Preoperative or postoperative use of retinoic acid can reduce the appearance of milia, but because retinoic acid can prolong wound healing and may cause irritation, it is only recommended after the post-operative erythema has subsided. Milia usually resolve spontaneously and should be treated only when the patient has a strong desire to do so. Milia can usually be punctured with a fine needle or scraped off with a curette and can also be treated with a laser.

6.5.12 Telangiectasia

Pre-existing telangiectasia may visually become darker and more pronounced due to depigmentation of the skin after chemical peels. In addition, phenol peels can exacerbate telangiectasia. Patients need to be informed of this risk before treatment. If the patient needs treatment, intense pulsed light or pulsed dye laser treatment can be used.

6.5.13 Sensitive skin

Peeling agents act on the skin barrier in a process of first damage and then repair. Repeated application of low concentrations of alpha-hydroxy acids (5% glycolic acid, 5% lactic acid), which induces exfoliation without increasing trans-epidermal water loss (TEWL), stimulates ceramide biosynthesis, and increases the number and secretion of lamellar bodies, can improve the skin barrier function (Kim *et al* 2001). However, the temporary impairment of the skin barrier function after chemical peels can lead to increased skin sensitivity. Before treatment, patients should be informed to use moisturizing products with simple ingredients during treatment, avoid using skin care products containing alcohol, spices, etc, and avoid using too many types of skin care products. Strict sun protection is required during the treatment period and the patient must try to avoid going out. When it is necessary to go out, you should strengthen sun protection, mainly physical sun protection, such as using wide-brimmed sun hats, sunglasses, sun protection masks or masks, and sun umbrellas.

6.5.14 Allergic reactions

Contact dermatitis is uncommon, but can occur with any chemical peel, especially when chemical peels are performed frequently, at inappropriate concentrations, or with the wrong choice of chemical peel. Resorcinol, salicylic acid, kojic acid and lactic acid are more likely to cause allergic contact dermatitis than other peeling agents, and some patients may also have adverse reactions such as contact urticaria (Anitha 2010). If an allergic reaction occurs, antihistamine treatment can be given in time. Anaphylaxis is indistinguishable from erythema and edema directly caused by chemical peels, and antihistamines should be given prophylactically if the patient has a history of allergic reaction to any peeling agent.

6.5.15 Pruritis

Pruritus is due to re-epithelialization of the skin and usually begins 2 weeks after treatment and lasts for about a month. More common after medium to deep chemical peels. If the patient is troubled by pruritus, oral antihistamines and topical glucocorticoids can be given, but to avoid adverse reactions such as skin atrophy or telangiectasia, long-term use of glucocorticoids should be avoided.

References

Abdel-Motaleb A A, Abu-Dief E E and Hussein M 2017 Dermal morphological changes following salicylic acid peeling and microdermabrasion *J. Cosmet. Dermatol.* **16** e9–e14

Anitha B 2010 Prevention of complications in chemical peeling *J. Cutan. Aesthet. Surg.* **3** 186–8

Brody H J, Monheit G D, Resnik S S and Alt T H 2000 A history of chemical peeling *Dermatol. Surg.* **26** 405–9

Butler P E, Gonzalez S, Randolph M A, Kim J, Kollias N and Yaremchuk M J 2001 Quantitative and qualitative effects of chemical peeling on photo-aged skin: an experimental study *Plast. Reconstr. Surg.* **107** 222–8

Costa I M C *et al* 2017 Review in peeling complications *J. Cosmet. Dermatol.* **16** 319–26

Dainichi T *et al* 2006 Chemical peeling by SA-PEG remodels photo-damaged skin: suppressing p53 expression and normalizing keratinocyte differentiation *J. Invest. Dermatol.* **126** 416–21

Ditre C M, Griffin T D, Murphy G F, Sueki H, Telegan B, Johnson W C, Yu R J and Van Scott E J 1996 Effects of alpha-hydroxy acids on photoaged skin: a pilot clinical, histologic, and ultrastructural study *J. Am. Acad. Dermatol.* **34** 187–95

He L *et al* 2020 Update of the treatments for skin photoaging *China J. Lepr. Skin Dis.* **36** 687–9 696

Imayama S, Ueda S and Isoda M 2000 Histologic changes in the skin of hairless mice following peeling with salicylic acid *Arch. Dermatol.* **136** 1390–5

Kim I H, Kim H K and Kye Y C 1996 Effects of tretinoin pretreatment on TCA chemical peel in guinea pig skin *J. Korean Med. Sci.* **11** 335–41

Kim T H *et al* 2001 The effects of topical alpha-hydroxyacids on the normal skin barrier of hairless mice *Br. J. Dermatol.* **144** 267–73

Lu J, Cong T, Wen X, Li X, Du D, He G and Jiang X 2019 Salicylic acid treats acne vulgaris by suppressing AMPK/SREBP1 pathway in sebocytes *Exp. Dermatol.* **28** 786–94

Maity N *et al* 2011 Bioactive compounds from natural resources against skin aging *Phytomedicine* **19** (special issue)

Marczyk B, Mucha P, Budzisz E and Rotsztejn H 2014 Comparative study of the effect of 50% pyruvic and 30% salicylic peels on the skin lipid film in patients with acne vulgaris *J. Cosmet. Dermatol.* **13** 15–21

National Toxicology Program 2007 Photocarcinogenic study of glycolic acid and salicylic acid in SKH-1 ice *Technical Report* TR 524 National Toxicology Program, Research Triangle Park, NC

Obagi S 2020 *Procedures in Cosmetic Dermatology Series: Chemical Peels* 3rd edn (Amsterdam: Elsevier) pp 39–40

Scholz Brooks G J *et al* 1994 Fruit acid extracts, a fresh approach to skin renewal *Int. J. Cosmet. Sci.* **16** 265–72

Usuki A, Ohashi A, Sato H, Ochiai Y, Ichihashi M and Funasaka Y 2003 The inhibitory effect of glycolic acid and lactic acid on melanin synthesis in melanoma cells *Exp. Dermatol.* **12** 43–50

Van Scott E J and Yu R J 1984 Hyperkeratinization, corneocyte cohesion, and alpha hydroxy acids *J.Am. Acad. Dermatol.* **11** 867–79

Vemula S *et al* 2018 Assessing the safety of superficial chemical peels in darker skin: a retrospective study *J. Am. Acad. Dermatol.* **79** 508–513.e2

Weirich E G, Longauer J K and Kirkwood A H 1976 Dermatopharmacology of salicylic acid. III. Topical contra-inflammatory effect of salicylic acid and other drugs in animal experiments *Dermatologica* **152** 87–99

Xiao Y and Li Q 2003 Treatment of skin photoaging using chemical peelings *Chin. J. Aesthetic Plast. Surg.* **014** 49–50

Yang R and Jiang X 2019 Chinese experts' consensus on clinical application of chemical peeling *J. Pract. Dermatol.* **12** 257–62

IOP Publishing

Skin Photoaging (Second Edition)

Rui Yin, Yang Xu and Chengfeng Zhang

Chapter 7

Microneedling for skin rejuvenation

Rui Yin

Microneedling is a pioneering approach in the field of facial rejuvenation that stimulates cutaneous collagen production, which helps to reduce the appearance of fine lines and improve skin texture. Patients seeking a more youthful complexion have increasingly gravitated towards this procedure due to its subtle yet perceptible results and the relatively short recovery time it entails. Widely recognized for its minimal invasiveness and low risk of adverse effects, this technology offers an alternative solution for managing the signs of photoaging on the face.

7.1 Introduction

Microneedling (MN), also known as percutaneous collagen induction therapy, is a minimally invasive procedure that has gained significant attention in dermatology. It emerged in the 1990s as a treatment for scars, striae, and skin laxity (Aslam and Alster 2014). This technique involves the use of fine needles to roll or glide across the skin, create micro-injuries to the skin, reaching both the dermal papillary layer and the reticular layer through purely mechanical means, which stimulates the body's natural scarless wound healing cascade process, leading to increased collagen and elastin production (Bonati *et al* 2017). MN offers an excellent safety profile across all skin types, without causing thermal injury or necrosis of chromophore targets, unlike lasers or light modalities (e.g. non-ablative lasers, fractional lasers and intense pulsed light). It has already been established as the treatment of choice for skin rejuvenation, acne scarring, wrinkles, surgical scars, pigmentation issues, melasma, enlarged pores, and transdermal drug delivery (Alster and Graham 2018). The reported high efficacy, safety, and minimal post-treatment recovery rates associated with MN have led to increased patient satisfaction and greater recognition of its popularity among clinicians. MN has now become an integral component of conventional treatment regimens, significantly altering the approach to improving photoaged/aged skin and acne scarring (Alster and Graham 2018).

doi:10.1088/978-0-7503-5112-6ch7

7.2 Mechanism of action

Microneedling works on the principle of controlled skin injury, which initiates a cascade of biological responses leading to skin regeneration and remodeling. The process begins with the creation of micro-punctures or micro-channels in the skin by the fine needles of a microneedling device. These micro-injuries are deep enough to reach the dermis, where they stimulate fibroblasts, the skin's primary collagen-producing cells. In the early 2000s, a drum-shaped microneedle device was developed, equipped with minuscule needles designed to create micro-wounds on the skin's surface for the treatment of facial wrinkles and laxity (Fernandes 2002, Fernandes 2005). Studies have shed light on the mechanism behind microneedling, revealing that the mechanical injury inflicted triggers a cascade of wound healing responses in the dermis, including inflammation, proliferation, and remodeling. This cascade results in the release of various growth factors, such as platelet-derived growth factor, fibroblast growth factor, and transforming growth factors.

7.2.1 Activation of fibroblasts and collagen synthesis

The primary mechanism of microneedling is the activation of fibroblasts. These cells, upon receiving signals from the micro-injuries, become more metabolically active and begin to produce collagen, the main structural protein of the skin's extracellular matrix. Following the skin injury, the increase in collagen synthesis leads to a fibronectin network formed, providing a scaffold for the deposition of collagen type III, which is later replaced by type I collagen. This transition can take place over several weeks to months, leading to the thickening of the dermis and improved skin texture and elasticity (Orentreich and Orentreich 1995).

7.2.2 Release of growth factors

Another critical aspect of microneedling's action is the release of growth factors. The micro-injuries trigger the release of various growth factors, including platelet-derived growth factor (PDGF), fibroblast growth factor (FGF), transforming growth factor beta (TGF-β), vascular endothelial growth factor, FGF-7, epidermal growth factor (Aust *et al* 2010). These factors play a pivotal role in the wound healing process by promoting cell migration, proliferation, and the synthesis of extracellular matrix components such as collagen and elastin (Fernandes 2002, Aust *et al* 2008, 2010). Among them, the increased expression of TGFβ3 (a cytokine that prevents abnormal scarring) and the enhanced expression of type I collagen genes play an important role in promoting regeneration and healing without scarring (Bonati *et al* 2017).

7.2.3 Angiogenesis and neovascularization

Microneedling also induces angiogenesis, the formation of new blood vessels, which is essential for the nourishment of the newly formed tissue during the healing process. This leads to improved blood supply and oxygenation to the skin, further enhancing the skin's health and appearance (Zeitter *et al* 2014).

7.2.4 Elastin production and remodeling

In addition to collagen, microneedling also stimulates the production of elastin, another vital component of the skin's extracellular matrix. The synthesis of new elastin fibers contributes to the skin's elasticity and the reduction of fine lines and wrinkles. Over time, the newly formed elastin and collagen fibers reorganize, leading to a more youthful and firmer skin appearance (El-Domyati *et al* 2015).

7.2.5 Skin tightening and texture improvement

The cumulative effect of increased collagen and elastin production, along with the reorganization of the extracellular matrix, results in skin tightening and improved texture. The skin's surface becomes smoother, and the appearance of scars and wrinkles is diminished (Ablon 2018, Fabbrocini *et al* 2012).

7.2.6 Melanin regulation and pigmentation changes

Microneedling has also been shown to influence melanin production and distribution in the skin. By affecting melanocyte activity and melanin dispersion, microneedling can help address pigmentary disorders such as melasma and post-inflammatory hyperpigmentation (Busch *et al* 2018).

7.2.7 Enhanced permeability for topical agents

The micro-channels created by microneedling increase the skin's permeability, allowing for better absorption of topically applied agents. This property can be harnessed to enhance the effectiveness of treatments such as serums, creams, and other topical formulations containing active ingredients for skin rejuvenation (Fabbrocini *et al* 2012).

7.2.8 Immune modulation and anti-inflammatory effects

Lastly, microneedling has been suggested to have immune-modulatory effects. The procedure can lead to a transient immune response that may contribute to the observed skin improvements. Additionally, there is evidence to suggest that microneedling may have anti-inflammatory effects, which could be beneficial in conditions such as acne vulgaris (Lee *et al* 2016).

Furthermore, microneedling facilitates the delivery of topical products into the dermis, enhancing their penetration and effects, such as platelet-rich plasma and human stem cell conditioned medium, which contribute additional growth factors to the process.

7.3 Device specifics

The effectiveness of microneedling therapy is significantly influenced by the device used to create the micro-injuries on the skin. Various microneedling devices have been developed, each with unique features designed to optimize treatment outcomes.

These devices can range from simple manual rollers to more complex, automated, and motorized systems.

7.3.1 Manual rollers

Manual derma rollers are among the most commonly used microneedling devices. They typically consist of a cylindrical barrel covered with fine needles of a specific length, which can vary from 0.25 to 3 mm in diameter (Alster and Graham 2018). The needles are arranged in a uniform pattern, and the roller is rolled across the skin to create micro-punctures. The simplicity of these devices makes them accessible and easy to use for at-home treatments. However, the effectiveness of manual rollers can be limited by the evenness of pressure applied and the consistency of needle depth (Fabbrocini *et al* 2009, El-Domyati *et al* 2015).

7.3.2 Motorized devices

To address the limitations of manual rollers, motorized microneedling devices have been developed. These devices are battery-operated and allow for adjustable needle depths and speeds. The motorization ensures a more consistent application of pressure and needle penetration, which can lead to more uniform and effective treatment outcomes (Aust *et al* 2010, Zeitter *et al* 2014).

7.3.3 Automated microneedling systems

Automated systems take microneedling precision a step further. These devices use computer-controlled movements to create a precise pattern of micro-injuries. Some systems are capable of varying the depth and density of the needles, allowing for a customized treatment plan based on the patient's specific needs and skin condition (Alster and Graham 2018, Fabbrocini *et al* 2012).

7.3.4 Radiofrequency-coupled microneedling

Innovative advancements in microneedling technology have led to the development of devices that combine microneedling with radiofrequency energy. These devices use the microneedling process to create channels in the skin, which are then targeted with radiofrequency energy to stimulate additional collagen production and skin tightening. This combination therapy offers the potential for enhanced aesthetic results (El-Domyati *et al* 2015, Hong *et al* 2020).

7.3.5 Single-use needle cartridges

To ensure hygiene and prevent cross-contamination, many modern microneedling devices utilize single-use, sterile needle cartridges. These cartridges can be easily attached to the device and are disposed of after a single treatment, providing a safe and effective treatment option.

In conclusion, the choice of microneedling device depends on various factors, including the patient's skin condition, the desired depth of treatment, and the practitioner's preference for a particular device. Advances in microneedling device technology continue to enhance the safety, precision, and effectiveness of this popular aesthetic procedure.

7.4 Microneedling for wrinkles and skin rejuvenation

In the treatment of photoaging, microneedling has been widely applied and has been proven to effectively ameliorate facial rhytides (Alster and Graham 2018, Aust *et al* 2011, Lima Eva *et al* 2013). It primarily stimulates the production of new collagen through micro-injuries, filling and smoothing out wrinkles. These micro-injuries induce angiogenesis, promoting the formation of new blood vessels that nourish the skin and enhance its vitality, resulting in a more youthful and radiant complexion. The creation of micro-wounds triggers the release of growth factors that are crucial for the synthesis and deposition of collagen and elastin in the dermis; hence, the facial skin becomes more elastic and firmer after microneedling treatment (Doddaballapur 2009). Studies have indicated a reduction in the severity of perioral wrinkles, along with a significant increase in collagen types I, III, and VII, and elevated levels of tropoelastin (El-Domyati *et al* 2015). The reorganization of existing collagen fibers and the concurrent increase in new structural dermal components account for the observed skin tightening. The enhancement of dermal collagen and elastic fibers contributes to the reduction and softening of wrinkles post-microneedling treatment. Clinical studies suggest that at least 6–8 weeks are required for the production of dermal collagen to become clinically apparent. Consequently, patients typically undergo several microneedling sessions biweekly or monthly to achieve optimal improvement in rhytides and overall skin rejuvenation (El-Domyati *et al* 2015).

Additionally, microneedling treatment can enhance transdermal drug delivery, assisting topical agents used for skin rejuvenation to achieve better percutaneous absorption. This includes topical retinoids, antioxidants, vitamins, amino acids, and peptides, thereby amplifying the efficacy of these components within the skin.

7.5 Side effects and complications

7.5.1 Common temporary side effects

The most frequently reported side effects of microneedling are generally mild and short-lived. These include the following.

Erythema and discomfort: The skin may appear red and flushed following the procedure, which typically resolves within a few hours to a day. This reaction is a natural response to the micro-injuries inflicted by the microneedling process.

Edema (swelling): Some patients may experience localized swelling, which is also a part of the body's healing response. This usually subsides within 24–48 h with cold compresses and anti-inflammatory measures (Aust *et al* 2008).

Pinpoint bleeding: Given the nature of the treatment, which involves creating micro-punctures, it is common to observe minimal bleeding at the treatment sites. It is self-limiting and resolves quickly after the procedure with gentle pressure and ice water-soaked gauze.

Desquamation (peeling): Similar to the peeling that occurs after other skin resurfacing procedures, microneedling can cause mild skin peeling. This is a sign of the skin's renewal process and usually resolves within a few days.

7.5.2 Less common but potentially more serious side effects

While rare, there are more serious side effects that can occur if microneedling is not performed correctly or if post-treatment care is not adhered to.

Infection: As with any procedure that involves breaking the skin, there is a risk of infection. However, this can be minimized with proper sterile technique and post-treatment care, including the use of antibiotics and avoiding the use of contaminated products.

Hyperpigmentation: Some individuals, particularly those with darker skin types, may experience hyperpigmentation. This is usually preventable with the use of skin lightening agents and strict sun protection (El-Domyati *et al* 2015, Aust *et al* 2008, Pahwa *et al* 2012).

Hypopigmentation: Conversely, in some cases, hypopigmentation may occur, where the skin may lose some of its color. This is typically temporary and resolves over time.

Granuloma formation: There have been reports of granuloma formation following microneedling, particularly when used in conjunction with certain topical products. The use of topical medications during or immediately after microneedling may increase the incidence of adverse effects due to the creation of channels within the epidermis and dermis, serving as gateways for immunogenic particles and potentially triggering immune responses (Kontochristopoulos *et al* 2016). This underscores the importance of using medically approved products and following post-treatment guidelines (Soltani-Arabshahi *et al* 2014).

Scarring: Although microneedling is designed to promote collagen production without causing scarring, improper technique or excessive needle depth can potentially lead to scarring. 'Tram track scarring' has been associated with high-pressure needling and the use of long needles over bony prominences (Pahwa *et al* 2012).

Allergic reactions: Local and systemic hypersensitivity have also been reported, attributed to the use of unsuitable topical products. It is essential to advise against the use of non-prescribed skincare products post-microneedling and for physicians to be cautious with topical agents to prevent complications (Soltani-Arabshahi *et al* 2014).

7.5.3 Risk aversion

To mitigate the risk of adverse reactions, it is crucial to select a qualified and experienced practitioner who adheres to best practices for the procedure. Patients should also be provided with comprehensive pre- and post-treatment instructions, including:

- Avoiding sun exposure and using high-SPF sunscreens.
- Keeping the treatment area clean and avoiding the use of potentially irritating skincare products.
- Avoiding strenuous activities that could increase blood flow to the treated area.

While microneedling is generally considered safe with a low risk of adverse effects, it is not without potential complications. By understanding these risks and taking appropriate precautions, both practitioners and patients can maximize the benefits of microneedling while minimizing any undesirable outcomes.

7.6 Conclusion

Microneedling is a safe, minimally invasive, and effective form of cosmetic treatment for a variety of skin conditions, including wrinkles and stretch marks. This technique stimulates neogenesis of skin components offering a therapeutic approach to correct aesthetic discrepancies attributable to injury, disease, or senescence (Ramaut *et al* 2018). Distinguished by its low risk of dyspigmentation and scarring, microneedling emerges as a preferable alternative to more aggressive modalities such as ablative laser therapy and chemical peels. Due to its simplicity and low complication rate, microneedling has been widely used in clinical practice. Published clinical studies, including randomized controlled studies, prospective clinical studies, and controlled clinical trials, provide evidence for the use of microneedling alone and in combination. Nonetheless, the absence of standardized evidence-based guidelines remains an unresolved issue. In addition, the methodology of microneedling use, as well as the choice of drugs and/or products used in combination, will also produce different clinical outcomes. In the future, there are still many unknowns that need to be explored in depth in the field of rejuvenation. Each study's methodology was critically assessed, and findings were summarized to determine the existing evidence and identify areas requiring further research.

References

Ablon G 2018 Safety and effectiveness of an automated microneedling device in improving the signs of aging skin *J. Clin. Aesthet. Dermatol.* **11** 29–34

Alster T S and Graham P M 2018 Microneedling: a review and practical guide *Dermatol. Surg.* **44** 397–404

Aslam A and Alster T S 2014 Evolution of laser skin resurfacing: from scanning to fractional technology *Dermatol. Surg.* **40** 1163–72

Aust M C *et al* 2008 Percutaneous collagen induction: minimally invasive skin rejuvenation without risk of hyperpigmentation—fact or fiction? *Plast. Reconstr. Surg.* **122** 1553–63

Aust M C *et al* 2010 Percutaneous collagen induction. Scarless skin rejuvenation: fact or fiction? *Clin. Exp. Dermatol.* **35** 437–9

Aust M C *et al* 2011 Percutaneous collagen induction—regeneration in place of cicatrisation? *J. Plast. Reconstr. Aesthet. Surg.* **64** 97–107

Bonati L M, Epstein G K and Strugar T L 2017 Microneedling in all skin types: a review *J. Drugs Dermatol.* **16** 308–13

Busch K H *et al* 2018 Medical needling: effect on skin erythema of hypertrophic burn scars *Cureus* **10** e3260

Doddaballapur S 2009 Microneedling with dermaroller *J. Cutan. Aesthet. Surg.* **2** 110

El-Domyati M, Barakat M, Awad S, Medhat W, El-Fakahany H and Farag H 2015 Multiple microneedling sessions for minimally invasive facial rejuvenation: an objective assessment *Int. J. Dermatol.* **54** 1361–9

Fabbrocini G *et al* 2012 Collagen induction therapy for the treatment of upper lip wrinkles *J. Dermatolog. Treat.* **23** 144–52

Fabbrocini G, Fardella N, Monfrecola A *et al* 2009 Acne scarring treatment using skin needling *Clin. Exp. Dermatol.* **34** 874–9

Fernandes D 2002 Percutaneous collagen induction: an alternative to laser resurfacing *Aesthet. Surg. J.* **22** 307–9

Fernandes D 2005 Minimally invasive percutaneous collagen induction *Oral Maxillofac. Surg. Clin. North Am.* **17** 51–63

Hong J Y, Kwon T R, Kim J H *et al* 2020 Prospective, preclinical comparison of the performance between radiofrequency microneedling and microneedling alone in reversing photoaged skin *J. Cosmet. Dermatol.* **19** 1105–9

Kontochristopoulos G, Kouris A, Platsidaki E, Markantoni V, Gerodimou M and Antoniou C 2016 Combination of microneedling and 10% trichloroacetic acid peels in the management of infraorbital dark circles *J. Cosmet. Laser Ther.* **18** 289–92

Lee J C, Daniels M A and Roth M Z 2016 Mesotherapy, microneedling, and chemical peels *Clin. Plast. Surg.* **43** 583–95

Lima Eva L M T D 2013 Microneedling: experimental study and classification of the resulting injury *Surg. Cosmet. Dermatol.* **5** 110–4

Orentreich D S and Orentreich N 1995 Subcutaneous incisionless (subcision) surgery for the correction of depressed scars and wrinkles *Dermatol. Surg.* **21** 543–9

Pahwa M, Pahwa P and Zaheer A 2012 'Tram track effect' after treatment of acne scars using a microneedling device *Dermatol. Surg.* **38** 1107–8

Ramaut L, Hoeksema H, Pirayesh A, Stillaert F and Monstrey S 2018 Microneedling: where do we stand now? A systematic review of the literature *J. Plast. Reconstr. Aesthet. Surg.* **71** 1–14

Soltani-Arabshahi R, Wong J W, Duffy K L and Powell D L 2014 Facial allergic granulomatous reaction and systemic hypersensitivity associated with microneedle therapy for skin rejuvenation *JAMA Dermatol.* **150** 68–72

Zeitter S *et al* 2014 Microneedling: matching the results of medical needling and repetitive treatments to maximize potential for skin regeneration *Burns* **40** 966–73

IOP Publishing

Skin Photoaging (Second Edition)

Rui Yin, Yang Xu and Chengfeng Zhang

Chapter 8

Mesotherapy for skin rejuvenation

Qi Luan

Mesotherapy involves injecting various active substances into the skin for facial rejuvenation, including absorbable and non-absorbable ingredients. Targeted injections stimulate collagen production, improving skin texture, hydration, and elasticity while reducing signs of aging. It addresses wrinkles, fine lines, age spots, and enhances skin tone and lip volume. Mesotherapy offers a safe, effective non-surgical option for comprehensive facial rejuvenation.

8.1 Introduction and definition

The concept of mesotherapy originates from the Greek words 'mesos' (middle) and 'therapeia' (to treat medically) and continues to be utilized today (Lee *et al* 2016). Mesotherapy is a minimally invasive therapeutic technique involving the intradermal (ID) or subcutaneous (SQ) microinjection of beneficial compounds, including pharmaceutical and homeopathic medications, plant extracts, vitamins, hyaluronic acid (HA), and other bioactive substances, into the dermis and/or subcutaneous fat (Atiyeh *et al* 2008, Khalili *et al* 2022). The topical application of these active substances is often ineffective due to limitations in penetrating the lipid barrier and the stratum corneum.

In the context of treating photoaging, mesotherapy microinjections have been employed to create an optimal physiological environment that stimulates fibroblast activity and collagen production. HA is utilized for maintaining or restoring youthful skin characteristics, such as firmness, brightness, and moisture content. Mesotherapy facilitates the delivery of active ingredients that can reverse the processes associated with photoaging, including elastin degradation and continuous transepidermal water loss (Plachouri and Georgiou 2019).

Mesotherapy has experienced significant growth in popularity in recent years due to its ability to provide convenient and targeted drug delivery to specific areas of interest with minimal training required for physicians (Plachouri and Georgiou 2019). It is widely utilized in aesthetic dermatology for various purposes, including

doi:10.1088/978-0-7503-5112-6ch8

the treatment of cellulite, localized fat deposits, alopecia, skin rejuvenation, acne scars, dark circles under the eyes, and melasma. However, further clinical studies are necessary to validate the efficacy of mesotherapy for skin rejuvenation, including wrinkle reduction, increased skin elasticity, and improved pigmentation. Mesotherapy utilizes various substances such as minerals, antioxidants, and vitamins, all of which must be sterile, water-soluble, and suitable for intradermal injection. It is important to use separate syringes for different substances and avoid mixing them within the same syringe (Atiyeh and Abou Ghanem 2021). This chapter aims to comprehensively review all mesotherapy-related applications, safety considerations, and reported complications documented in the literature thus far, providing valuable insights into this technique.

8.2 Application in photoaging

The efficacy of mesotherapy is largely determined by the primary ingredients injected into the skin. Accordingly, mesotherapy substances can be categorized into three main types based on their ingredients. The first type involves a judicious combination and allocation of known beneficial ingredients, forming a cocktail formula that includes vitamins, minerals, homeopathic remedies, and proteins. The second type comprises injectable products containing ingredients with well-defined efficacy, such as poly-l-lactic acid (PLLA) and polydeoxyribonucleotide (PDRN). The third type consists of component ingredients derived from purified autologous human blood, which are then reinjected into the body. Overall, mesotherapy products fall into these three categories, all aimed at facial rejuvenation and reversing the effects of photoaging.

8.2.1 Platelet-rich plasma (PRP) and platelet-rich fibrin (PRF)

Platelet-rich plasma (PRP) is a concentration of platelets and plasma rich in monocytes, macrophages, growth factors, and anti-inflammatory factors, playing a pivotal role in inflammation reduction and tissue regeneration (Maisel-Campbell *et al* 2020). Platelet-rich fibrin (PRF) stands as a newer alternative to PRP, representing the second generation of platelet concentrates (Mijiritsky *et al* 2021). Platelet-rich plasma (PRP) therapies have been employed for over three decades to treat a variety of conditions, harnessing the regenerative potential of autologous PRP in the field of regenerative medicine. Recently, PRP has gained significant traction in medical cosmetology. The core mechanism of PRP therapy involves the injection of concentrated platelets into targeted areas, which can initiate tissue repair through the release of numerous biologically active factors, including growth factors, cytokines, lysosomes, and adhesion proteins. These substances facilitate processes such as angiogenesis, neocollagenesis, and adipogenesis (Choi *et al* 2022).

During the various stages of tissue repair, a complex interplay of growth factors, cytokines, and locally acting regulators is essential for enhancing beneficial cellular functions. These interactions occur through multiple signaling mechanisms, including the endocrine, paracrine, autocrine, and intracrine pathways. Each mechanism contributes to the modulation of cellular activities, facilitating effective tissue regeneration and repair. The coordinated action of these bioactive molecules ensures the precise regulation of healing processes, optimizing outcomes in tissue recovery and regeneration (Giusti *et al* 2020). A notable advantage of PRP is its autologous nature, resulting in no known adverse reactions, unlike other commonly used products. However, there remains a lack of clear regulations regarding the formulation and composition of injectable PRP (Beitzel *et al* 2015), leading to substantial variations in platelet content, white blood cell (WBC) count, red blood cell (RBC) contamination, and platelet growth factor (PGF) concentrations among PRP compositions (Everts *et al* 2006). Additionally, the integration of autologous PRP injections with other treatment modalities, such as chemical peels, laser therapies, and other microneedling and radiofrequency devices, has demonstrated substantial enhancements in skin rejuvenation outcomes. This combination approach leverages the regenerative properties of PRP to amplify the effectiveness of these treatments, resulting in improved skin texture, tone, and overall appearance. The synergistic effects observed from these combined therapies indicate a more comprehensive and effective strategy for achieving optimal skin rejuvenation (Emer 2019).

Platelet-rich plasma (PRP) is prepared through a method known as differential centrifugation, wherein the centrifugal force is adjusted to separate cellular components according to their specific gravities. Various commercially available PRP systems and kits exist, each with protocols that vary by brand and treatment indication. Traditionally, 10–22 ml of whole blood is drawn via venipuncture and mixed with an anticoagulant. The blood is then centrifuged to separate it into red blood cells (RBCs), platelet-poor plasma (PPP), and the desired PRP layer. Subsequent centrifugation steps isolate and collect the PRP while discarding the RBCs and PPP. The concentrated PRP can be activated with calcium chloride or thrombin to release alpha granules and growth factors, although activation is often unnecessary for common dermatological and cosmetic applications (Emer 2019). Activation of growth factors in PRP begins rapidly, with initial activation occurring within 10 min. This process progresses efficiently, reaching nearly complete activation—approximately 100%—within 60 min. This rapid activation timeline is critical for ensuring the effectiveness of PRP in therapeutic applications, allowing for the prompt release of the bioactive molecules necessary for tissue regeneration and repair. The duration of viability for activated growth factors in PRP remains uncertain, as factors such as pH and temperature fluctuations can impact their stability within hours after collection (Emer 2019).

Various classification systems have been proposed for different PRP formulations based on PRP terminologies and product descriptions (Rossi *et al* 2019). The objective of classifying PRP systems is to establish standardized protocols for the production, definition, and formulation of PRP for clinical use. PRP preparations are typically divided into four categories: pure platelet-rich plasma (P-PRP), leukocyte platelet-rich plasma (L-PRP), pure platelet-rich fibrin matrix (P-PRF), and leukocyte and platelet-rich fibrin matrix (L-PRF). In the field of aesthetic dermatology, P-PRP, which involves minimal leukocyte collection, is most commonly employed due to its specific advantages in enhancing skin rejuvenation

(Leo *et al* 2015). Pure platelet-rich fibrin (P-PRF) preparations, characterized by a lower concentration of platelets and the inclusion of fibrin, effectively bind and trap growth factors, facilitating their gradual release over several days. This sustained release mechanism makes P-PRF particularly suitable for applications such as fat grafting procedures, where a prolonged release of growth factors can enhance the survival and integration of grafted fat cells. The gradual delivery of bioactive molecules supports ongoing tissue regeneration and repair, optimizing the outcomes of such aesthetic interventions (Sclafani 2011).

The absolute PRP content, purity, and biological properties of PRP and related products vary considerably, impacting biological effects and clinical trial outcomes. The choice of PRP preparation device introduces the first critical variable. In clinical practice, physicians may employ two distinct PRP preparation devices and methods, including standard blood cell separators and gravitational centrifugation techniques and devices. The method and process of PRP procurement result in significant differences in yields, concentration, purity, viability, and activation status (Fadadu *et al* 2019). Furthermore, ongoing developments in clinical PRP product technologies and scientific data suggest the need for different PRP formulations to address various indications under specific conditions. Therefore, we anticipate that parameters and variables for ideal PRP production will continue to evolve in the future.

The reporting of PRP preparation protocols in clinical studies has been highly inconsistent, making it challenging to compare PRP products and their associated therapy outcomes. Progress has been made in understanding the variations in PRP formulations, concentrations, and delivery routes influencing tissue repair and regeneration for treating skin photoaging (Chahla *et al* 2017). However, further research is necessary to establish a consensus regarding PRP terminologies related to PRP bioformulations to effectively and safely treat skin photoaging.

Currently, PRP is characterized, and its therapeutic effect is determined by its absolute platelet concentration, shifting from the initial definition of PRP, which consisted of a platelet concentration above baseline values, to a minimum platelet concentration of more than $1 \times 10^6\ \mu l^{-1}$ or approximately a five-fold increase in platelets from baseline (Haunschild *et al* 2020). Unfortunately, the failure to report PRP concentration in most clinical reports has been a significant limitation. In two studies, concentrations of at least $1 \times 10^6\ \mu l^{-1}$ were reported, which is the suggested minimum for effective treatment (Cameli *et al* 2017, Hui *et al* 2017). The concentration range was wide (524–3760 × 109 platelets ml^{-1}), with the maximum concentration being seven times the lowest. While some studies have suggested that excessively high platelet concentrations are unnecessary for effectiveness, there was one study that showed no correlation between platelet concentration and skin improvement (Kang *et al* 2014). Until a consensus on standardized PRP preparations and formulations is reached, PRP concentration should adhere to a clinical PRP recipe to contribute to significant tissue repair mechanisms and progressive clinical outcomes.

In addition to dose-dependency, the effects of PRP on cell activity appear to be highly time-dependent. Short-term exposure to human platelet lysate stimulates bone cell proliferation and chemotaxis, while long-term exposure to PRP results in decreased levels of alkaline phosphatase and mineral formation (Soffer *et al* 2004).

PRP has emerged as a significant therapeutic approach in aesthetic medicine. Various factors and active cells within PRP promote neocollagenesis, angiogenesis, and overall proliferation of dermal fibroblasts, leading to soft tissue remodeling. The top evidence-based cosmetic indications for PRP include skin rejuvenation, dermal augmentation, and improvements in striae distensae. Recent research findings on the effects of PRP on cell senescence and aging are promising, making it reasonable to recommend its use in treating photoaging. The mechanisms behind skin rejuvenation involve notable increases in collagen density and dermal elastic fibers when PRP is used.

A prospective, single-center, single-dose, open-label, non-randomized controlled clinical study was conducted to evaluate the efficacy and safety of intradermal PRP injection in facial rejuvenation. In this study, PRP was injected into the right infra-auricular area and the entire face, while saline was injected into the left infra-auricular area as a control. Histopathological examinations were performed before PRP treatment, 28 days after PRP treatment, and after saline treatments. Twenty women aged 40–49 years participated in the study. The results showed that the mean optical density (MOD) of collagen fibers was significantly higher on the PRP-treated side, with a PRP-to-saline improvement ratio of 1.93:1 (89.05%–46.01%). No serious side effects were detected. The study suggested that PRP increased dermal collagen and could be considered an effective and safe procedure for facial skin rejuvenation (Abuaf *et al* 2016).

Another split-face study was designed to compare the efficacy and safety of PRP with ready-made growth factors in skin rejuvenation. Twenty adult females with Fitzpatrick skin types III–IV and Glogau photoaging types II and III were enrolled. They underwent a split-face therapy where each side was randomly assigned to treatment with either ready-made growth factors or autologous PRP. All patients received six sessions at 2 week intervals. Evaluation was carried out using the Global Aesthetic Improvement Scale (GAIS) and optical coherence tomography (OCT), with patients followed up for 6 months. Results showed significant improvement in both GAIS (skin turgor and overall vitality) and OCT (epidermal and dermal thickness) assessments for both procedures. A significant negative correlation was found between patients' age, sun exposure, and GAIS. Burning sensation was significantly higher in areas with ready-made growth factors, while patient satisfaction was significantly higher in areas with PRP. Improvement was more sustained in areas treated with PRP during follow-up. The study suggested that PRP is effective and safe for skin rejuvenation, comparable to ready-made growth factors but with noticeably higher longevity (Gawdat *et al* 2017).

Both of these aforementioned studies are randomized controlled trials evaluating PRP for facial rejuvenation in 40 patients, strongly demonstrating the efficacy and safety of PRP. Additionally, the objective of another study was to evaluate the efficacy and safety of a single autologous PRP intradermal injection for the treatment of facial wrinkles and rejuvenation. Twenty subjects with different types of facial wrinkles were included. All subjects received a single PRP intradermal injection and were clinically assessed before and after treatment for 8 weeks using the Wrinkle Severity Rating Scale (WSRS), Skin Homogeneity and Texture (SHnT) Scale, Physician Assessment Scale, and Subject Satisfaction Scale. The most significant results were observed in younger subjects with mild and moderate wrinkles of the nasolabial folds. Fourteen of seventeen subjects with nasolabial folds showed more than 25% improvement in their appearance. The study suggested that single PRP intradermal injection is well-tolerated and capable of rejuvenating the face and significantly correcting wrinkles, especially in the nasolabial folds (Elnehrawy *et al* 2017).

Several studies have demonstrated improvements in skin elasticity, texture, hydration, wrinkles, and micropigmentation following PRP injection. These improvements were observed 1–3 months after treatment (Hui *et al* 2017, Soffer *et al* 2004) and lasted for 6 months and lasted 6 months (Abuaf *et al* 2016) after 3 PRP treatments, with gradual return to baseline (Cameli *et al* 2017, Everts *et al* 2019, Redaelli *et al* 2010).

Combinations of PRP with laser resurfacing, botulinum toxin, and HA injection are effective and popular in treating skin photoaging. A split-face randomized clinical trial was conducted to compare the efficacy of PRP in 'fluid' versus 'gel' form combined with fractional CO_2 laser in treating atrophic acne scars. Twenty-seven patients with atrophic acne scars were enrolled in the study. Treatment with fractional CO_2 laser plus plasma fluid/gel was randomly assigned to the right/left sides of the face. Clinical and optical coherence tomography (OCT) assessments were conducted at baseline, 1 month, and 3 months after the last session. The results revealed a significant improvement in clinical assessment scores at the third-month follow-up on the plasma-gel- and plasma-fluid-treated sides compared to those at the first-month follow-up. Scar depth decreased significantly at the third-month follow-up compared to baseline on both plasma gel- and plasma fluid-treated sides. The pain score was significantly lower on the plasma fluid-treated side compared to the plasma gel-treated side. This study concluded that the use of PRP in combination with fractional CO_2 laser, in both fluid and gel forms, yielded significant results in treating atrophic acne scars. Patients reported a more noticeable effect but experienced more intense pain with plasma gel than with fluid-treated sides (Gawdat *et al* 2022).

Another combined study of PRP and fractional CO_2 laser was designed to make clear whether the combination of PRP and ultra-pulsed fractional CO_2 laser had a synergistic effect on facial rejuvenation. Totally, 13 facial aging females were treated with ultra-pulsed fractional CO_2 laser. One side of the face was randomly selected as the experimental group and injected with autologous PRP, the other side acted as the control group and was injected with physiological saline at the same dose. Comprehensive assessment of clinical efficacy was performed by satisfaction scores, dermatologists' double-blind evaluation and the VISIA skin analysis system. After treatment for 3 months, subjective scores of facial wrinkles, skin texture, and skin elasticity were higher than that in the control group. Similarly, improvement of skin wrinkles, texture, and tightness in the experimental group was better compared with the control group. Additionally, the total duration of erythema, edema, and crusting was decreased, in the experimental group compared with the control group. PRP combined with ultra-pulsed fractional CO_2 laser had a synergistic effect on facial rejuvenation, shortening duration of side effects, and promoting better therapeutic effect (Hui *et al* 2017).

Generally speaking, PRP injections are safe and may offer modest benefits for aging skin, with the most convincing evidence supporting improvement in facial skin aging. PRP can be easily harvested from patients' own whole blood using various commercially available systems, making it a safe and convenient procedure. Furthermore, combining PRP with other treatment modalities, such as laser therapies, microneedling, and dermal fillers, has demonstrated synergistic effects, enhancing overall anti-aging outcomes. The dermatologic community emphasizes the need for more studies to further standardize and define PRP protocols in skin rejuvenation.

8.2.2 Polydeoxyribonucleotide (PDRN)

Mesotherapy involves delivering various beneficial substances directly into the dermis and subcutaneous layer of affected areas through several injecting methods, including aminophylline, enzymes, minerals, L-carnitine, and more recently, polydeoxyribonucleotides (PDRN). PDRN is an active compound extracted from salmon trout sperm and purified to contain a high percentage of DNA, promoting and stimulating wound healing by enhancing angiogenesis and increasing fibroblast growth rates (Veronesi *et al* 2017). Combining PDRN with other therapies has become a popular cosmetic procedure for skin rejuvenation, with a growing interest reflected in numerous experimental and clinical studies conducted with this DNA-derived drug (Hwang *et al* 2018).

PDRN is a proprietary and registered DNA-derived drug, comprising a mixture of deoxyribonucleotides with molecular weights ranging from 50–1500 kDa, derived from *Oncorhynchus mykiss* (salmon trout) or *Oncorhynchus keta* (chum salmon) sperm DNA (Valdatta *et al* 2004). The most common molecular weight falls between 80 and 200 kDa, with a peak of the Gaussian distribution at 132 kDa. PDRN is extracted and purified at high temperatures, a process yielding $> 95\%$ pure active substance with inactivated proteins and peptides. Pharmacokinetics of PDRN have been evaluated after a single intraperitoneal administration of $8\ \mathrm{mg\,kg^{-1}}$ in animal models. Measurable levels of PDRN were observed 15 min post-injection, peaking at 1 h after administration, with a bioavailability of 90%. Subsequently, PDRN levels gradually decreased, remaining measurable even 6 h post-injection. Its half-life is approximately 3 h and is unaffected by dosage. The pharmacodynamic effects of PDRN may exceed its plasma half-life significantly, as the drug initiates a cascade of events involving numerous transduction effectors that last much longer than its plasma half-life (Squadrito *et al* 2017).

PDRN functions by stimulating cell growth and repair processes in the body. It achieves this by delivering nucleotides, the building blocks of DNA, to the cells. These nucleotides activate the cells and aid in increasing the production of new skin cells, resulting in an improvement in skin appearance. Additionally, PDRN possesses anti-inflammatory properties, which can help reduce damage caused by UV radiation and enhance the appearance of aging skin (Gennero *et al* 2013).

Numerous clinical studies have explored the use of PDRN in treating skin photoaging. These studies have demonstrated that PDRN can significantly enhance the appearance of photoaged skin, reducing fine lines and wrinkles, and improving skin texture and tone. In some instances, PDRN has proven to be as effective as other skin rejuvenation treatments, such as topical retinoids and laser therapy (Squadrito *et al* 2017). As a treatment for skin photoaging, PDRN, an A2A receptor agonist, exerts angiogenic effects by augmenting vascular endothelial growth factor (VEGF) (Bitto *et al* 2008a, Bitto *et al* 2008b) and tissue repair effects by stimulating fibroblasts (Galeano *et al* 2008, Sini *et al* 1999). Activation of the A2A receptor induces an anti-inflammatory effect by inhibiting several pro-inflammatory mediators (Bitto *et al* 2013, Chan *et al* 2007). Additionally, the salvage pathway is implicated in PDRN's stimulatory effect on accelerated tissue regeneration and wound healing (Thellung *et al* 1999). PDRN may also shield cells from UV-induced DNA damage, likely through priming the salvage pathway. Following exposure of human dermal fibroblasts to ultraviolet B radiation, which leads to the accumulation of hazardous photoproducts, the addition of PDRN to the cell culture immediately activates the p53 protein and enhances DNA repair (Belletti *et al* 2007). Thus, collectively, these *in vitro* findings support the notion that PDRN, with its dual anti-inflammatory and collagen synthesis effects, could be beneficial in skin rejuvenation when administered to the affected areas.

PDRN enhanced the skin repair process and increased wound-breaking strength in diabetic animals, indicating its potential importance in skin rejuvenation. This effect was supported by a significant rise in the expression of VEGF, a key regulator of angiogenesis that is often impaired in diabetes-related wound disorders. A study assessed the effects of PDRN on diabetes-related healing defects using an incisional skin-wound model created on the backs of female diabetic mice (db+/db+) and their normal littermates (db+/+m). Animals received daily treatment for 12 days with PDRN (8 $mg\,kg^{-1}\,ip^{-1}$) or its vehicle (100 μl 0.9% NaCl). Mice were euthanized 3, 6, and 12 days after skin injury to measure VEGF mRNA expression and protein synthesis, assess angiogenesis and tissue remodeling through histological evaluation, and study CD31, angiopoietin-1, and transglutaminase-II. The study found that PDRN injection in diabetic mice led to increased VEGF mRNA expression and protein content in wounds on day 6. PDRN injection improved impaired wound healing and increased wound-breaking strength in diabetic mice. Furthermore, PDRN significantly increased CD31 immunostaining and induced expression of transglutaminase-II and angiopoietin-1 (Galeano *et al* 2008).

Thermal injury is also characterized by a deficient skin repair process and impaired angiogenesis. The effects of PDRN were investigated in mice with deep-dermal second-degree burn injuries. Treatment with PDRN enhanced burn wound re-epithelialization and reduced the time to final wound closure. The wound healing properties of PDRN may result from stimulating the altered cell-cycle machinery, which is deeply impaired in several conditions. In a diabetic setting, the drug stimulated the proliferation of granulation tissue by activating cyclins that drive cell-cycle progression and suppressing the cell-cycle negative regulators p15 and p27 (Bitto *et al* 2008a).

PDRN is typically administered via skin injection by a trained medical professional, and the treatment is usually completed in one session with minimal downtime. It is generally considered safe for clinical use, with mild swelling and redness at the injection site being the most common reported side effects, typically resolving within a few days. Serious side effects are rare, but as with any medical treatment, there is a risk of side effects and potential complications (Squadrito *et al* 2017).

As a promising treatment for skin rejuvenation, PDRN has been supported by several clinical studies demonstrating its efficacy in improving the appearance of aging skin. Administered via injection, it is generally regarded as safe for clinical use. Further research is necessary to fully comprehend the potential benefits and limitations of PDRN for skin rejuvenation, but it presents a promising new option for individuals seeking to enhance their skin's appearance (Sun *et al* 2023).

8.2.3 Exosomes

Exosomes are small vesicles, typically 30–150 nm in diameter, released by cells to facilitate cell-to-cell communication. They contain various biomolecules such as proteins, lipids, and RNA, which can be transferred to other cells, influencing their behavior (Doyle and Wang 2019). Exosomes play roles in immune regulation, tissue repair, and cancer progression, constituting a rapidly growing field with potential applications in disease diagnosis and treatment (Pegtel and Gould 2019).

The structure of exosomes comprises a lipid bilayer enclosing the inner contents of the vesicle. This lipid bilayer consists of phospholipids, cholesterol, and other lipids derived from the endoplasmic reticulum and Golgi apparatus of the parent cell. Additionally, exosomes contain a small number of proteins, including those involved in vesicle formation, sorting, and fusion, as well as regulators of their contents. In essence, the size, shape, and density of exosomes are primarily determined by the specific protein, lipid, enzymatic, and mineral content of each individual exosome, making them highly variable (Booth *et al* 2006, Fang *et al* 2007). Exosome analytics encompasses the study of exosome composition, function, and biological significance. While achieving accuracy in exosome measurements remains challenging, unbiased cryo–electron microscopy experiments have provided valuable insights into the range of exosomes and other extracellular vehicles (EVs) circulating in human biofluids (Arraud *et al* 2014, Sódar *et al* 2016). This field has seen rapid growth in recent years, with the development of numerous technologies and methods for exosome isolation, characterization, and analysis. Key applications of exosome analytics include studying exosome biogenesis, identifying exosomal biomarkers for disease diagnosis, and developing exosome-based therapies for various diseases.

Exosomes contain a wide range of transmembrane proteins, lipid-anchored membrane proteins, peripherally associated membrane proteins, and soluble proteins within the exosome lumen, playing a crucial role in regulating exosome formation and function. Some of the most commonly found exosomal proteins include the following.

Tetraspanins: A family of proteins that play a role in regulating exosome formation, as well as in the interaction of exosomes with target cells. Tetraspanins do not contain catalytic activities of their own, but rather, they facilitate the trafficking, function, stability, and oligomerization of other membrane proteins (Hemler 2003).

Rab GTPases: These proteins regulate vesicle formation and trafficking. The inner membrane of exosomes is rich in acylated, lipid-anchored proteins, including prenylated small GTPases (Rabs, Ras, Rho, etc), myristoylated signaling kinases (e.g. Src), and palmitoylated membrane proteins (Hsu *et al* 2010, Ostrowski *et al* 2010). Virally encoded, acylated Gag proteins can dominate the composition of exosomes released by infected cells within this protein class (Fang *et al* 2007).

Heat shock proteins (HSPs): These proteins protect cells from stress and are implicated in various diseases. Exosomal HSPs were first reported in 1995, noting that exosomal HSP70 tightly associates with exosomal transferrin receptor (TfR) during vesicular secretion from reticulocytes (Mathew *et al* 1995).

Alix: A protein involved in exosome and vesicle formation. Alix represents another class of exosomal scaffolding protein, binding to syntenin, tumor suppressor gene 101 protein (TSG101), and charged multivesicular body protein 4 (CHMP4) of the endosomal sorting complexes required for transport (ESCRTs) (Radulovic and Stenmark 2018).

TSG101: A protein involved in sorting proteins into exosomes, emerging as a commonly used marker of exosomes (Nabhan *et al* 2012).

CD63: A transmembrane protein regulating exosome formation and function. CD63, an exosomal tetraspanin, interacts complexly with syntenin. High syntenin expression has been shown to retain CD63 at the plasma membrane by masking its constitutive endocytosis signal (Latysheva *et al* 2006).

Calnexin: A protein involved in quality control in the endoplasmic reticulum (Kozlov and Gehring 2020).

CD9: A transmembrane protein involved in exosome formation and regulation of exosomal RNA. CD9 is commonly used as an exosomal marker protein (Hemler 2003).

In addition to these well-known exosomal proteins, growing evidence suggests that other proteins and signaling pathways are also involved in exosome formation and function.

Moreover, exosomes are known to harbor a variety of glycoconjugates and lipids crucial for their function and stability (Jeppesen *et al* 2019). Glycoconjugates are molecules composed of a carbohydrate (sugar) chain linked to a protein or lipid (Gabius and Kayser 2014). Exosomes feature a diverse array of glycoconjugates on their surface, including glycoproteins, glycolipids, and proteoglycans. These molecules play roles in exosome biogenesis, cargo sorting, and intercellular communication (Phuyal *et al* 2014). Lipids are a major component of exosomes, comprising up to 50% of their total mass. Exosomes contain a variety of lipids, including phospholipids, sphingolipids, cholesterol, and ceramides (Egea-Jimenez and

Zimmermann 2020). These lipids play important roles in maintaining the structural integrity of exosomes and in regulating exosome formation and cargo sorting (Donoso-Quezada *et al* 2021). Ceramide, a specific lipid type, has been demonstrated to be crucial in exosome biogenesis. Ceramide, a bioactive lipid, regulates diverse cellular processes including cell death, proliferation, and differentiation. It participates in exosome formation and the sorting of specific cargo molecules into exosomes (Arya *et al* 2022). In summary, exosomes host a variety of glycoconjugates and lipids vital for their function and stability. Research into these molecules is pivotal, shedding light on the mechanisms governing exosome biogenesis, cargo sorting, and their potential diagnostic and therapeutic applications.

Exosomes are recognized to house a variety of RNA molecules, including messenger RNA (mRNA), microRNA (miRNA), and other non-coding RNAs. These exosomal RNAs participate in various cellular processes, such as gene expression, cell-to-cell communication, and immune regulation. Generated within the parent cell, exosomal RNAs are sorted into the exosome through diverse mechanisms, including the endosomal sorting complex required for transport (ESCRT) machinery and tetraspanin-mediated sorting (Wei *et al* 2021). Once enclosed within the exosome, these RNA molecules are shielded from degradation by extracellular RNases and can be transferred to recipient cells, where they regulate gene expression and other cellular processes. MicroRNAs are extensively studied among exosomal RNA types. These small non-coding RNA molecules regulate gene expression and are implicated in various physiological and pathological processes, including cancer, immune regulation, and neurodegenerative diseases. Exosomal miRNAs participate in intercellular communication, modulating gene expression in recipient cells and altering cellular behavior. In addition to miRNAs, exosomes also harbor other RNA molecules, including messenger RNAs and long non-coding RNAs (Kitagawa *et al* 2022, Yang *et al* 2019). These RNA molecules are involved in various cellular processes, such as gene expression regulation, cell differentiation, and immune modulation (Kilchert *et al* 2016). The evolutionarily conserved RNA exosome is a multisubunit ribonuclease complex responsible for processing and/or degrading numerous RNAs. Recently, mutations in genes encoding both structural and catalytic subunits of the RNA exosome have been associated with human disease (Fasken *et al* 2020).

Exosomes are also recognized to harbor various types of DNA, including mitochondrial DNA (mtDNA), chromosomal DNA fragments, and viral DNA (Munagala *et al* 2021). Exosomal DNA is believed to originate from the parent cell and is packaged into exosomes through diverse mechanisms, including the ESCRT machinery and tetraspanin-mediated sorting (Sharma and Johnson 2020). Once enclosed within the exosome, exosomal DNA can be shielded from degradation and transferred to recipient cells. One extensively studied type of exosomal DNA is mtDNA, released from cells in response to various stressors like oxidative stress. It can be taken up by recipient cells, where it activates innate immune signaling pathways and induces inflammation. Exosomal mtDNA is implicated in various pathological conditions, including cancer, cardiovascular disease, and neurodegenerative diseases (Arance *et al* 2021). Research into exosomal RNAs and DNA is pivotal, shedding light on the mechanisms governing intercellular communication and the potential diagnostic and therapeutic applications of exosomal RNAs and DNA.

Exosomes have been demonstrated to fulfill crucial roles in various biological processes, encompassing intercellular communication, immune regulation, and tissue repair. Below are some of the biological roles of exosomes.

Intercellular communication: Exosomes facilitate the transfer of diverse biomolecules, including proteins, RNAs, and lipids, between cells. This transfer enables cells to communicate and regulate each other's behavior, such as inducing cell differentiation, promoting cell survival, or inhibiting cancer cell proliferation (Shi *et al* 2020).

Immune regulation: Exosomes modulate the immune response by transferring proteins and RNAs to immune cells. Exosomes derived from immune cells can also serve as antigen-presenting vesicles, initiating or inhibiting immune responses (Li *et al* 2021).

Tissue repair and regeneration: Exosomes contribute to tissue repair and regeneration by delivering growth factors and other molecules to damaged cells. They also promote blood vessel growth, crucial for tissue regeneration (Huang *et al* 2021).

Disease pathogenesis: Exosomes have been implicated in the pathogenesis of various diseases, including cancer, neurodegenerative diseases, and viral infections. They can promote cancer cell proliferation, metastasis, and immune evasion. In neurodegenerative diseases, exosomes can transfer toxic proteins, such as tau and alpha-synuclein, between cells, contributing to the spread of pathology (D'Anca *et al* 2019, Miranda and Di Paolo 2018). In viral infections, exosomes can facilitate viral replication and immune evasion (Crenshaw *et al* 2018).

Therapeutic applications: The ability of exosomes to deliver biomolecules to target cells renders them an attractive therapeutic tool. They can be engineered to carry therapeutic molecules, such as drugs, siRNAs, or CRISPR/Cas9, to specific cells or tissues for treating various diseases (He *et al* 2018).

Exosomes have garnered increasing attention as a potential therapeutic tool in regenerative medicine, including skin rejuvenation. Exosomes derived from mesenchymal stem cells (MSCs) have demonstrated the ability to promote skin regeneration and mitigate signs of aging (Bian *et al* 2022). When applied topically, exosomes can traverse the skin barrier and deliver their cargo of bioactive molecules, including growth factors and RNAs, to target cells (Yang *et al* 2021). They stimulate collagen and elastin production, crucial components of the skin's extracellular matrix responsible for skin elasticity and firmness. Additionally, exosomes promote blood vessel growth, vital for skin regeneration (Wu *et al* 2018).

Clinical studies have yielded promising results regarding the use of exosomes for skin rejuvenation. For instance, a recent clinical trial assessed the efficacy of exosomes derived from MSCs in enhancing skin quality in patients with photoaging. The findings revealed that topical application of exosomes significantly enhanced skin elasticity, thickness, and hydration, while diminishing the appearance of fine lines and wrinkles (Zhang *et al* 2020). Another potential avenue for exosome application in skin rejuvenation involves combining them with other cosmetic treatments, such as microneedling or laser therapy (Bagheri *et al* 2018). Exosomes can augment the efficacy of these procedures by fostering skin regeneration and mitigating inflammation (Mi *et al* 2023, Su *et al* 2021).

In summary, the utilization of exosomes in skin rejuvenation represents a burgeoning area of research with considerable promise. Although further investigation is warranted to comprehensively grasp the mechanisms of action and establish optimal treatment protocols, exosomes present a non-invasive and potentially efficacious approach to enhancing skin quality and alleviating the signs of aging.

8.2.4 Others

Mesotherapy presents a non-invasive and potentially efficacious cosmetic approach involving the injection of small amounts of various substances, such as vitamins, minerals, and other bioactive compounds, into the skin to enhance its quality and appearance (Bifarini *et al* 2022). Alongside PRP, PDRN, and exosomes, polycomponent mesotherapy solutions such as new cellular treatment factors (NCTF) have been widely and effectively utilized for years to address skin aging (Braccini and Dohan Ehrenfest 2010, Iranmanesh *et al* 2022). Comprising 55 active ingredients, including vitamins, minerals, amino acids, and hyaluronic acid, NCTF is administered via skin injection, with its blend of active components designed to enhance skin quality and diminish signs of aging. Hyaluronic acid, a natural skin component, aids in moisture retention and skin elasticity. Injecting hyaluronic acid improves skin hydration, reduces wrinkle appearance, and enhances skin firmness (Hu *et al* 2020). Other NCTF active ingredients aim to boost collagen production, refine skin texture and tone, and alleviate inflammation. These components encompass vitamins A, B, C, and E, boasting antioxidant properties safeguarding the skin from free radical damage. They also include essential minerals such as zinc, copper, and magnesium, vital for optimal skin function. Amino acids such as proline and lysine, pivotal for collagen and elastin synthesis, are incorporated to stimulate protein production. Moreover, NCTF features nucleotides, DNA, and RNA building blocks fostering skin repair and regeneration. Overall, NCTF constitutes a comprehensive mesotherapy solution enhancing skin quality and counteracting signs of aging. It is commonly employed to address fine lines, wrinkles, and loss of skin elasticity. However, consulting a qualified and experienced practitioner before undergoing any cosmetic procedure, including NCTF injections, is crucial.

Currently, an increasing number of single-functional-component mesotherapeutic products are being utilized in clinical treatments for skin aging, such as hyaluronic acid (HA), PLLA, PCL, and CaHA.

In contrast to the crosslinked hyaluronic acid commonly used in dermal fillers, non-crosslinked HA remains in its natural, fluid state without chemical modification or crosslinking. In mesotherapy, non-crosslinked HA is delicately injected into the skin using a fine needle, delivering small amounts of the hyaluronic acid directly to the targeted area. This process aids in skin hydration, plumping, improving texture and tone, and reducing the appearance of fine lines and wrinkles.

A clinical study published in 2018 assessed changes in skin biophysical parameters and appearance following pneumatic injections of non-cross-linked hyaluronic acid for facial aging treatment. Twenty-eight healthy female volunteers received pneumatic injections of non-cross-linked hyaluronic acid into the face for five consecutive weeks. Skin biophysical parameter assessment and clinical evaluation were conducted using the CK Multi-Probe Adapter and VISIA system. The study concluded that pneumatic injections of non-crosslinked hyaluronic acid were both safe and effective in improving skin parameters such as transepidermal water loss (TEWL), texture, pores, and wrinkles, with results maintained for up to 3 months post-treatment (Cheng *et al* 2018).

Non-crosslinked HA is commonly utilized either as a standalone treatment for skin rejuvenation or in combination with other mesotherapy ingredients, such as vitamins, amino acids, and peptides. Due to its natural presence in the skin, non-crosslinked HA is generally well-tolerated and rarely causes adverse reactions or allergies.

Another study published in 2021 investigated the efficacy and safety of non-cross-linked hyaluronic acid filler combined with L-carnosine in treating horizontal neck wrinkles in 13 female subjects. The study concluded that this combination was safe and effective for treating horizontal neck wrinkles, with improvements lasting up to 6 months post-treatment. Adverse events were rare, with local complications considered common reactions to the filler injection procedure (Wang *et al* 2021).

Furthermore, a study published in 2022 evaluated the effectiveness of a non-cross-linked hyaluronic acid-based soft tissue filler in correcting lateral canthal lines and periorbital lines in 59 female Caucasian patients. Significant increases in skin firmness and viscoelasticity were observed in the lateral canthal and perioral region. Although not statistically significant, skin hydration increased in both areas after 8 weeks. The study concluded that non-crosslinked hyaluronic acid injections were safe and effective in reducing the appearance of perioral wrinkles, with improvements observed for up to 3 months post-treatment (Sulovsky *et al* 2022).

These studies collectively suggest that non-crosslinked hyaluronic acid injections represent a safe and effective option for treating various signs of skin aging, including fine lines, wrinkles, and skin laxity.

Poly-L-lactic acid (PLLA) is a synthetic polymer commonly employed in cosmetic procedures, including mesotherapy for skin rejuvenation. It is a biocompatible and biodegradable substance with a history of medical use, such as in dissolvable sutures (Fitzgerald *et al* 2018). When injected into the skin, PLLA stimulates collagen production, improving skin texture and diminishing the appearance of fine lines and wrinkles. PLLA administration involves injection and typically necessitates a series of treatments over several months to achieve desired outcomes. A 2018 study assessed the safety and efficacy of injected PLLA in improving rhytids and crepiness of the décolletage in 25 healthy female volunteers aged 40–70 years with moderate-to-severe crepiness and wrinkling. The findings demonstrated PLLA's safety and efficacy in enhancing rhytids and skin quality of the photodamaged décolletage (Wilkerson and Goldberg 2018).

Although PLLA application to the face is well-documented, there are relatively fewer publications on its use in other body regions. A study aimed to extend prior findings by evaluating PLLA's efficacy and safety in treating contour (including lifting) deformities of the buttock region. This prospective, multicenter (three sites), single-cohort, open-label clinical trial involved thirty female subjects treated with PLLA in the bilateral buttocks over three sessions, each spaced 1 month apart, and followed for 6 months post-treatment. Various safety and clinical efficacy parameters were assessed at each visit, including the Global Assessment of Improvement Scale (GAIS), subject satisfaction, and skin health parameters. The study concluded that PLLA is safe and effective for buttock contouring and improving skin health parameters (Kwon *et al* 2019).

One advantage of PLLA is its long-lasting results, with effects lasting up to 2 years. Results are natural-looking, as PLLA gradually increases collagen production, providing subtle and gradual skin quality improvement. PLLA injections are commonly used on noticeable facial aging areas like cheeks, chin, and temples, as well as on other body areas like the décolletage. Another study provided recommendations on PLLA use for skin laxity treatment in non-facial areas such as the neck, décolletage, arms, abdomen, buttocks, and thighs, including patient selection, product preparation, and injection techniques. PLLA offers a promising method for skin laxity treatment in bodily areas, enhancing body contour and appearance. Further research is warranted to better comprehend PLLA efficacy and durability in non-facial indications, ensuring optimal patient outcomes (Haddad *et al* 2019). PLLA is a safe and effective option for skin aging treatment, commonly combined with other mesotherapy products such as hyaluronic acid, vitamins, and antioxidants to augment effectiveness.

Polycaprolactone (PCL) is a synthetic polymer utilized in mesotherapy for skin aging. It is a biocompatible and biodegradable substance with a history of medical use, including in surgical sutures and medical implants (Kwon *et al* 2019). When injected into the skin, PCL stimulates collagen production, enhancing skin texture and reducing the appearance of fine lines and wrinkles. PCL administration involves injection and typically entails a series of treatments over several months to achieve desired outcomes. Given its unique characteristics and versatility in treating various areas, this paper presents expert recommendations focusing on indications, treatment areas, procedures, and injection techniques. These recommendations serve as guidelines for physicians seeking to perform safe and effective PCL collagen stimulator treatments for facial rejuvenation and volume augmentation (de Melo *et al* 2017).

One advantage of PCL is its long-lasting results, with effects lasting up to 18 months. The results are natural-looking, as PCL gradually increases collagen production, providing subtle and gradual skin quality improvement (Hong *et al* 2021). PCL injections are commonly used on noticeable facial aging areas such as the cheeks, chin, and temples, and can also enhance skin quality in other body areas. A 2022 study examined the efficacy of PCL filler on enlarged facial pores and skin texture improvement. Seven participants with enlarged facial pores underwent deep-dermal injection of PCL-based filler. Skin quality measurements, including skin evenness, red areas,

UV spots, wrinkles, and pore numbers, were evaluated with an automated aesthetic camera before and 3 months after the injection session. While skin evenness, red areas, UV spots, and wrinkles showed no significant improvement post-filler injection, no serious adverse events were reported. However, the PCL-based filler demonstrated notable effectiveness in improving enlarged facial pores, particularly in moderate to severe cases (Marefat *et al* 2022).

PCL is a safe and effective option for treating skin aging and is often combined with other mesotherapeutic products such as hyaluronic acid, vitamins, and antioxidants to enhance its effectiveness. These studies collectively suggest that PCL injections are a safe and effective option for addressing various signs of skin aging, including wrinkles and skin laxity.

Calcium hydroxylapatite (CaHA) is a biocompatible and biodegradable substance commonly utilized in anti-aging treatments. It is a synthetic variant of the mineral hydroxyapatite, naturally present in the body and serving as a vital component of bones and teeth. Upon injection into the skin, CaHA functions as a scaffold, prompting collagen production and fostering the growth of new collagen fibers (Nowag *et al* 2023). This mechanism aids in enhancing skin texture and diminishing the visibility of fine lines and wrinkles. The administration of CaHA involves injection and typically necessitates a series of treatments spanning several months to achieve optimal results.

A 2018 study examined the combined application of microfocused ultrasound with visualization (MFU-V) and diluted calcium hydroxyl appetite for addressing neck and décolletage issues in 47 subjects with moderate to severe lines in these areas. Baseline and 90 day photographs were evaluated by two independent, blinded assessors using three scales: the Merz Aesthetics décolleté wrinkles scale, Fabi–Bolton chest wrinkle scale, and Allergan transverse neck lines scale. The study concluded that the combination of MFU-V with 1:1 diluted CaHA proved effective in ameliorating the appearance of neck and décolletage lines and wrinkles (Casabona and Nogueira Teixeira 2018).

One of the benefits of CaHA is its ability to deliver long-lasting results, with the effects of the treatment lasting for up to 2 years (Wollina and Goldman 2020). Additionally, the outcomes appear natural as CaHA functions by gradually boosting collagen production, resulting in a subtle and gradual enhancement in skin quality. CaHA injections are commonly utilized to address noticeable signs of aging in facial areas such as the cheeks, chin, and temples (Guida *et al* 2021, Moradi *et al* 2021), and they can also enhance skin quality in other body regions such as the hands (Lim and Mulcahy 2017). CaHA stands as a safe and effective choice for combating skin aging and is frequently employed alongside other mesotherapeutic products like hyaluronic acid, vitamins, and antioxidants to bolster its efficacy (Fakih-Gomez and Kadouch 2022).

Both hyaluronic acid (HA) and calcium hydroxyapatite (CaHA) fillers play a role in facial rejuvenation, yet they induce distinct effects concerning volume restoration and dermal biostimulation. A study set out to evaluate clinical and ultrasonographic enhancements in facial skin laxity through a method combining HA and CaHA injections. A 120 day follow-up quasi-experimental study was conducted, involving

15 women with mild facial flaccidity scores who received subcutaneous injections of up to 3 ml of HA followed by 3 ml of 1:1 diluted CaHA. All participants expressed high satisfaction with the results and reported significant improvement. Ultrasonography revealed an 11.1% increase in dermal thickness and enhanced dermal uniformity. Mild and transient local adverse effects were observed. In summary, the combined technique employing HA and CaHA fillers was well-tolerated and led to substantial satisfaction, along with safe improvement in facial skin laxity and dermal thickness in women with mild midface aging (Galmés-Truyols *et al* 2011).

These findings suggest that CaHA injections present a secure and efficacious approach for addressing various manifestations of skin aging, encompassing wrinkles, volume loss, and hand rejuvenation.

8.3 Safety and complications

Mesotherapy is a procedure involving the injection of active substances into the dermis and subcutaneous tissue to address various local medical and cosmetic conditions. While generally regarded as a safe and minimally invasive approach to treating skin aging, it carries the potential for adverse reactions due to its broad application and lack of standardized processes. It is crucial to recognize that the risks and side effects of mesotherapy can vary depending on the specific substances and techniques employed, as well as the individual's skin type and medical history. Nevertheless, like any medical procedure, mesotherapy entails potential side effects and risks to consider. Some potential side effects of mesotherapy for skin aging may include: discomfort or pain at the injection site, swelling, redness, or bruising, itching or rash, infection, allergic reactions, formation of lumps or nodules under the skin, skin discoloration, scarring, or tissue damage. All the aforementioned complications are categorized into infectious and non-infectious agents (Plachouri and Georgiou 2019).

8.3.1 Infectious complications

The utilization of needles and injections in mesotherapy carries the potential for infectious complications if proper hygiene and safety protocols are not adhered to. Some possible infectious complications of mesotherapy may include the following.

Bacterial infections: These can arise if bacteria are introduced into the skin through the injection site, manifesting symptoms such as redness, swelling, pain, and pus. Bacterial infections pose a risk when mesotherapy procedures lack appropriate hygiene and safety measures. Symptoms of a bacterial infection may entail redness, swelling, pain, and the presence of pus at the injection site. In severe instances, the infection may disseminate to other body parts, resulting in fever, chills, and systemic symptoms. Various types of bacterial infections can emerge as complications of mesotherapy. Common bacterial infections associated with mesotherapy include *Staphylococcus aureus*, *Streptococcus pyogenes*, *Pseudomonas aeruginosa*, and *Mycobacterium fortuitum*. Treating a mycobacterial infection can prove challenging due to its resistance to multiple antibiotics (Galmés-Truyols *et al* 2011). Therapy should be tailored following an antibiogram assessment and typically involves at least two antibiotic agents to mitigate the risk of induced resistance (Wongkitisophon *et al* 2011).

To mitigate the risk of bacterial infections in mesotherapy, it is imperative to select a qualified and experienced practitioner who adheres to proper safety and hygiene protocols. This may involve employing sterile needles and equipment, prepping the skin before injection, and refraining from reusing needles or syringes. Additionally, informing the practitioner of any allergies, medical conditions, or medications beforehand can help reduce the likelihood of adverse reactions or complications. Should you experience any unusual or concerning symptoms following mesotherapy treatment, promptly contact your practitioner. In cases of suspected bacterial infection, swift medical intervention with antibiotics may be necessary to prevent further complications stemming from the spread of infection.

Viral infections: While bacterial infections are a prevalent complication of mesotherapy, viral infections are less common but still possible. Viral infections can be introduced into the skin through contaminated needles, syringes, or other equipment used during mesotherapy. Some of the most common viral infections associated with mesotherapy include herpes simplex virus (HSV), human papillomavirus (HPV), and hepatitis B/C. HSV is a common virus that can cause cold sores or genital herpes. It can be transmitted through contaminated needles or by contact with a cold sore or genital herpes lesion during the procedure. Symptoms of a viral infection following mesotherapy may include redness, swelling, pain, and the presence of blisters or lesions at the injection site. In severe cases, the infection can spread to other parts of the body, resulting in fever, fatigue, and other systemic symptoms.

In addition to bacterial and viral infections, other types of infections, such as fungal infections, may also arise as complications of mesotherapy. Prompt medical attention should be sought if any symptoms of infection develop after mesotherapy, such as redness, swelling, pain, fever, or drainage at the injection site. Your healthcare provider can conduct diagnostic tests and prescribe appropriate treatment to prevent the infection from spreading and causing further complications.

8.3.2 Non-infectious complications

In addition to potential infectious complications, there are also non-infectious complications that can arise from mesotherapy. Some of the possible non-infectious complications of mesotherapy may include the following.

Granuloma: Granulomatous foreign body reactions have been reported to occur after the mesotherapy application of agents such as phosphatidylcholine, deoxycholate, buflomedil, silica, or carnitine, as well as a large number of several other different substances. These reactions occur when the body reacts to a foreign substance, such as a filler or other injection material, by forming granulomas, which are small, hard nodules that can be felt under the skin.

Late-onset immune-mediated adverse effects associated with the use of poly-l-lactic acid can appear years after the initial injection with a reported range of 6–60 months

(Alijotas-Reig *et al* 2009). These effects include inflammatory nodules, papules, and edema. Previous reports of adverse effects associated with poly-l-lactic acid identify a maximum of 60 months before the development of a subcutaneous nodule (Alijotas-Reig *et al* 2009). In one case study, a 48-year-old immunocompetent woman presented with a solitary visible noninflammatory nodule on her right temple, which at the time of presentation had been present for 2 months. She received injections of botulinum toxin A and HA after her last PLLA treatment and before the development of the temporal nodule. However, none of these other treatments were on or near the temporal regions. Histopathologic analysis of the lesion on the temple, including analysis under polarized light, showed a subcutaneous foreign body granuloma compatible with foreign body reaction to poly-l-lactic acid. Although she was treated with an intralesional steroid, it was unclear whether or to what extent this actually provided clinical benefit in terms of resolution, especially given that she also had other granulomas that later resolved without any treatment. This case of the longest yet reported latency period for the development of subcutaneous nodules with foreign body reaction observed with poly-l-lactic acid serves to raise awareness of a broader expected time course for complications associated with poly-l-lactic acid injection procedures (Storer *et al* 2016).

These examples highlight the importance of choosing a qualified and experienced practitioner for mesotherapy treatment, and of carefully considering the risks and benefits of any injection material before undergoing treatment. If you experience any unusual symptoms after mesotherapy, such as lumps or nodules, it's important to contact your practitioner right away to discuss possible treatment options.

Pain or discomfort: Injection of the mesotherapy solution can cause temporary pain, discomfort, and bruising at the injection site.

Swelling or redness: Some people may experience mild to moderate swelling or redness at the injection site, which usually subsides within a few days.

Allergic reactions: Rarely, mesotherapy can cause allergic reactions to the injected substances, which may manifest as itching, hives, or difficulty breathing.

Skin discoloration: Injection of mesotherapy solution can sometimes cause temporary skin discoloration or hyperpigmentation.

Scarring or tissue damage: In rare cases, mesotherapy can cause scarring or tissue damage, particularly if the injection is not performed correctly or the solution is injected into the wrong area.

8.4 Conclusion

Mesotherapy, a minimally invasive procedure, entails injecting various substances such as hyaluronic acid and peptides into the skin for cosmetic purposes, including facial rejuvenation. These injections target specific facial areas, such as the cheeks and forehead, to stimulate collagen production, enhance skin hydration and elasticity, and diminish signs of aging such as fine lines and wrinkles. Mesotherapy also improves skin texture and tone, addresses age spots and sun damage, and enhances lip volume and contour. It offers a safe and effective non-surgical option for comprehensive facial rejuvenation.

References

Abuaf O K *et al* 2016 Histologic evidence of new collagen formulation using platelet rich plasma in skin rejuvenation: a prospective controlled clinical study *Ann. Dermatol.* **28** 718–24

Alijotas-Reig J, Garcia-Gimenez V and Vilardell-Tarres M 2009 Late-onset immune-mediated adverse effects after poly-L-lactic acid injection in non-HIV patients: clinical findings and long-term follow-up *Dermatology* **219** 303–8

Arance E *et al* 2021 Determination of exosome mitochondrial DNA as a biomarker of renal cancer aggressiveness *Cancers* **14** 199

Arraud N *et al* 2014 Extracellular vesicles from blood plasma: determination of their morphology, size, phenotype and concentration *J. Thromb. Haemost.* **12** 614–27

Arya S B, Chen S, Jordan-Javed F and Parent C A 2022 Ceramide-rich microdomains facilitate nuclear envelope budding for non-conventional exosome formation *Nat. Cell Biol.* **24** 1019–28

Atiyeh B S and Abou Ghanem O 2021 An update on facial skin rejuvenation effectiveness of mesotherapy EBM V *J. Craniofac. Surg.* **32** 2168–71

Atiyeh B S, Ibrahim A E and Dibo S A 2008 Cosmetic mesotherapy: between scientific evidence, science fiction, and lucrative business *Aesthet. Plast. Surg.* **32** 842–9

Bagheri H S *et al* 2018 Low-level laser irradiation at a high power intensity increased human endothelial cell exosome secretion via WNT signaling *Lasers Med. Sci.* **33** 1131–45

Beitzel K *et al* 2015 US definitions, current use, and FDA stance on use of platelet-rich plasma in sports medicine *J. Knee Surg.* **28** 29–34

Belletti S, Uggeri J, Gatti R, Govoni P and Guizzardi S 2007 Polydeoxyribonucleotide promotes cyclobutane pyrimidine dimer repair in UVB-exposed dermal fibroblasts *Photodermatol. Photoimmunol. Photomed.* **23** 242–9

Bian D, Wu Y, Song G, Azizi R and Zamani A 2022 The application of mesenchymal stromal cells (MSCs) and their derivative exosome in skin wound healing: a comprehensive review *Stem Cell Res. Ther.* **13** 24

Bifarini B *et al* 2022 Intradermal therapy (mesotherapy): the lower the better *Clin. Ter.* **173** 79–83

Bitto A *et al* 2008a Polydeoxyribonucleotide improves angiogenesis and wound healing in experimental thermal injury *Crit. Care Med.* **36** 1594–602

Bitto A *et al* 2008b Polydeoxyribonucleotide (PDRN) restores blood flow in an experimental model of peripheral artery occlusive disease *J. Vasc. Surg.* **48** 1292–300

Bitto A *et al* 2013 Adenosine receptor stimulation by polynucleotides (PDRN) reduces inflammation in experimental periodontitis *J. Clin. Periodontol.* **40** 26–32

Booth A M *et al* 2006 Exosomes and HIV Gag bud from endosome-like domains of the T cell plasma membrane *J. Cell Biol.* **172** 923–35

Braccini F and Dohan Ehrenfest D M 2010 Advantages of combined therapies in cosmetic medicine for the treatment of face aging: botulinum toxin, fillers and mesotherapy *Rev. Laryngol. Otol. Rhinol.* **131** 89–95

Cameli N *et al* 2017 Autologous pure platelet-rich plasma dermal injections for facial skin rejuvenation: clinical, instrumental, and flow cytometry assessment *Dermatol. Surg.* **43** 826–35

Casabona G and Nogueira Teixeira D 2018 Microfocused ultrasound in combination with diluted calcium hydroxylapatite for improving skin laxity and the appearance of lines in the neck and décolletage *J. Cosmet. Dermatol.* **17** 66–72

Chahla J *et al* 2017 A call for standardization in platelet-rich plasma preparation protocols and composition reporting: a systematic review of the clinical orthopaedic literature *J. Bone Joint Surg. Am.* **99** 1769–79

Chan E S, Fernandez P and Cronstein B N 2007 Adenosine in inflammatory joint diseases *Purinergic Signal* **3** 145–52

Cheng H Y, Chen Y X, Wang M F, Zhao J Y and Li L F 2018 Evaluation of changes in skin biophysical parameters and appearance after pneumatic injections of non-cross-linked hyaluronic acid in the face *J. Cosmet. Laser Ther.* **20** 454–61

Choi S Y, Kim S and Park K M 2022 Initial healing effects of platelet-rich plasma (PRP) gel and platelet-rich fibrin (PRF) in the deep corneal wound in rabbits *Bioengineering* **9** 405

Crenshaw B J, Gu L, Sims B and Matthews Q L 2018 Exosome biogenesis and biological function in response to viral infections *Open Virol. J.* **12** 134–48

D'Anca M *et al* 2019 Exosome determinants of physiological aging and age-related neurodegenerative diseases *Front. Aging Neurosci.* **11** 232

De Melo F *et al* 2017 Recommendations for volume augmentation and rejuvenation of the face and hands with the new generation polycaprolactone-based collagen stimulator (Ellansé(®)) *Clin. Cosmet. Investig. Dermatol.* **10** 431–40

Donoso-Quezada J, Ayala-Mar S and González-Valdez J 2021 The role of lipids in exosome biology and intercellular communication: function, analytics and applications *Traffic* **22** 204–20

Doyle L M and Wang M Z 2019 Overview of extracellular vesicles, their origin, composition, purpose, and methods for exosome isolation and analysis *Cells* **8** 727

Egea-Jimenez A L and Zimmermann P 2020 Lipids in exosome biology *Handb. Exp. Pharmacol.* **259** 309–36

Elnehrawy N Y, Ibrahim Z A, Eltoukhy A M and Nagy H M 2017 Assessment of the efficacy and safety of single platelet-rich plasma injection on different types and grades of facial wrinkles *J. Cosmet. Dermatol.* **16** 103–11

Emer J 2019 Platelet-rich plasma (PRP): current applications in dermatology *Skin Ther. Lett.* **24** 1–6

Everts P A *et al* 2006 Differences in platelet growth factor release and leucocyte kinetics during autologous platelet gel formation *Transfus Med.* **16** 363–8

Everts P A, Pinto P C and Girão L 2019 Autologous pure platelet-rich plasma injections for facial skin rejuvenation: biometric instrumental evaluations and patient-reported outcomes to support antiaging effects *J. Cosmet. Dermatol.* **18** 985–95

Fadadu P P, Mazzola A J, Hunter C W and Davis T T 2019 Review of concentration yields in commercially available platelet-rich plasma (PRP) systems: a call for PRP standardization *Reg. Anesth. Pain Med.* **44** 652–9

Fakih-Gomez N and Kadouch J 2022 Combining calcium hydroxylapatite and hyaluronic acid fillers for aesthetic indications: efficacy of an innovative hybrid filler *Aesthet. Plast. Surg.* **46** 373–81

Fang Y *et al* 2007 Higher-order oligomerization targets plasma membrane proteins and HIV gag to exosomes *PLoS Biol.* **5** e158

Fasken M B *et al* 2020 The RNA exosome and human disease *Methods Mol. Biol.* **2062** 3–33

Fitzgerald R, Bass L M, Goldberg D J, Graivier M H and Lorenc Z P 2018 Physiochemical characteristics of poly-L-lactic acid (PLLA) *Aesthet. Surg. J.* **38** S13–s7

Gabius H J and Kayser K 2014 Introduction to glycopathology: the concept, the tools and the perspectives *Diagn. Pathol.* **9** 4

Galeano M *et al* 2008 Polydeoxyribonucleotide stimulates angiogenesis and wound healing in the genetically diabetic mouse *Wound Repair Regen.* **16** 208–17

Galmés-Truyols A *et al* 2011 An outbreak of cutaneous infection due to *Mycobacterium abscessus* associated to mesotherapy *Enferm. Infecc. Microbiol. Clin.* **29** 510–4

Gawdat H I, El-Hadidy Y A, Allam R and Abdelkader H A 2022 Autologous platelet-rich plasma 'fluid' versus 'gel' form in combination with fractional CO_2 laser in the treatment of atrophic acne scars: a split-face randomized clinical trial *J. Dermatolog. Treat.* **33** 2654–63

Gawdat H I, Tawdy A M, Hegazy R A, Zakaria M M and Allam R S 2017 Autologous platelet-rich plasma versus readymade growth factors in skin rejuvenation: a split face study *J. Cosmet. Dermatol.* **16** 258–64

Gennero L *et al* 2013 Protective effects of polydeoxyribonucleotides on cartilage degradation in experimental cultures *Cell Biochem. Funct.* **31** 214–27

Giusti I, D'ascenzo S, Macchiarelli G and Dolo V 2020 *In vitro* evidence supporting applications of platelet derivatives in regenerative medicine *Blood Transfus* **18** 117–29

Guida S *et al* 2021 Hyperdiluted calcium hydroxylapatite for the treatment of skin laxity of the neck *Dermatol. Ther.* **34** e15090

Haddad A *et al* 2019 Recommendations on the use of injectable poly-L-lactic acid for skin laxity in off-face areas *J. Drugs Dermatol.* **18** 929–35

Haunschild E D *et al* 2020 Platelet-rich plasma augmentation in meniscal repair surgery: a systematic review of comparative studies *Arthroscopy* **36** 1765–74

He C, Zheng S, Luo Y and Wang B 2018 Exosome theranostics: biology and translational medicine *Theranostics* **8** 237–55

Hemler M E 2003 Tetraspanin proteins mediate cellular penetration, invasion, and fusion events and define a novel type of membrane microdomain *Annu. Rev. Cell Dev. Biol.* **19** 397–422

Hong J Y *et al* 2021 *In vivo* evaluation of novel particle-free polycaprolactone fillers for safety, degradation, and neocollagenesis in a rat model *Dermatol. Ther.* **34** e14770

Hsu C *et al* 2010 Regulation of exosome secretion by Rab35 and its GTPase-activating proteins TBC1D10A-C *J. Cell Biol.* **189** 223–32

Hu L, Zhao K and Song W M 2020 Effect of mesotherapy with nanochip in the treatment of facial rejuvenation *J. Cosmet. Laser Ther.* **22** 84–9

Huang J *et al* 2021 Cell-free exosome-laden scaffolds for tissue repair *Nanoscale* **13** 8740–50

Hui Q, Chang P, Guo B, Zhang Y and Tao K 2017 The clinical efficacy of autologous platelet-rich plasma combined with ultra-pulsed fractional CO_2 laser therapy for facial rejuvenation *Rejuvenation Res.* **20** 25–31

Hwang K H, Kim J H, Park E Y and Cha S K 2018 An effective range of polydeoxyribonucleotides is critical for wound healing quality *Mol. Med. Rep.* **18** 5166–72

Iranmanesh B, Khalili M, Mohammadi S, Amiri R and Aflatoonian M 2022 Employing hyaluronic acid-based mesotherapy for facial rejuvenation *J. Cosmet. Dermatol.* **21** 6605–18

Jeppesen D K *et al* 2019 Reassessment of exosome composition *Cell* **177** 428–45

Kang B K, Shin M K, Lee J H and Kim N I 2014 Effects of platelet-rich plasma on wrinkles and skin tone in Asian lower eyelid skin: preliminary results from a prospective, randomised, split-face trial *Eur. J. Dermatol.* **24** 100

Khalili M, Amiri R, Iranmanesh B, Zartab H and Aflatoonian M 2022 Safety and efficacy of mesotherapy in the treatment of melasma: a review article *J. Cosmet. Dermatol.* **21** 118–29

Kilchert C, Wittmann S and Vasiljeva L 2016 The regulation and functions of the nuclear RNA exosome complex *Nat. Rev. Mol. Cell Biol.* **17** 227–39

Kitagawa M, Wu P, Balkunde R, Cunniff P and Jackson D 2022 An RNA exosome subunit mediates cell-to-cell trafficking of a homeobox mRNA via plasmodesmata *Science* **375** 177–82

Kozlov G and Gehring K 2020 Calnexin cycle—structural features of the ER chaperone system *FEBS J.* **287** 4322–40

Kwon T R *et al* 2019 Biostimulatory effects of polydioxanone, poly-d, l lactic acid, and polycaprolactone fillers in mouse model *J. Cosmet. Dermatol.* **18** 1002–8

Latysheva N *et al* 2006 Syntenin-1 is a new component of tetraspanin-enriched microdomains: mechanisms and consequences of the interaction of syntenin-1 with CD63 *Mol. Cell. Biol.* **26** 7707–18

Lee J C, Daniels M A and Roth M Z 2016 Mesotherapy, microneedling, and chemical peels *Clin. Plast. Surg.* **43** 583–95

Leo M S, Kumar A S, Kirit R, Konathan R and Sivamani R K 2015 Systematic review of the use of platelet-rich plasma in aesthetic dermatology *J. Cosmet. Dermatol.* **14** 315–23

Li J, Liu M, Hu Y and Guo G 2021 Research progress on exosome-mediated immune regulation of acute lung injury *Zhonghua Wei Zhong Bing Ji Jiu Yi Xue* **33** 118–21

Lim A and Mulcahy A 2017 Hand rejuvenation: combining dorsal veins foam sclerotherapy and calcium hydroxylapatite filler injections *Phlebology* **32** 397–402

Maisel-Campbell A L *et al* 2020 A systematic review of the safety and effectiveness of platelet-rich plasma (PRP) for skin aging *Arch. Dermatol. Res.* **312** 301–15

Marefat A, Dadkhahfar S, Tahvildari A and Robati R M 2022 The efficacy of polycaprolactone filler injection on enlarged facial pores *Dermatol. Ther.* **35** e15600

Mathew A, Bell A and Johnstone R M 1995 Hsp-70 is closely associated with the transferrin receptor in exosomes from maturing reticulocytes *Biochem. J.* **308** 823–30

Mi P *et al* 2023 Stem cell-derived exosomes for chronic wound repair *Cell Tissue Res.* **391** 419–23

Mijiritsky E *et al* 2021 Use of PRP, PRF and CGF in periodontal regeneration and facial rejuvenation—a narrative review *Biology* **10** 317

Miranda A M and Di Paolo G 2018 Endolysosomal dysfunction and exosome secretion: implications for neurodegenerative disorders *Cell Stress* **2** 115–8

Moradi A *et al* 2021 Effectiveness and safety of calcium hydroxylapatite with lidocaine for improving jawline contour *J. Drugs Dermatol.* **20** 1231–8

Munagala R *et al* 2021 Exosome-mediated delivery of RNA and DNA for gene therapy *Cancer Lett.* **505** 58–72

Nabhan J F, Hu R, Oh R S, Cohen S N and Lu Q 2012 Formation and release of arrestin domain-containing protein 1-mediated microvesicles (ARMMs) at plasma membrane by recruitment of TSG101 protein *Proc. Natl Acad. Sci. USA* **109** 4146–51

Nowag B, Casabona G, Kippenberger S, Zöller N and Hengl T 2023 Calcium hydroxylapatite microspheres activate fibroblasts through direct contact to stimulate neocollagenesis *J. Cosmet. Dermatol.* **22** 426–32

Ostrowski M *et al* 2010 Rab27a and Rab27b control different steps of the exosome secretion pathway *Nat. Cell Biol.* **12** 19–30

Pegtel D M and Gould S J 2019 Exosomes *Annu. Rev. Biochem.* **88** 487–514

Phuyal S, Hessvik N P, Skotland T, Sandvig K and Llorente A 2014 Regulation of exosome release by glycosphingolipids and flotillins *FEBS J.* **281** 2214–27

Plachouri K M and Georgiou S 2019 Mesotherapy: safety profile and management of complications *J. Cosmet. Dermatol.* **18** 1601–5

Radulovic M and Stenmark H 2018 ESCRTs in membrane sealing *Biochem. Soc. Trans.* **46** 773–8

Redaelli A, Romano D and Marcianó A 2010 Face and neck revitalization with platelet-rich plasma (PRP): clinical outcome in a series of 23 consecutively treated patients *J. Drugs Dermatol.* **9** 466–72

Rossi L A *et al* 2019 Classification systems for platelet-rich plasma *Bone Joint J.* **101-b** 891–6

Sclafani A P 2011 Safety, efficacy, and utility of platelet-rich fibrin matrix in facial plastic surgery *Arch. Facial Plast. Surg.* **13** 247–51

Sharma A and Johnson A 2020 Exosome DNA: critical regulator of tumor immunity and a diagnostic biomarker *J. Cell. Physiol.* **235** 1921–32

Shi Z Y, Yang X X, Malichewe C, Li Y S and Guo X L 2020 Exosomal microRNAs-mediated intercellular communication and exosome-based cancer treatment *Int. J. Biol. Macromol.* **158** 530–41

Sini P *et al* 1999 Effect of polydeoxyribonucleotides on human fibroblasts in primary culture *Cell Biochem. Funct.* **17** 107–14

Sódar B W *et al* 2016 Low-density lipoprotein mimics blood plasma-derived exosomes and microvesicles during isolation and detection *Sci. Rep.* **6** 24316

Soffer E, Ouhayoun J P, Dosquet C, Meunier A and Anagnostou F 2004 Effects of platelet lysates on select bone cell functions *Clin. Oral Implants Res.* **15** 581–8

Squadrito F *et al* 2017 Pharmacological activity and clinical use of PDRN *Front. Pharmacol.* **8** 224

Storer M, Euwer R, Calame A and Kourosh A S 2016 Late-onset granuloma formation after poly-L-lactic acid injection *JAAD Case Rep.* **2** 54–6

Su N *et al* 2021 Mesenchymal stromal exosome-functionalized scaffolds induce innate and adaptive immunomodulatory responses toward tissue repair *Sci. Adv.* **7** eabf7207

Sulovsky M *et al* 2022 A prospective open-label, multicentre study evaluating a non-cross-linked hyaluronic acid based soft-tissue filler in the correction of lateral canthal and perioral lines *J. Cosmet. Dermatol.* **21** 191–8

Sun Y *et al* 2023 A chitosan derivative-crosslinked hydrogel with controllable release of polydeoxyribonucleotides for wound treatment *Carbohydr. Polym.* **300** 120298

Thellung S, Florio T, Maragliano A, Cattarini G and Schettini G 1999 Polydeoxyribonucleotides enhance the proliferation of human skin fibroblasts: involvement of A2 purinergic receptor subtypes *Life Sci.* **64** 1661–74

Valdatta L, Thione A, Mortarino C, Buoro M and Tuinder S 2004 Evaluation of the efficacy of polydeoxyribonucleotides in the healing process of autologous skin graft donor sites: a pilot study *Curr. Med. Res. Opin.* **20** 403–8

Veronesi F *et al* 2017 Polydeoxyribonucleotides (PDRNs) from skin to musculoskeletal tissue regeneration via adenosine A(2A) receptor involvement *J. Cell. Physiol.* **232** 2299–307

Wang S *et al* 2021 Clinical efficacy and safety of non-cross-linked hyaluronic acid combined with L-carnosine for horizontal neck wrinkles treatment *Aesthet. Plast. Surg.* **45** 2912–7

Wei D *et al* 2021 RAB31 marks and controls an ESCRT-independent exosome pathway *Cell Res.* **31** 157–77

Wilkerson E C and Goldberg D J 2018 Poly-L-lactic acid for the improvement of photodamage and rhytids of the décolletage *J. Cosmet. Dermatol.* **17** 606–10

Wollina U and Goldman A 2020 Long lasting facial rejuvenation by repeated placement of calcium hydroxylapatite in elderly women *Dermatol. Ther.* **33** e14183

Wongkitisophon P, Rattanakaemakorn P, Tanrattanakorn S and Vachiramon V 2011 Cutaneous *Mycobacterium abscessus* infection associated with mesotherapy injection *Case Rep. Dermatol.* **3** 37–41

Wu P, Zhang B, Shi H, Qian H and Xu W 2018 MSC-exosome: a novel cell-free therapy for cutaneous regeneration *Cytotherapy* **20** 291–301

Yang G H *et al* 2021 Overcome the barriers of the skin: exosome therapy *Biomater. Res.* **25** 22

Yang L *et al* 2019 Long non-coding RNA HOTAIR promotes exosome secretion by regulating RAB35 and SNAP23 in hepatocellular carcinoma *Mol. Cancer* **18** 78

Zhang K *et al* 2020 Topical application of exosomes derived from human umbilical cord mesenchymal stem cells in combination with sponge spicules for treatment of photoaging *Int. J. Nanomed.* **15** 2859–72

Chapter 9

Thread lift for photoaging

Zhuanli Bai

Facial aging encompasses complex changes in bone, soft tissue, and skin. As bones remodel and soft tissue weakens, skin thins and loses elasticity, leading to wrinkles and sagging. Thread lifting effectively addresses soft tissue sagging and contouring, often combined with minimally invasive techniques such as botulinum toxin and fillers for optimal rejuvenation. By using lifting, tightening, and biological stimulation, thread lifting achieves lifting, firming, and textural improvements. Its principle involves mechanical lifting and tissue response, promoting collagen synthesis and wound healing. Thread lifting, with its minimal invasiveness and immediate results, is increasingly favored in cosmetic surgery for facial rejuvenation.

9.1 Introduction

Facial aging is a composite, interrelated degenerating process involving changes to the bone, soft tissue, and skin. While each anatomical layer undergoes an aging process of its own, the facial bones are the framework for the attachment of overlying soft tissue, the bones recede and remodel with age, accompanying weakening of the supporting ligaments, resulting in repositioning and atrophied irregularly of fat pads. The skin thins and weakens as the dermis atrophies from aging. Aged skin slowly loses elasticity and develops wrinkles, which is manifested clinically as skin wrinkles, facial soft tissue sagging, mid-facial emptiness and flattening, and sunken local soft tissue. Thread lifting is an effective treatment for soft tissue sagging and contouring (Atiyeh *et al* 2021, Rezaee *et al* 2020). Combined with minimally invasive techniques such as botulinum toxin injection, photoelectric treatment, and filler injection, it can achieve effective rejuvenation treatment effects and is increasingly widely accepted (Fabi *et al* 2021, Moon *et al* 2021).

Thread lifting refers to a cosmetic surgical technique where different types of sutures are implanted at different layers in the subcutaneous soft tissue, such as fat or fascia layers, to treat facial aging (Kaya and Cakmak 2022). By using lifting, tightening, and biological stimulation methods, aesthetic effects can be achieved

doi:10.1088/978-0-7503-5112-6ch9

such as lifting and tightening of sagging tissues, filling of sunken tissues, improvement of skin texture, and shaping of facial contour (Tong and Rieder 2019). It is generally considered that the main principle of thread lifting is the mechanical lifting effect generated by barbed anchor and pulling (Fukaya 2018), as well as the tissue's response to the thread, such as adhesion and the collagen mechanical lifting effect. Subsequently, the mechanical force will be transformed into a chemical signal, such as growth factors (TGF-β1, β2, β3), which inhibit the expression of matrix metalloproteinases, reduce collagen protein degradation, and produce a wound healing effect and tissue response to the thread, resulting in effective tightening and lifting of the facial tissue by promoting cellular proliferation and neovascular formation (Ha *et al* 2022, Yoon *et al* 2019). In addition, granulation tissue formation stimulates fibroblasts, creating tension near the thread, which can also lift the skin tissue (Kim *et al* 2019). Thread lifting has small injuries and immediate results. Generally, there is no need for surgical incisions and sutures, only needle puncture wounds are left behind. Local anesthesia is commonly used, and the operation time is short with a fast recovery time. It has received increasing attention in the field of minimally invasive cosmetic surgery. Thread lift is also known as barbed suture lift, suture suspension, silhouette suture, etc. Sulamanidze first introduced the concept of using bidirectional sutures made of polypropylene to lift the face in the 1990s (Sulamanidze and Sulamanidze 2009). In 2002, he published an article describing his thread lifting technique, called 'featherlifting.' (Sulamanidze *et al* 2002). In 2004, the FDA approved the use of Aptos threads, a bidirectional barbed permanent suture (Aptos Ltd, Moscow, Russia) in the United States. Aptos threads work by lifting and securing sagging skin and soft tissue with barbed threads buried beneath the skin. Following Sulamanidze's invention of Aptos threads, Woffles Wu first designed the bidirectional barbed long thread lift in 2002 (Wu 2004). It is made of 2.0 polypropylene thread, 60 cm in length, with bidirectional barbs distributed from the center to both sides to allow the thread to fold within the temporal fascia without scalp detachment or knotting, aiding in fast recovery. The thread forms a fibrous capsule around it, increasing its strength and adhesive force to the soft tissue, and ultimately prolonging the lifting effect. In 2004, Woffles Wu published an article describing his barbed suture lift technique, which was subsequently named the 'Woffles Lift' (Wu 2004). This technique uses a bidirectional barbed suture made of polypropylene to provide long-term support for sagging soft tissue in the face. Wu has observed the results and complications of the Woffles Lift for over 20 years and has demonstrated its effectiveness in providing long-term support for sagging soft tissue. The complication rate is low, at about 1%–2%. Because the wires are non-absorbable polypropylene, they are easy to extract if the wires are improperly placed. In an independent study, Sasaki found that the Woffles Thread had the greatest lifting and pulling forces compared with all available barbed or cog threads (Sasaki *et al* 2008). Later, Iss thread, contour thread, silhouette thread, and various PDO threads were introduced. In 2006, a new type of polypropylene suture material was approved by the FDA for facial surgery (Kolster Methods, Inc., Corona, CA). This was the first suture material to introduce an absorbable component, combining permanent polypropylene suture material with unidirectional cones made of absorbable poly-L-lactic acid (PLLA) and polyglycolic acid (PLGA). There are currently two fully absorbable suture materials on the market in the United States that can be used for minimally invasive lifting techniques. The absorbable suture materials include PLLA/PLGA and polydioxanone (PDO). In 2015, the FDA granted a 501(k) clearance for a new suture for lifting, which is composed of 18% PLGA and 82% PLLA monofilament with bidirectional cones (Silhouette InstaLift; Sinclair Pharma, Irvine, CA). Compared with traditional barbed sutures, the cone on the Silhouette InstaLift suture provides strong lifting force without the risk of breaking. PLGA and PLLA have also been shown to stimulate the production of type I and III collagen. PDO is another widely used absorbable suture material that became popular in South Korea around 2012 and was approved for use in the United States in 2015. It is a colorless, crystalline, biodegradable synthetic polymer. Since then, thread lifting has made more progress in cosmetic surgical applications, but there are still some differences in the maintenance time and mechanism of action of thread lifting.

Facial aging is caused by multiple factors, such as skin and soft tissue laxity, deep fat atrophy, muscle relaxation and bone resorption, resulting in wrinkles, sagging of soft tissue, hollowing of the submalar area, jowling, nasolabial fold deepening, and so on. By reconstructing the three-dimensional facial contour to observe the facial features at different ages, it is found that the aging face is characterized by the imbalance, loss and accumulation of deep fat pad volume, as well as sagging of skin and soft tissue. The focus of thread lifting should be on the repositioning and lifting of sagging soft tissues in the middle and lower face. Although thread lifting is widely used in facial rejuvenation, it also has its indications. Currently recognized suitability criteria are: (i) having a low body mass index; (ii) mild to moderate soft tissue sagging; (iii) having significant bony contours; (iv) having good skin quality. Therefore, for mild to moderate soft tissue sagging, such as reduced fat in the midface, sagging cheeks, nasolabial folds, perioral wrinkles, sagging jawline, marionette lines, etc, selecting anchor points, thread materials, and thread paths based on anatomical diagnosis can often achieve good treatment results with thread lifting.

9.2 Classification

Thread lifting has a history of 30 years in facial rejuvenation, and the material of the thread has been constantly updated during the application process. With the continuous in-depth study of the mechanism of aging, there has been more personalized progress in the design of thread lifting plans. The placement of barbed sutures in various areas of the face and neck can reliably and permanently improve wrinkles and soft tissue folds and sagging, mainly due to the formation of fibrous connective tissue around the thread and the local contraction of myofibroblasts during the degradation process of the thread (Goldberg 2020). Therefore, the ideal thread should have a slow degrading and biocompatible structure that is initially strong enough to pull and anchor tissue, stimulate fibroblast migration and peripheral collagen regeneration to produce long-term effects. Various types of threads differ in their microstructure, tensile strength, elasticity, anchoring ability in human tissues, and biocompatibility. The classification of thread lifting is currently mainly based on the classification of the thread.

- *The duration of the suture material*: There are two types absorbable and non-absorbable suture materials. The non-absorbable suture material has a longer-lasting treatment effect, while the absorbable suture material is safe but has a shorter duration of effect.
- *The material of the thread*: Polypropylene, polydioxanone (PDO), polydioxanone-propanediol copolymer (PPDO), polylactic-co-glycolic acid (PLGA), poly-L-lactic acid (PLLA), polycaprolactone (PCL), poly(lactic acid-co-caprolactone) [P(LA/CL)], polyglycolic acid (PGA), polyglycolic acid-lactic acid (PGLA), and so on.
- *The appearance of the thread*: Including barbed suture, cones suture, smooth suture and spiral suture, etc. Barbed suture includes unidirectional barbed suture, bidirectional barbed suture, etc.
- *The length of the thread*: Long threads and short threads.

9.3 Clinical application

9.3.1 Indications

Indications include: mild to moderate sagging of the mid-facial skin and subcutaneous tissue, deepening of the nasolabial folds; mild to moderate sagging of the lower facial skin and subcutaneous tissue, accumulation of submental fat, indistinct jawline contour; accumulation of fat at the jowl and submandibular area; early sagging of the neck skin and platysma muscle, submental fat accumulation; mild tear trough and laxity of the lower eyelid skin and soft tissue; and static wrinkles and laxity of facial skin.

9.3.2 Contraindications

Absolute contraindications: Severe systemic diseases such as severe hypertension, diabetes, etc; impaired coagulation mechanisms or long-term oral anticoagulant therapy; tendency for hypertrophic scarring; autoimmune diseases; active infectious lesions on the facial skin.

Relative contraindications: Severe sagging of facial skin and subcutaneous tissue; facial subcutaneous tissue has received multiple and large amounts of hyaluronic acid filler treatments; excessive or insufficient fat in the facial subcutaneous tissue; excessive subcutaneous fibrosis or scar due to open facelift surgery, liposuction, etc; unknown filler injections in the facial threading area; menstruation in women; excessive expectations of minimally invasive cosmetic effects or severe psychological disorders; severe psychological disorders.

9.3.3 Preoperative assessment

Careful evaluation of the following facial conditions should be performed before surgery and scored: thickening of the temporal muscle; temporal depression;

prominent zygomatic bone; sunken cheeks; hypertrophy of the masseter muscle; marionette lines; cheek fat; submental fat; facial asymmetry; relaxation of the fat pad in the malar area; and nasolabial folds. The thread placement plan is then established based on the specific evaluation results.

9.3.4 Thread lifting relevant anatomy

The goal of facial rejuvenation is to reverse the anatomical changes that occur with aging. Facial soft tissues can usually be divided into five layers: skin, subcutaneous superficial fat layer, SMAS fascia layer, ligamentous interval layer, and periosteal layer. Thread lifting is often used in the mid-to-lower face, where sags of the superficial soft tissue due to gravity, loss of elasticity of skin and ligament septum, absorption of bone supporting the face, and loss of volume of deep fat pad, are considered important anatomical factors that contribute to facial aging. Therefore, thread lifting should set different thread lifting plans based on different anatomical evaluations, and thread lifting levels can be placed in the subcutaneous superficial fat layer or sagging deep fat pad, achieving a youthful effect of soft tissue lifting and repositioning.

9.3.5 Preoperative preparation and perioperative management

9.3.5.1 Preoperative preparation

It is very important to collect a detailed medical history of the patient before treatment. See table 9.1.

9.3.5.2 Preoperative management

Laboratory examination: Blood routine, coagulation profile, viral screening (hepatitis B, hepatitis C, syphilis, HIV etc), electrocardiogram, liver and kidney function, blood glucose, urine routine, etc.

Table 9.1. Medical history collection.

Information	Detailed
General medical history information	General systemic medical history, such as diabetes, hypertension, heart disease, bleeding disorders, chronic lung disease, immune system disorders, etc.
Facial surgery and injection history	Botulinum toxin injection, filler injection, face and neck liposuction, history of previous facial and neck thread lifting surgery, etc.
Medication history	Special drug usage: especially the use of blood-activating agents such as aspirin, danshen, etc, which must be discontinued for at least 1 week before arranging a thread lift surgery.
Mental state	Psychiatric disease, personality disorder, addiction disorder, depressive disorder, cognitive deviation, affective disorder, etc.

Preoperative photography: At least seven photos are taken, including standing frontal view, bilateral 45-degree view, bilateral 90-degree view, low head position, and upward view.

Preoperative special preparation: Including preoperative hair and facial cleansing, preparation of temporal hair, hair styling, and fixing with tape.

Surgical instrument preparation: Sterile surgical pack including four sterile treatment cloths, cloth forceps, hemostatic forceps, Adson forceps, surgical scissors, knife handle, No. 11 scalpel blade, etc.

Postoperative management and follow-up: After the operation, keep the face clean and dry and apply ice for 48 h. If necessary, take oral antibiotics and anti-swelling drugs, and apply antibiotic cream to the puncture holes. Try to avoid excessive facial expression and heavy massage after the operation. Facial energy-tightening equipment is prohibited for 6 months after the operation. Follow-up appointments should be scheduled for 1 month, 3 months, 6 months, and 1 year after the operation.

9.4 Treatment process

9.4.1 Anesthesia

Local infiltration anesthesia with lidocaine (with a concentration of 0.5%–1%, not exceeding a total amount of 400 mg) and epinephrine (with a concentration ratio of 1:100 000–1:200 000) is used. Point-like infiltration anesthesia is performed at the needle entry and knotting points, and tunnel infiltration anesthesia is performed with a long blunt needle or sharp needle for the buried suture design channel.

9.4.2 Design and implantation technique of various parts of the face

There are different lengths, distributions, and thicknesses of threads available in the market, and personalized operation should be carried out according to the design plan. The following are the key points of several common parts of facial thread implantation.

Mid-lower face lifting: This is commonly performed by lifting and tightening outward and upward. It mainly improves mild to moderate sagging of the mid-facial skin and shallow fat pad, reduces nasolabial folds, improves jowls and submandibular fat accumulation. The anchor point is often selected at the upper edge of the zygomatic arch within the temporal region, avoiding the superficial temporal artery. The distal end of the buried suture is distributed from top to bottom: near the nasolabial groove line between the nasal alar and the commissures of the mouth; near the midpoint of the marionette fold; near the intersection of the anterior margin of the masseter muscle and the mandibular margin. The number and distribution of the sutures can be adjusted according to individual differences, as well as the degree of sagging and depression in the mid-lower face. During the operation, the entry point is penetrated with a needle, and the depth of the needle is determined by the assistance of a hand pinching the feeling of the needle. The needle first reaches the temporal deep fascia layer, traverses the zygomatic arch ligament, and then the buried line is mostly located in the subcutaneous shallow fat layer. During the operation, attention should be paid to accurate layering to avoid damage to important facial tissue structures such as the facial nerve, blood vessels, and parotid duct. The operator should gently manipulate until the thread reaches the far end. Finally, the operating hand pulls the distal part of the thread, and the assisting hand performs gentle tissue reposition. After achieving satisfactory results, and the exposed thread ends are cut off close to the skin.

Lower facial lifting: This operation design is suitable for those who have obvious jowls outside the mouth, obvious submental sulcus, an unclear mandibular line, and significant sagging and accumulation of skin and subcutaneous fat above the mandibular border. The arrangement and fixation methods are basically the same as those of the mid-lower facial lifting. The lifting direction is mainly towards the lateral and posterior directions to tighten and improve the jowls outside the mouth and reshape the mandibular border. The anchoring points in the temporal area are mostly located within the hairline under the zygomatic arch and platysma auricular ligament behind the earlobe. The needle is inserted from the entry point and passes through the superficial subcutaneous fat layer. During the operation, attention should be paid to avoiding important facial tissues such as facial nerve and blood vessels until reaching the distant lifting point. After satisfactory soft tissue reposition, the exposed thread is cut off.

Cheek lift: The superficial malar fat pad is the location of the 'apple muscle' in the midface, and the volume loss of its deep supportive fat pad is an important cause of sagging in the midface with aging. By using special anatomical anchoring points, support can be provided for the soft tissues of the cheek, and the sagging superficial malar fat pad can be repositioned to its original anatomical position, making facial rejuvenation possible. APTOS, QUILL, and Happy Lift™ threads all have respective designs for cheek lifting. They use bidirectional barbed threads, with the zygomatic arch ligament and the maxillary ligament of the cheek as anchor points, and follow parallel curved lines to gather and lift sagging soft tissue. The threads are arranged in a 'fishing net' pattern within the soft tissue to support and resist gravitational pull, thus restoring the plump and round aesthetic effect of the midface.

Lower eyelid buried suture: Skin and soft tissue atrophy, tear trough and palpebromalar groove appearance, and lower eyelid bagging are the main signs of lower eyelid aging. Using smooth or threaded PDO thread buried in the superficial subcutaneous layer can improve and achieve a rejuvenation effect. The main principle is that the thread can stimulate the proliferation of new collagen. The patient is in an upright position, and markings are made along the palpebromalar groove. The entry point is 2 cm below the vertical line of the outer canthus, and local anesthesia is performed at the entry point. According to the designed thread, multiple 5.0 buried sutures are inserted parallel to the orbital groove, located in the subcutaneous layer and the anterior orbital septum space, and the tear trough ligament is separated as much as possible. After being completely buried, the sheath is gently removed.

In addition, because various threads such as PDO and PLLA can stimulate collagen production, they have a definite effect on static wrinkles and mild sagging of the skin. Smooth or spiral fine threads can be used in different parts of the face according to the needs, and can be effectively combined with long thread lifting to achieve better rejuvenation effects. For example, the placement of mesh smooth threads and spiral threads in the nasolabial fold and marionette lines area can play an auxiliary lifting and fixing role.

9.5 Combined treatment

9.5.1 Combined use with botulinum toxin

Botulinum toxin can regulate the contraction of the muscles and facial contour; it can be used in combination with thread lifting. Lower facial thread lifting is often used to correct jowls and sagging skin and subcutaneous fat accumulation above the lower jawline. Botulinum toxin injections in the lower jawline can also improve unclear jawlines and jaw pouches by relaxing the platysma muscle. Therefore, thread lifting and botulinum toxin injection in the platysma muscle can be used in combination to reduce the antagonistic force of the platysma muscle and make the lifting effect of the thread more obvious. In addition, enlarged masseter muscles can also degrade the jawline appearance, and combined use of botulinum toxin injection in the masseter muscle and thread lifting of the lower face can help establish a clear and smooth jawline.

9.5.2 Combined with hyaluronic acid

The placement level of thread lifting is mainly in the superficial subcutaneous fat layer, but for groove areas such as nasolabial folds, marionette lines, mostly accompanied with the volume loss of the deep fat pads located in the midface such as suborbicularis oculi fat and the deep medial cheek fat pad, hyaluronic acid can be combined with thread lifting to improve appearance. For patients with soft tissue sagging in the mid and lower face and depression under the cheekbones, an appropriate amount of hyaluronic acid can be filled at the depression site after mid-lower face thread lifting. Combined application can effectively enhance the treatment effect, but also control the filling volume and reduce the adverse reactions.

9.5.3 Combined use of thread lifting and autologous fat grafting

The combination of thread lifting and autologous fat grafting can achieve better soft tissue volume, contouring and skin texture. It is recommended to perform local liposuction before thread lifting surgery, such as for thickened nasolabial fat pads and the removal of excessive fat under the jawline and chin. The autologous fat grafting can then be performed after the thread lifting surgery. Thread lifting surgery can cause tissue trauma and anatomical displacement, increasing the risk of fat injection into blood vessels. The trauma from thread lifting surgery can also lead to changes in the vascular structure of the transplantation bed, affecting the survival rate of autologous fat grafting and the occurrence of nodules. Autologous fat grafting has different degrees of absorption rate, which can cause asymmetry. Preoperative communication with the patient is critical.

9.5.4 Combined with collagen injection

Collagen injection can provide nutrition for the tissue and supplement the volume loss. Combined with thread lifting, collagen injection is often used to improve the tear groove, middle buccal groove, and reduce dark area below the lower eyelashes.

9.5.5 Combined with micronized acellular dermal matrix (mADM)

After thread lifting, related facial tissue ligaments such as the zygomaticus major ligament and orbicularis retaining ligament are relaxed due to the tightening of the skin and subcutaneous tissue. At the same time, injecting regenerative materials such as mADM into the deep layer of the ligaments can effectively repair and enhance their function, achieving a synergistic therapeutic effect to repair and enhance ligament function.

9.5.6 Combined with skin mesotherapy

Skin mesotherapy can be combined with smooth and barbed threads to achieve treatment effects such as improving and brightening skin texture, reducing pore size, and reducing pigmentation.

9.5.7 Combination with photoelectric treatment

Combination with photoelectric instruments for facial rejuvenation and lifting such as radiofrequency and ultrasound equipment. The photoelectric treatment should be performed before thread lifting, as phototherapy treatments can promote collagen regeneration and tightening of subcutaneous tissues, which is beneficial for anchoring thread barbs during thread lifting and for enhancing surgical outcomes. If phototherapy is needed after thread lifting, it is recommended to wait for at least 3 months after surgery and preferably 6 months, when the thread body and tissue have formed a stable fibrous package, to avoid postoperative early photoelectric treatment which may cause uneven heating of the thread inside the tissue and lead to skin discoloration, burns, and other complications.

9.6 Complications

Thread lifting is a minimally invasive surgery compared to standard incisional surgery for facial rejuvenation, with minimal scarring, quick recovery, and fewer complications. Therefore, thread lifting is becoming increasingly popular among doctors and patients. The most common complaint about thread lifting is that the results do not meet the patient's expectations. The improvement in results is based on the patient's own anatomical basis, reasonable selection and design of thread materials, and superb surgical technique. Of course, the maintenance time of the lifting effect is another common problem. Therefore, it is important to clarify the patient's expectations before surgery, evaluate the anatomical basis and individual differences, and simulate the surgical effect. In addition, there are also some complications that can occur, such as swelling, bleeding, infection, damage to important blood vessels, nerves, and salivary glands, thread exposure, local unevenness, and facial asymmetry (Helling *et al* 2007, Li *et al* 2021, Paul 2008). Therefore, adequate preoperative communication and corresponding treatment after surgery are important for the prevention and treatment of complications.

9.6.1 Localized swelling

Swelling often occurs in the surgical area and occasionally affects the surrounding area. When the swelling is slight, it can generally disappear spontaneously in a short period of time. It is recommended to apply an ice compress and wear an elastic face mask after surgery to relieve the swelling. If necessary, short-term oral antibiotics or steroids can be taken.

9.6.2 Bleeding

Bleeding is often related to the bluntness and sharpness of the cannula needle, as well as the depth of needle insertion. Occasionally, it can occur due to the secondary damage of blood vessels caused by the protrusion of the thread or tissue traction after the procedure. Bleeding often occurs at the entry point of the thread and surrounding areas, and may be accompanied by pain, swelling, and bruising. Pressure is applied to stop bleeding and, after a few minutes, the entry point is observed for any continued bleeding. If necessary, the entry point can be changed to continue the procedure.

9.6.3 Infection

Enhancing preoperative preparation and strict sterile operation standards during surgery may prevent infections from occurring. If a local red and swollen infection occurs at the puncture site, it is mostly caused by the pulling of hair into the entry point by the barbs of the thread. If there is hair in the entry point, it must be removed. If the patient develops redness, swelling, pain, and elevated skin temperature after surgery, and the blood routine examination suggests suspicion of infection, it is recommended to start systemic anti-infective treatment as soon as possible. In addition, special infections such as non-tuberculous mycobacteria should also be taken seriously, and if infection is diagnosed, corresponding standardized treatment should be given in a timely manner.

9.6.4 Injury

Injury to important blood vessels, nerves, and salivary ducts is generally caused by unfamiliarity with the anatomical layer and lack of proficiency in surgical techniques. If this occurs, appropriate symptomatic treatment should be provided.

9.6.5 Thread exposure

Thread exposure can occur on the skin or in the oral cavity, and the exposed thread can be removed with a needle tip, usually without infection. Recently, manufacturers have developed flexible barbed wires that can be bent 180 degrees to prevent the thread tip from piercing the skin. Thread exposure in the oral cavity often occurs when exaggerated facial expressions or biting causes thread fracture and movement, eventually leading to exposure in the oral cavity and requiring removal from inside the mouth.

9.6.6 Local unevenness

The most common side effect of thread lifting is skin depressions or swelling, resulting in unevenness on the surface, which usually occurs during the early swelling period after the surgery. The cause for the unevenness can be attributed to the thread type and surgical method. It is generally caused by the uneven depth placement of the thread or the thread running in a too superficial layer, and if necessary, appropriate treatments such as sharp needle stripping or filling can be given. If a thread is hanging on the dermis layer, usually causing a dimple, it may be accompanied by pain and can be relieved by manual massage. In addition, nonuniform distribution of threads and different looseness of tissue can also cause uneven swelling and an uneven surface, but in general, it will self-relieve as the swelling goes down gradually.

9.6.7 Asymmetry

Asymmetry on both sides can be adjusted through precise preoperative design and fine-tuned surgical procedures. If necessary, combined injection and filling therapy can also be used.

9.7 Conclusion

Thread lifting has been widely used in the past 20 years. Successful and effective lifting treatment must be based on a comprehensive understanding of the facial anatomical changes caused by aging. In addition, effective thread lifting requires three elements: anchor points, direction, and lifting points. In clinical application, different types of threads should be selected and different thread lifting schemes should be designed based on the anatomical diagnosis and accurate assessment of different patients. Multiple threads can be used in combination, and thread lifting can also be combined with various rejuvenation treatments such as botulinum toxin injection, filler injection, phototherapy, and so on. Attention should also be paid to avoiding complications.

References

Atiyeh B S, Chahine F and Ghanem O A 2021 Percutaneous thread lift facial rejuvenation: literature review and evidence-based analysis *Aesthet. Plast. Surg.* **45** 1540–50

Fabi S G, Weiss R and Weinkle S H 2021 Absorbable suspension sutures: recommendations for use in a multimodal nonsurgical approach to facial rejuvenation *J. Drugs Dermatol.* **20** 23–9

Fukaya M 2018 Two mechanisms of rejuvenation using thread lifting *Plast. Reconstr. Surg. Glob. Open* **6** e2068

Goldberg D J 2020 Stimulation of collagenesis by poly-L-lactic acid (PLLA) and -glycolide polymer (PLGA)-containing absorbable suspension suture and parallel sustained clinical benefit *J. Cosmet. Dermatol.* **19** 1172–8

Ha Y I, Kim J H and Park E S 2022 Histological and molecular biological analysis on the reaction of absorbable thread: polydioxanone and polycaprolactone in rat model *J. Cosmet. Dermatol.* **21** 2774–82

Helling E R, Okpaku A, Wang P T and Levine R A 2007 Complications of facial suspension sutures *Aesthet. Surg. J.* **27** 155–61

Kaya K S and Cakmak O 2022 Facelift techniques: an overview *Facial Plast. Surg.* **38** 540–5

Kim C M, Kim B Y, Hye S D, Lee S J, Moon H R and Ryu H J 2019 The efficacy of powdered polydioxanone in terms of collagen production compared with poly-L-lactic acid in a murine model *J. Cosmet. Dermatol.* **18** 1893–8

Li Y L, Li Z H, Chen X Y, Xing W S and Hu J T 2021 Facial thread lifting complications in China: analysis and treatment *Plast. Reconstr. Surg. Glob. Open* **9** e3820

Moon H, Fundaro S P, Goh C L, Hau K C, Paz-Lao P and Salti G 2021 A review on the combined use of soft tissue filler, suspension threads, and botulinum toxin for facial rejuvenation *J. Cutan. Aesthet. Surg.* **14** 147–55

Paul M D 2008 Complications of barbed sutures *Aesthet. Plast. Surg.* **32** 149

Rezaee K S, Nabie R and Aalipour E 2020 Outcomes in thread lift for face, neck, and nose: a prospective chart review study with APTOS *J. Cosmet. Dermatol.* **19** 2867–76

Sasaki G H, Komorowska-Timek E D, Bennett D C and Gabriel A 2008 An objective comparison of holding, slippage, and pull-out tensions for eight suspension sutures in the malar fat pads of fresh-frozen human cadavers *Aesthet. Surg. J.* **28** 387–96

Sulamanidze M and Sulamanidze G 2009 APTOS suture lifting methods: 10 years of experience *Clin. Plast. Surg.* **36** 281–306

Sulamanidze M A, Fournier P F, Paikidze T G and Sulamanidze G M 2002 Removal of facial soft tissue ptosis with special threads *Dermatol. Surg.* **28** 367–71

Tong L X and Rieder E A 2019 Thread-lifts: a double-edged suture ? A comprehensive review of the literature *Dermatol. Surg.* **45** 931–40

Wu W T 2004 Barbed sutures in facial rejuvenation *Aesthet. Surg. J.* **24** 582–7

Yoon J H, Kim S S, Oh S M, Kim B C and Jung W 2019 Tissue changes over time after polydioxanone thread insertion: an animal study with pigs *J. Cosmet. Dermatol.* **18** 885–91

IOP Publishing

Skin Photoaging (Second Edition)

Rui Yin, Yang Xu and Chengfeng Zhang

Chapter 10

Laser treatment for skin photoaging

Zhen Zhang and Xiaojin Wu

Minimally invasive procedures in cutaneous laser surgery, particularly ablative and fractional ablative laser systems, have gained popularity for treating various skin conditions. Ablative lasers like CO_2 and Er:YAG target water as their chromophore, with CO_2 providing effective skin rejuvenation but limited by wide thermal damage. Er:YAG, introduced to counter the drawbacks of CO_2, offers minimal thermal damage and quicker recovery. Fractional ablative lasers, a newer development, create microthermal treatment zones (MTZ) for controlled ablation and collagen remodeling. Both types have clinical implications, with traditional ablative lasers suited for severe photoaging, while fractional ablative lasers offer shorter downtimes. However, caution is needed for darker skin tones due to increased risk of side effects such as hyperpigmentation. Contraindications include keloids, active infections, and systemic diseases. Complications such as prolonged erythema and acneiform eruptions are manageable, with fractional ablative lasers generally having milder side effects compared to fully ablative ones.

10.1 Introduction

With the increasing demands of patients undergoing cosmetic surgery, minimally invasive procedures have gained popularity. There are numerous applications for cutaneous laser surgery, including destruction of vascular and pigmented lesions, striae, as well as dermal remodeling for treatment of photodamage (Tanzi and Alster 2003). Most lasers work through selective photothermolysis where controlled destruction of a chromophore occurs without damage to the surrounding normal tissue (Anderson and Parrish 1983). Ablative and non-ablative laser systems have been successfully used in the treatment of photodamage, wrinkles and for increasing collagen production; the appeal of the process is that it is less invasive and requires less downtime compared to traditional surgery (Lolis and Goldberg 2012, Tanzi *et al* 2003).

doi:10.1088/978-0-7503-5112-6ch10

10.2 Ablative laser systems

10.2.1 Fully/confluent ablative lasers

10.2.1.1 Mechanism

Traditional ablative lasers, mainly consisted of CO_2 lasers and Er:YAG lasers, were generally used in the 1990s to treat skin photoaging. These skin resurfacing systems select water as their target chromophore. Apart from the approximately 100 mm deep contiguous ablation, they also generate a <200 mm thermal coagulation zone. The latter could cause heat-induced collagen contraction and dermal remodeling and was thought to be the key impeller for skin tightening (Kauvar 2014).

10.2.1.2 Classification

The CO_2 laser was one of the first lasers to be used in medicine. Its 10 600 nm wavelength enables it to be properly absorbed by water in the epidermis and dermis. In the 1960s, continuous-wave CO_2 lasers were used in surgery to cut neoplasm and coagulate at the same time. However, due to its much-too-wide thermal coagulation zone, its use in skin rejuvenation was limited. First replaced by pulsed and scanning CO_2 lasers in the early 1990s and then by high-energy pulsed and slash-scanned CO_2 lasers in the late 1990s, continuous-wave CO_2 lasers are no longer used in skin resurfacing. The ultra-pulsed CO_2 laser, as its pulse width is narrower than the thermal relaxation time of the skin, provides greater control over the depth of injury. Also, because of its ability to induce neocollagenesis and neoelastogenesis, it was commonly used in the 1990s for skin rejuvenation in lighter skin phototype (Adrian 1999, Koch 2002, Loesch *et al* 2014). The erbium:yttrium aluminum garnet (Er: YAG) laser was introduced as the second generation of ablative resurfacing system to counter the drawbacks of CO_2 lasers. The wavelength of Er:YAG lasers is 2940 nm, matching the peak absorption of water perfectly (3000 nm). As this wavelength has ten times the water absorption coefficient of CO_2 lasers and a very short extinction, vaporization of intracellular water can occur with more efficient tissue ablation. As a result, minimal thermal damage, shorter recovery times, and lower incidence of side effects could be achieved. However, the reduction of its thermal damage also leads to less tissue tightening effect than achieved by CO_2 lasers. Later, longer-pulsed Er:YAG lasers and then variable-pulsed Er:YAG lasers and 2790 nm erbium:yttrium scandium gallium garnet (Er:YSGG) lasers were developed in attempts to improve the tissue tightening effect while maintaining the low side-effect rate. However, CO_2 lasers still proved to be more effective in terms of skin rejuvenation and deep rhytides (Adrian 1999, Alster and Lupton 2001, Ross *et al* 2001, Weniger *et al* 2020). Recently, a novel carbon monoxide (CO) laser with a wavelength of 5500 nm was developed. It was found to generate deeper ablation craters and wider thermal coagulation zones than the traditional CO_2 laser at a normalized ablation threshold. Researchers hope this new development could benefit those who seek more effective rejuvenation by laser resurfacing. However, further *in vivo* studies must be performed to confirm its effect on dermal remodeling (Ha *et al* 2020).

10.2.1.3 Clinical implications

A traditional ablative laser requires a longer downtime along with causing more side effects due to its wide ablative area and fewer healthy skin islands. Therefore, fractional ablative lasers are now more commonly used for skin rejuvenation. However, traditional ablative lasers are still favored for those with severe rhytides/ photoaging and realistic expectations of treatment outcomes. The reduction of rhytides was reported to vary from 50% to 90% according to previous research. However, patients should be aware of and accept the longer post-operative downtime. Furthermore, in subjects with higher Fitzpatrick skin phototypes, the consensus among experts is to advise against traditional laser ablation to avoid permanent hypo/hyperpigmentation and scarring. Finally, it should be noted that patients' dietary habits may play a role in the outcome of ablative laser treatment. One recent study found that rejuvenation effect of ultra-pulsed CO_2 lasers in vegetarians is worse than that of the omnivores. Longer epidermal repair time and post-procedure erythema were also observed, probably due to the reduced collagen synthesis of vegetarians after CO_2 laser resurfacing (Alexiades-Armenakas *et al* 2008, Chen *et al* 2022, Fusano *et al* 2021, Hu *et al* 2022).

Contraindications to traditional ablative laser treatment include:

1. Patients with high expectations for treatment or who are unable to cooperate.
2. Patients with a history of keloids.
3. Patients with infection or skin diseases characterized by the Koebner phenomenon (vitiligo\psoriasis …).
4. Patients having undergone isotretinoin therapy within 1 year.
5. Patients having undergone a face lift procedure within 6 months.
6. Patients with dermatologic conditions that could result in reduction of adnexal structures (radiation therapy/scleroderma).
7. Patients with higher Fitzpatrick skin phototypes.
8. Patients with severe systemic diseases.
9. Pregnant or lactating patients (Alexiades-Armenakas *et al* 2008).

To minimize the side effects of facial resurfacing by fully ablative lasers, researchers set up general recommendations for Er:YAG laser depth settings (Weniger *et al* 2020).

1. The perioral area is one of the most suitable areas for laser resurfacing. However, severe complications often occur in that area. The recommended depth is less than 600 μm for the upper lip, 400 μm in the prejowls, and 300 μm on the chin. Note that depth should be adjusted during the procedure according to the immediate reaction. If much bleeding occurs before the predetermined depth is reached, the treatment should be stopped. On the other hand, more laser passes are required without obvious pinpoint bleeding or reduction of rhytids.
2. Most commonly, 100 μm of ablation and 50 μm of coagulation are used for cheeks. They could be treated with up to a maximum of 200 μm of ablation with 50 μm of coagulation in the most severe cases.

3. The forehead is most commonly treated with 150 μm of ablation and 50 μm of coagulation; 200 μm of ablation can also be used when needed. Coagulation mode in this area does appear to cause slight elevation in the brows. Laser settings do not need to be decreased when simultaneous forehead lifting surgery in the subgaleal or subperiosteal plane is performed.
4. The rich sebaceous glands in the skin of most noses allows for aggressive resurfacing settings. Two passes of 200 μm of ablation and 50 μm of coagulation can be used except in the thinnest skin. The use of coagulation is very helpful with even deeper resurfacing of this area such as for rhinophyma, as bleeding is lessened.
5. For the periorbital area, most commonly, 80 μm of ablation and 50 μm of coagulation are used. It is observed that the skin retraction caused by laser resurfacing could match a 6 mm skin-only resection in the upper eyelids. Therefore, for patients who lack support or with pre-existing malposition in the periorbital area, settings could be reduced to 50 μm of ablation and 25 μm of coagulation. In severely lax patients, a canthoplasty may be required before any resurfacing.

10.2.1.4 Complications

1. *Immediate post-operative reactions*: Edema, oozing, pain, scabs, etc, usually occur within 1–10 d after laser resurfacing.
2. *Prolonged erythema*: This inevitably occurs during the wound healing process, but the duration of erythema after traditional laser ablation is long, and prolonged erythema is expected for up to 6 months.
3. *Post-inflammatory hypo/hyperpigmentation (PIH)*: Individuals with darker skin types are more prone to PIH due to higher melanin deposits in the epidermis. Topical treatment with hydroquinone or tranexamic acid might help. Avoidance of darker skin toned patients, precise control over the ablation depth during the procedure and strict sun protection afterwards might help to reduce the occurrence of PIH.
4. *Acne and milia*: These most commonly occur in patients with a history of acne 1–2 weeks after the laser procedure. Acneiform eruptions and milia can disappear spontaneously. However, regular anti-acne therapy might be helpful. Careful nursing of the post-procedure area could help to reduce the acne and milia.
5. *Infection*: The routine use of antibacterial prophylaxis and post-procedure steroids might lead to more fungal infections. Due to the high incidence of latent HSV infection in the general population, any patient (regardless of prior known HSV history) planning to undergo full-face or perioral resurfacing should be given oral antiviral prophylaxis.
6. *Scarring and ectropion formation*: Excessive ablation, post-procedure infection and removal of immature skin crust are all risk factors for scarring and ectropion. Scars are most often seen in the perioral area, chin, jaw, and neck (Li *et al* 2018, Weniger *et al* 2020).

10.2.2 Fractional ablative laser (FAL)

10.2.2.1 Mechanism

The FAL was introduced in the early new millennium to minimize the side effects and downtime of laser facial rejuvenation. Its basic notion is called fractional photothermolysis (FP), a milestone theory proposed by Dieter Mainstein *et al* (Manstein *et al* 2004).

Fractionated systems deliver pixelated laser energy in a grid-like pattern creating microscopic vertical columns known as microthermal treatment zones (MTZs). As the distribution of thermal excitation is proportional to the level of the local optical energy density times the local optical absorption coefficient, the diameter and depth of the MTZs depend on the settings of the fractional laser. An *in vivo* study demonstrated that, at a fluence of 50, 100, and 300 mJ/microbeam, ablation reached the superficial dermis, mid-dermis, and deep dermis, respectively. The study also pointed out that the ablation width increased from 224 mm to 261 mm and 398 mm. The pattern and size of the FAL spot can be adjusted to meet the clinical need. The spared skin adjacent to the MTZs provides a reservoir of viable tissue that allows for migration of keratinocyte and fibroblast to the ablated epidermis and dermis. As the proportion of MTZs is generally less than 40%, re-epithelialization can be ensured within 24 h. The rapid repair of the epidermal skin barrier allows for higher energy and thus deeper penetrance can be achieved without severe adverse effects (Chen *et al* 2022, Grunewald *et al* 2011, Laubach *et al* 2006).

Genetic expression profiling of *ex vivo* skin from Asian patients 24 h after FAL demonstrated upregulation of genes encoding Wnt5a, CYR61 and heat shock protein (HSP)90—all three play a role in the cell migration and differentiation during the wound healing process. Remarkably increased expression of HSP-70 and HSP-47 was found in a histologic study. In the study, continued formation of new and remodeled collagen and elastic fibers was observed for up to 6 months after the procedure (Kim *et al* 2013, Xu *et al* 2011).

10.2.2.2 Classification

Similar to traditional ablative lasers, the FALs consist mainly of 2940 nm Er:YAG laser, 2790 nm Er:YSGG laser and 10 600 nm CO_2 laser. Compared to non-ablative lasers, FALs have a higher absorption coefficient by water. Thus, a FAL can create a real MTZ by vaporization of the epidermis and part of dermis. As energy from a 2940 nm Er:YAG laser and 2790 nm Er:YSGG laser are absorbed by water more efficiently than a CO_2 laser, the penetration depth is reduced. The penetration depth is 20–30 $\mu m\ J^{-1}\ cm^{-2}$ for a CO_2 laser versus 1–3 $\mu m\ J^{-1}\ cm^{-2}$ for an Er:YAG laser. A better water absorption coefficient also results in reduced collateral thermal damage and less coagulation for Er:YAG lasers (Alexiades-Armenakas *et al* 2012, Chen *et al* 2022).

Theoretically, deeper penetration of the CO_2 laser and a wider coagulation zone adjacent to the MTZ should result in more neocollagenesis and improved collagen remodeling. The effect of a CO_2 laser on facial rejuvenation is indeed better than an Er:YAG laser in terms of a traditional ablative laser. However, two split-face randomized controlled trials compared the effect of fractional CO_2 and Er:YAG lasers on skin wrinkle reduction. They came to the consistent conclusion that, although the fractional CO_2 laser showed slightly better efficacy than the Er:YAG laser during the first 2 months after the procedure, no significant differences were noted between the two lasers. Their side effects were also comparable. The fractional Er:YAG laser caused more bleeding during the procedure due to its reduced coagulation effect. However, the fractional CO_2 laser was more painful and generated a more prominent crust during the post-procedure recovery. Differences between the two FALs were found in other dermatologic indications. A split scar randomized controlled trial conducted by Choi *et al* demonstrated superior efficacy of the fractional CO_2 laser on hypertrophic scars, especially in improving the pliability of the scars (Choi *et al* 2014, Karsai *et al* 2010, Robati and Asadi 2017).

10.2.2.3 Clinical implications

Traditional ablative lasers have been proven to be effective for facial rejuvenation and scarring in light skin toned individuals. However, the high incidence rate of infection, hyperpigmentation, scarring and its long recovery time have limited its use. In contrast, FALs achieve a shorter downtime and fewer complications through the intervening areas of healthy skin tissue around MTZs. In 2013, El-Domyati *et al* compared the effect on facial rejuvenation between a single-session full ablative Er:YAG laser and four sessions of a fractional ablative Er:YAG laser. Skin biopsies of treated areas were obtained to make an objective comparison. In the end, both of the above lasers induced dermal neocollagenesis without significant differences. The result suggested that the efficacy of the multiple fractional ablative Er:YAG laser is comparable to that of a single-session full ablative Er:YAG laser. The former was favored for its shorter recovery time and higher safety profile (El-Domyati *et al* 2013).

However, the use of FAL is still controversial for the darker skin toned population. A comprehensive review written by Wat *et al* in 2016 confirmed the therapeutic effect of FAL on photoaging in Asian people. The literature proved its effect on rhytides and wrinkles, skin elasticity, skin pore size and texture. However, prolonged erythema and PIH were frequently seen after FAL in Asians. One study reported a more than 12% incidence rate of PIH on patients with darker skin phototypes (Fitzpatrick III–VI) after FAL. Therefore, they suggested practitioners be cautious and save FAL only for severely photodamaged Asian individuals. For those in their 40s–60s, skin dyspigmentation is generally more prominent than facial rhytides. Therefore, a combined therapy with FAL and non-ablative fractional laser may achieve better results (Chan *et al* 2007, Wat *et al* 2017).

Contraindications for an FAL are similar to those for a traditional ablative laser:

1. Patients with high expectations for treatment or who are unable to cooperate.
2. Patients with a history of keloids.
3. Active infection in the areas to be treated (particularly herpes infection).
4. Patients in an active stage of skin diseases characterized by the Koebner phenomenon (vitiligo\psoriasis…).

5. Whether patients who have undergone isotretinoin therapy should be excluded is still controversial.
6. Patients with dermatologic conditions that could result in reduction of adnexal structures (radiation therapy/scleroderma).
7. Patients with severe systemic diseases.
8. Pregnant or lactating individuals (Alexiades-Armenakas *et al* 2008, Rubenstein *et al* 1986).

10.2.2.4 Complications

The types of adverse reactions associated with FAL are virtually the same as those associated with conventional laser resurfacing, including prolonged erythema, acneiform eruption and milia, hyper/hypopigmentation, infection, and scarring. However, the side effects are usually milder and persist for a shorter time than for fully ablative lasers (Metelitsa and Alster 2010).

10.3 Nonablative laser systems

10.3.1 Nonfractionated near-infrared (NIR) lasers

10.3.1.1 Mechanism

Near infrared refers to the 750–1400 nm part of the electromagnetic spectrum. The absorption coefficient of melanin and hemoglobin in this laser spectrum is decreasing, while the absorption coefficient of water is gradually increasing. Therefore, the main target chromophore of NIR lasers is water. In practice, thermal injury caused by dermal water absorption of laser energy results in neocollagenesis. As NIR lasers are far less effectively absorbed by water than ablative lasers, vaporization of the skin tissue is spared. Deeper penetration and fewer side effects are also expected. In fact, the downtime of NIR lasers could be limited to hours. However, a lower thermal coagulation effect is also related to lower dermal remodeling and fewer cosmetic effects than the ablative laser systems. Moreover, NIR laser penetration and thermal stimulation are also based on the energy level. Therefore, a surface cooling system must be equipped to avoid epidermal burns (Aslam and Alster 2014).

10.3.1.2 Classification

Commonly used pulsed NIR lasers include the 1320 nm Nd:YAG laser, 1064 nm Nd:YAG laser and 1450 nm diode laser.

The relatively weaker water absorption of the 1320 nm laser enables it to penetrate deeper into the dermal layer. Histologic studies on a 1320 nm Nd:YAG laser confirmed the increased papillary collagen density and neocollagenisis of type I and III collagen after treatment. However, it has been reported that the clinical improvement of a 1320 nm Nd:YAG laser on facial rejuvenation is relatively mild. Multiple treatments and long-term follow-ups are needed to make appropriate evaluation (Goldberg 2000, Trelles *et al* 2001, El-Domyati *et al* 2011).

Histological examination was carried out immediately after 1450 nm diode laser treatment and showed that the depth of the thermal damage area reached 644 μm in the dermis. Biopsy at 2 months post-procedure showed that the dermal fibrosis reached 420 μm deep in the dermis. Like the 1320 nm laser, despite the dermal collagen remodeling observed from histological studies, the clinical effect of the 1450 nm diode laser on wrinkle improvement is quite subtle. However, thanks to the thermal effect of the 1450 nm diode laser on sebaceous glands and associated tissues, studies have found it quite effective in the treatment of active acne as well as acne scars (Hardaway *et al* 2002, Jih *et al* 2006).

The target chromophores of the 1064 nm Nd:YAG laser include melanin, hemoglobin and water. The thermal effect on the skin dermis could be induced by the absorption of laser energy by the above chromophores. The energy of the 1064 nm Nd:YAG laser could reach deep into the dermis thanks to the relatively weak absorption coefficient of its chromophores. Studies have also confirmed its diffuse heating effect on dermal tissue. In two histological studies (one on hairless mice and the other on a porcine skin model), dermal collagen formation was observed. An increased level of TGF-β expression was also found in a hairless mouse model after the laser procedure. Meanwhile, the quasi-long pulse and q-switched 1064 nm Nd:YAG lasers were also found to be effective in the treatment of skin photoaging because of their photomechanical and photothermal effects. The rejuvenation effect of picosecond 1064 nm Nd:YAG lasers will be discussed in a later section (Lee *et al* 2012, Lee *et al* 2009, Dayan *et al* 2003, Dayan *et al* 2003).

10.3.1.3 Clinical implications

The major advantage of NIR lasers over ablative laser systems lies in their protection of the epidermis, which results in shorter downtime and fewer complications. Additionally, NIR laser rejuvenation could be applied safely to areas where the prerequisite pilosebaceous units for re-epithelization are scarce. Therefore, NIR lasers are ideal for those who seek rapid recovery or rejuvenation for special areas such as the neck and hands. It is worth noting that, despite the histological findings of neocollagenesis, the clinical effect of the NIR laser is not that evident. Repeated sessions and a more than 3 month treatment period are expected. Most reported clinical improvement rates range from 30% to 50% after several treatments. Patients should be fully informed before treatment to avoid complaints afterwards (Aslam and Alster 2014, Sadick and Schecter 2004).

10.3.1.4 Complications

The most common side effects of NIR lasers were pain during the procedure and edema and erythema immediately after the procedure. The edema and erythema generally resolve spontaneously within 24 h. Insufficient cooling of the epidermis may lead to the formation of blisters afterwards. Post-inflammatory hyperpigmentation could probably occur after blisters. Fortunately, long-term sequelae such as scars or keloids are very rarely seen (Alster and Lupton 2002).

10.3.2 Nonablative fractional lasers (NAFL)

10.3.2.1 Mechanism

NAFL also targets water containing tissues in a pixelated pattern. In contrast to AFL devices, NAFLs carry out thermal damage to the dermis without vaporization of the epidermis. The damage to the dermal–epidermal junction is also minimized so that complications become milder, and downtime is shortened. The MTZs in NAFL are conical zones of thermally coagulated collagen instead of real tunnels of vaporized tissue. Treatment depth depends on three factors, the energy levels used during the treatment, the passes and laser density. Generally, a higher energy power results in deeper penetration. The density of the treatment refers to the density of MTZs and correlates with the severity of inflammation. Too much overlap of the MTZs might cause ablative-like laser resurfacing (Chen *et al* 2022, Manstein *et al* 2004).

Several molecular studies demonstrated a significant increase in pro-inflammatory cytokines after NAFL. The changes include an elevated level of tumor necrosis factor α (TNFα) and interleukin β (ILβ) followed by the upregulation of HSP-70 and several MMPs. The latter was found to play a pivotal role in the induction of collagen remodeling. Gene expression of collagen I and III was initially decreased after NAFL but gradually increasing several days after. The wound repairing process observed at the molecular level further confirmed the effect of NFAL on facial collagen stimulation (Orringer *et al* 2010, Said *et al* 2021).

10.3.2.2 Classification

NAFL also consists of near-infrared lasers whose wavelengths range from 700 to 1400 nm. Mainstream NAFLs include an 1540/1550 nm erbium glass laser, 1927 nm thulium fiber laser, 1440 nm diode laser and 1440/1320 nm Nd:YAG laser. Moreover, as the use of a picosecond q-switched Nd:YAG laser and 755 nm Alex-andr-ite laser in NAFL is gaining more attention, we will discuss them later in the section on fractionated picosecond lasers.

As discussed in the above sections, the penetration depth of NAFLs is positively correlated to the energy level and inversely proportional to the absorption coefficient of water. The water absorption peaks at 1440 and 1927 nm results in their relatively shallow penetration. At maximum energy level, the penetration depth of 1440 and 1927 nm lasers is 390 and 200 μm, respectively. In contrast, low water absorption of 1540 and 1550 nm lasers enables them to penetrate up to 1400 μm at maximum energy level. The popularity of fractionated 1540 or 1550 nm erbium-doped fiber lasers may be contributed to partly by its penetration depth as well as its safety and low complication rate. Its incidence rate of PIH was reported to be 15.3%, far lower than the 34.7% PIH rate of diode lasers. The authors of the study ascribed the high PIH rate to cryogen-induced injury and high fluence during treatment. The fractionated 1927 nm thulium laser is often used to treat photoaging patients with superficial and diffuse pigmentation. It can also be used in combination with the 1550 nm laser so that both superficial and deep photoaging lesions can be targeted (Chen *et al* 2022, Lee *et al* 2013, Polder *et al* 2012, Hu *et al* 2022).

10.3.2.3 Clinical implications

NAFL is proved to be effective in wrinkle reduction, skin tightening and improvement of enlarged pores, atrophic scars, and skin texture. Since NAFL targets mainly water molecules instead of melanin, it is safer to be used in individuals with darker skin types. The fact that the epidermis remains intact during NAFLs also contributes to their relatively lower complication rate compared to traditional ablative lasers and AFLs. In contrast to non-fractional non-ablative NIR lasers, NAFLs avoid bulk heating by protecting adjacent skin tissue next to MTZs. Before the advent of the fractional picosecond laser, some experts considered NAFLs the first-line laser treatment for wrinkle reduction in Asian populations (Chen *et al* 2022, Wat *et al* 2017).

Using appropriate parameters, NAFL may achieve clinical effects comparable to AFL in terms of orbital and forehead wrinkle reduction. A randomized controlled study compared the efficacy of a 1550 nm fractional Yb/Er laser (Finescan, TNC Spectronics, Bangkok, Thailand) and 2940 nm Variable Square Pulse (VSP) Er: YAG laser (Dualis SP, Fotona, Ljubljana, Slovenia) in the treatment of periorbital wrinkles in Thai patients. Both kinds of lasers proved to be effective with no significant difference between the two lasers. However, erythema, edema and the burning sensation after 2940 nm Er:YAG lasted longer, and the incidence of post-operative PIH was also higher. Another RCT conducted in Korea compared the clinical effect of a NAFL device (1550 nm erbium glass laser: Fraxel Xena, Solta Medical) with an AFL device (fractional Er:YAG laser: Avvio, Won Tech, Daejeon, Korea) on photoaging. NAFL proved to be more effective in wrinkle reduction while AFL performed better in improving uneven skin tone and skin texture. Both devices achieved statistically significant improvements. However, it was noted that the NAFL group had higher patient satisfaction and clinician assessment scores. Continued improvement was observed at 3 months follow-up after NAFL, while the effect of AFL only lasted for 1 month. Nevertheless, the pain scores and incidence of acneiform eruptions were higher in NAFL group. The results lead to the assumption that the treatment parameters of AFL might not be high enough to trigger effective rejuvenation (Wattanakrai *et al* 2012, Moon *et al* 2015).

In recent years, melasma has increasingly been considered as a photoaging skin disorder in genetically predisposed patients. As NAFL attracted much attention thanks to its high safety profile for photoaging, several studies focused on its effect on melasma. One pilot study investigated the effect of a 1550 nm erbium-doped laser (Fraxel SR, Solta Medical) on melasma. Treatments were given at 1–2 weeks intervals. A clinically significant effect was observed at 3 months follow-up. However, later studies found the results of the 1550 nm NAFL to be transient and possibly cause a rebound afterwards. One study treated 25 melasma patients with four sessions of 1550 nm NAFL (Fraxel SR, Solta Medical). Twenty-four percent of patients were graded as 'definitely improved' at 1 month after laser sessions. However, at 6 months follow-up, only 17% maintained the improvement while 13% of patients were graded as 'lesion aggravated'. The same problem was also encountered by 1927 nm thulium fiber NAFL in the treatment of melasma. Therefore, NAFL should be considered as a last resort choice for recalcitrant melasma or used in combination with other topical or systemic treatments (Wat *et al* 2017, Passeron and Picardo 2018, Rokhsar and Fitzpatrick 2005, Lee *et al* 2009, Ho *et al* 2013, Niwa *et al* 2013).

10.3.2.4 Complications

Its high safety profile and short downtime are two major advantages of NAFL. In a retrospective study of 362 patients who underwent 856 sessions of laser therapy by a 1550 nm erbium or 1927 nm thulium fiber laser, only 5% experienced short-term adverse events. No long-term complications were observed (Lee *et al* 2014).

The most common side effects after NAFL are erythema and edema, but the duration is usually shorter than for AFL. The erythema after NAFL usually resolves 2–7 d after laser therapy. The duration is even shorter with low-energy diode lasers. High energy or density of NAFL may result in a superficial crust which generally lasts 3–7 d. In this case, post-procedural erythema may last several weeks. After NAFL, acneiform eruption and HSV infection can also be induced, and the former usually lasts 2–4 weeks and resolves spontaneously (Wat *et al* 2017, Wattanakrai *et al* 2012).

The treatment density and energy of NAFL, especially density, play a major role in the PIH of Asian individuals. A histological study confirmed the significant inflammation and granulation caused by high-density fractional photothermolysis. The assumed pigment incontinence in the dermal–epidermal junction induced by intense inflammation may be the cause of PIH. One study compared the effect of more sessions of low-density NAFL with that of fewer sessions of high-density NAFL and found that the incidence rate of PIH rose from 6% to 18%. Avoidance of suntan before NAFL, appropriate setting and cooling during the procedure, and careful nursing of the affected area after laser therapy could greatly reduce the risk of complications, especially PIH (Chan *et al* 2007, Chan *et al* 2010, Kono *et al* 2007).

10.3.3 Fractional picosecond laser (FPL)

10.3.3.1 Mechanism

A picosecond laser destructs its target chromophore mainly through its photomechanical or photoacoustic effect. When equipped with a diffractive or holographic lens array, the laser beam can be divided into fractionated microbeams. High fluence was concentrated in the center of the low fluence background, adjacent tissue was spared, just as for a traditional fractional laser (Weiss *et al* 2017).

The tissue effect of FPL was mainly realized by a process called laser-induced optical breakdown (LIOB). To be specific, a chromophore in the epidermis, most commonly melanin, absorbs high energy from FPL and releases a free electron (also called a seed electron) which triggers the formation of localized ionized plasma through collision and activation of new electrons. The keratinocytes are then vaporized by the heat released by plasma and leave intraepidermal and/or dermal vacuoles. Histologic findings confirm the vacuoles formed immediately after FPL. Adjacent tissues are intact. Over the next 24 h, the vacuoles are filled with melanin and other cellular debris. Vacuolar contents are excreted through the epidermis 3–7 d after FPL. The thermal and photoacoustic effects of plasma might stimulate a cell signaling process and activate a cascade reaction which results in the dermal remodeling of collagen and elastin. It should be mentioned that most previous studies have found the epidermis intact after FPL, but two recent studies have demonstrated that FPL may also be ablative when the energy is high enough (Tanghetti 2016, Tanghetti and Jennings 2018, O'Connor *et al* 2021, Dai *et al* 2020).

10.3.3.2 Classification

There are two major types of FPL, namely the diffractive lens array for a 755 nm picosecond alexandrite laser and the holographic lens array for a 532/1046 nm picosecond Nd:YAG Laser. Both the above FPLs can create intraepidermal vacuoles and induce dermal neocollagenesis and neoelastinogenesis through the LIOB effect. Nevertheless, the different wavelengths may result in different clinical effects and side effects in individuals with different Fitzpatrick skin types.

One study compared the histopathological differences between 532, 1064 and 755 nm FPLs. They found that the 755 nm picosecond alexandrite laser was less prone to cause vascular damage than the fractional 532/1046 nm picosecond Nd:YAG laser. Based on their calculation, the absorption ratio of melanin to hemoglobin was 54:1 for the 755 nm laser, far higher than that for the 1064 nm laser (16:1) and 532 nm laser (2.4:1). The more prominent energy absorption by melanin for 755 nm FPL may contribute to less collateral damage, especially to blood vessels. In addition, the authors noted in their daily practice and from the histological findings that vascular damage was most often caused by the 1064 nm FPL. Speculation is that the higher energy threshold of the 1064 nm FPL to produce LIOB might lead to a collateral thermal effect on blood vessels. On the other hand, the 1064 nm FPL may be preferred when vascular damage is also present in the therapeutic target (Tanghetti and Jennings 2018).

Another split-face RCT compared the effect of a single-wavelength 1064 nm FPL with a dual-wavelength 532/1064 nm FPL on facial photoaging. No significant difference was observed between them in terms of efficacy scores and satisfaction ratings. The histological changes were also similar: intraepidermal vacuole formation, dermal vascular damage, dermal neocollagenesis and neoelastinogenesis were observed in both groups. The result indicate that the 1064 nm laser may play a major role in the treatment of photoaging by a dual-wavelength FPL (Zhang *et al* 2021).

10.3.3.3 Clinical implications

A systematic review of picosecond lasers in 2020 summarized studies on FPL facial rejuvenation and rated its evidence level as 2a, indicating that several high-quality prospective left-to-right comparison trials supported its efficacy, but high-quality RCT studies were still lacking. Nevertheless, the use of FPL in wrinkle reduction has been cleared by the US FDA and is gaining increasing attention because of its high safety profile. It is also worth mentioning that FPL plays a unique role in the treatment of dark-skinned individuals whose laser options may be limited due to the high PIH rate. As the preferred chromophore for most FPLs is melanin, for individuals with high FST, a high concentration of melanin in the epidermis might absorb most of the FPL energy emitted and leaves the collateral vessels intact. In a histological study, none of the subjects with the highest FST showed any sign of vascular damage after either 755, 532, or 1064 nm FPL. In addition, FPLs also achieve good results in terms of long-term efficacy. A recent long-term follow-up study of FPL found that its effect on facial photoaging could persist for up to 3 years (Tanghetti and Jennings 2018, Wu *et al* 2021, Yu *et al* 2021).

Compared with other NAFL, FPL's more precise targeting of the chromophore helps it to reduce collateral damage. However, whether FPL performs significantly better than NAFL in terms of clinical effect remains undecided. To date, there have been five published trials comparing the clinical efficacy between FPL and NAFL, including four split-face controlled studies and one left-to-right comparison trial. Among the five studies, two demonstrated better efficacy of FPL over NAFL while the other three showed no significant difference between the two. Interestingly, the FPLs used in the above five studies were all 1064/532 nm FPL (Kwon *et al* 2020, Yim *et al* 2020, Shi *et al* 2021, Chayavichitsilp *et al* 2020, Wu *et al* 2021).

In contrast, most practitioners believe that FPL is safer and has fewer side effects compared with AFL but is not as effective. However, a randomized split-face trial on FST III–V Asians showed a similar effect of FPL to CO_2 AFL on atrophic acne scars. The incidence rate of PIH was significantly reduced in the FPL group. It is worth noting that the treatment setting of CO_2 AFL in this study was quite conservative, while the energy setting of FPL was at the peak. Moreover, the efficacy was compared after only a single session. More high-quality controlled studies may help to reach more convincing conclusions (Sirithanabadeekul *et al* 2021).

10.3.3.4 Complications

The complications are also comparable between FPL and NAFL, including erythema, edema, purpura, acneiform eruption and HSV infection. The usually transient adverse reactions after FPL contribute to its relatively short downtime. The most common side-effect is post-procedural erythema and edema, which usually resolves spontaneously within 1–2 d. The incidence rate of purpura after 1064 nm FPL is higher than for 532 and 755 nm FPL, which also disappears spontaneously within a few days. However, PIH and other long-term side effects after FPL are very rare and are mainly related to excessive energy or density settings.

References

Adrian R M 1999 Pulsed carbon dioxide and erbium-YAG laser resurfacing: a comparative clinical and histologic study *J. Cutan. Laser Ther.* **1** 29–35

Alexiades-Armenakas M R, Dover J S and Arndt K A 2008 The spectrum of laser skin resurfacing: nonablative, fractional, and ablative laser resurfacing *J. Am. Acad. Dermatol.* **58** 719–37 738–40

Alexiades-Armenakas M R, Dover J S and Arndt K A 2012 Fractional laser skin resurfacing *J. Drugs Dermatol.* **11** 1274–87

Alster T S and Lupton J R 2001 Erbium:YAG cutaneous laser resurfacing *Dermatol. Clin.* **19** 453–66

Alster T S and Lupton J R 2002 Are all infrared lasers equally effective in skin rejuvenation *Semin. Cutan. Med. Surg.* **21** 274–9

Anderson R R and Parrish J A 1983 Selective photothermolysis: precise microsurgery by selective absorption of pulsed radiation *Science* **220** 524–7

Aslam A and Alster T S 2014 Evolution of laser skin resurfacing: from scanning to fractional technology *Dermatol. Surg.* **40** 1163–72

Chan H H *et al* 2007 The prevalence and risk factors of post-inflammatory hyperpigmentation after fractional resurfacing in Asians *Lasers Surg. Med.* **39** 381–5

Chan N P *et al* 2010 The use of non-ablative fractional resurfacing in Asian acne scar patients *Lasers Surg. Med.* **42** 710–5

Chayavichitsilp P *et al* 2020 Comparison of fractional neodymium-doped yttrium aluminum garnet (Nd:YAG) 1064-nm picosecond laser and fractional 1550-nm erbium fiber laser in facial acne scar treatment *Lasers Med. Sci.* **35** 695–700

Chen S X *et al* 2022 Review of lasers and energy-based devices for skin rejuvenation and scar treatment with histologic correlations *Dermatol. Surg.* **48** 441–8

Choi J E *et al* 2014 Ablative fractional laser treatment for hypertrophic scars: comparison between Er:YAG and CO_2 fractional lasers *J. Dermatolog. Treat.* **25** 299–303

Dai Y X *et al* 2020 Efficacy and safety of ablative resurfacing with a high-energy 1,064 Nd:YAG picosecond-domain laser for the treatment of facial acne scars in Asians *Lasers Surg. Med.* **52** 389–95

Dayan S H *et al* 2003 Nonablative laser resurfacing using the long-pulse (1064-nm) Nd:YAG laser *Arch. Facial Plast. Surg.* **5** 310–5

Dayan S *et al* 2003 Histological evaluations following 1,064-nm Nd:YAG laser resurfacing *Lasers Surg. Med.* **33** 126–31

El-Domyati M *et al* 2011 Effects of the Nd:YAG 1320-nm laser on skin rejuvenation: clinical and histological correlations *J. Cosmet. Laser Ther.* **13** 98–106

El-Domyati M *et al* 2013 Fractional versus ablative erbium:yttrium-aluminum-garnet laser resurfacing for facial rejuvenation: an objective evaluation *J. Am. Acad. Dermatol.* **68** 103–12

Fusano M *et al* 2021 Ultrapulsed CO_2 resurfacing of photodamaged facial skin in vegan and omnivore patients: a multicentric study *Lasers Surg. Med.* **53** 1370–5

Goldberg D J 2000 Full-face nonablative dermal remodeling with a 1320 nm Nd:YAG laser *Dermatol. Surg.* **26** 915–8

Grunewald S *et al* 2011 *In vivo* wound healing and dermal matrix remodelling in response to fractional CO_2 laser intervention: clinicopathological correlation in non-facial skin *Int. J. Hyperthermia* **27** 811–8

Ha L *et al* 2020 First assessment of a carbon monoxide laser and a thulium fiber laser for fractional ablation of skin *Lasers Surg. Med.* **52** 788–98

Hardaway C A *et al* 2002 Non-ablative cutaneous remodeling with a 1.45 microm mid-infrared diode laser: phase I *J. Cosmet. Laser Ther.* **4** 3–8

Ho S G *et al* 2013 A retrospective study of the management of Chinese melasma patients using a 1927 nm fractional thulium fiber laser *J. Cosmet. Laser Ther.* **15** 200–6

Hu S, Atmakuri M and Rosenberg J 2022 Adverse events of nonablative lasers and energy-based therapies in subjects with Fitzpatrick skin phototypes IV to VI: a systematic review and meta-analysis *Aesthet. Surg. J.* **42** 537–47

Jih M H *et al* 2006 The 1450-nm diode laser for facial inflammatory acne vulgaris: dose-response and 12-month follow-up study *J. Am. Acad. Dermatol.* **55** 80–7

Karsai S *et al* 2010 Ablative fractional lasers (CO and Er:YAG): a randomized controlled double-blind split-face trial of the treatment of peri-orbital rhytides *Lasers Surg. Med.* **42** 160–7

Kauvar A N 2014 Fractional nonablative laser resurfacing: is there a skin tightening effect? *Dermatol. Surg.* **40** S157–63

Kim J E *et al* 2013 Gene profiling analysis of the early effects of ablative fractional carbon dioxide laser treatment on human skin *Dermatol. Surg.* **39** 1033–43

Koch R J 2002 Laser skin resurfacing *Otolaryngol. Clin. North Am.* **35** 119–33

Kono T *et al* 2007 Prospective direct comparison study of fractional resurfacing using different fluences and densities for skin rejuvenation in Asians *Lasers Surg. Med.* **39** 311–4

Kwon H H *et al* 2020 Comparison of a 1064-nm neodymium-doped yttrium aluminum garnet picosecond laser using a diffractive optical element vs a nonablative 1550-nm erbium-glass laser for the treatment of facial acne scarring in Asian patients: a 17-week prospective, randomized, split-face, controlled trial *J. Eur. Acad. Dermatol. Venereol.* **34** 2907–13

Laubach H J *et al* 2006 Skin responses to fractional photothermolysis *Lasers Surg. Med.* **38** 142–9

Lee H M *et al* 2013 Split-face study using a 1,927-nm thulium fiber fractional laser to treat photoaging and melasma in Asian skin *Dermatol. Surg.* **39** 879–88

Lee H S *et al* 2009 Treatment of melasma in Asian skin using a fractional 1,550-nm laser: an open clinical study *Dermatol. Surg.* **35** 1499–504

Lee M C *et al* 2009 Skin rejuvenation with 1,064-nm Q-switched Nd:YAG laser in Asian patients *Dermatol. Surg.* **35** 929–32

Lee S M *et al* 2014 Adverse events of non-ablative fractional laser photothermolysis: a retrospective study of 856 treatments in 362 patients *J. Dermatolog. Treat.* **25** 304–7

Lee Y B *et al* 2012 Effects of long-pulsed 1,064-nm neodymium-doped yttrium aluminum garnet laser on dermal collagen remodeling in hairless mice *Dermatol. Surg.* **38** 985–92

Li D, Lin S B and Cheng B 2018 Complications and posttreatment care following invasive laser skin resurfacing: a review *J. Cosmet. Laser Ther.* **20** 168–78

Loesch M M *et al* 2014 Skin resurfacing procedures: new and emerging options *Clin. Cosmet. Investig. Dermatol.* **7** 231–41

Lolis M S and Goldberg D J 2012 Radiofrequency in cosmetic dermatology: a review *Dermatol. Surg.* **38** 1765–76

Manstein D *et al* 2004 Fractional photothermolysis: a new concept for cutaneous remodeling using microscopic patterns of thermal injury *Lasers Surg. Med.* **34** 426–38

Metelitsa A I and Alster T S 2010 Fractionated laser skin resurfacing treatment complications: a review *Dermatol. Surg* **36** 299–306

Moon H R *et al* 2015 A prospective, randomized, double-blind comparison of an ablative fractional 2940-nm erbium-doped yttrium aluminum garnet laser with a nonablative fractional 1550-nm erbium-doped glass laser for the treatment of photoaged Asian skin *J. Dermatolog. Treat.* **26** 551–7

Niwa M A *et al* 2013 Treatment of melasma with the 1,927-nm fractional thulium fiber laser: a retrospective analysis of 20 cases with long-term follow-up *Lasers Surg. Med.* **45** 95–101

O'Connor K, Cho S B and Chung H J 2021 Wound healing profile After 1064- and 532-nm picosecond lasers with microlens array of *in vivo* human skin *Lasers Surg. Med.* **53** 1059–64

Orringer J S *et al* 2010 Molecular mechanisms of nonablative fractionated laser resurfacing *Br. J. Dermatol.* **163** 757–68

Passeron T and Picardo M 2018 Melasma, a photoaging disorder *Pigment Cell Melanoma Res.* **31** 461–5

Polder K D *et al* 2012 Treatment of macular seborrheic keratoses using a novel 1927-nm fractional thulium fiber laser *Dermatol. Surg.* **38** 1025–31

Robati R M and Asadi E 2017 Efficacy and safety of fractional CO_2 laser versus fractional Er:YAG laser in the treatment of facial skin wrinkles *Lasers Med. Sci.* **32** 283–9

Rokhsar C K and Fitzpatrick R E 2005 The treatment of melasma with fractional photothermolysis: a pilot study *Dermatol. Surg.* **31** 1645–50

Ross E V *et al* 2001 One-pass CO_2 versus multiple-pass Er:YAG laser resurfacing in the treatment of rhytides: a comparison side-by-side study of pulsed CO_2 and Er:YAG lasers *Dermatol. Surg.* **27** 709–15

Rubenstein R *et al* 1986 Atypical keloids after dermabrasion of patients taking isotretinoin *J. Am. Acad. Dermatol.* **15** 280–5

Sadick N and Schecter A K 2004 Utilization of the 1320-nm Nd:YAG laser for the reduction of photoaging of the hands *Dermatol. Surg.* **30** 1140–4

Said T M, Konz M and Paasch U 2021 Comparison of heat shock protein 70 expression in response to different non-ablative lasers: an *in vitro* study *Photobiomodul. Photomed. Laser Surg.* **39** 221–8

Shi Y *et al* 2021 Comparison of fractionated frequency-doubled 1,064/532 nm picosecond Nd:YAG lasers and non-ablative fractional 1,540 nm Er: glass in the treatment of facial atrophic scars: a randomized, split-face, double-blind trial *Ann. Transl. Med.* **9** 862

Sirithanabadeekul P *et al* 2021 Comparison of fractional picosecond 1064-nm laser and fractional carbon dioxide laser for treating atrophic acne scars: a randomized split-face trial *Dermatol. Surg.* **47** e58–65

Tanghetti E A 2016 The histology of skin treated with a picosecond alexandrite laser and a fractional lens array *Lasers Surg. Med.* **48** 646–52

Tanghetti M E and Jennings J 2018 A comparative study with a 755 nm picosecond Alexandrite laser with a diffractive lens array and a 532 nm/1064 nm Nd:YAG with a holographic optic *Lasers Surg. Med.* **50** 37–44

Tanzi E L and Alster T S 2003 Single-pass carbon dioxide versus multiple-pass Er:YAG laser skin resurfacing: a comparison of postoperative wound healing and side-effect rates *Dermatol. Surg.* **29** 80–4

Tanzi E L, Lupton J R and Alster T S 2003 Lasers in dermatology: four decades of progress *J. Am. Acad. Dermatol.* **49** 1–31 31–34

Trelles M A, Allones I and Luna R 2001 Facial rejuvenation with a nonablative 1320 nm Nd:YAG laser: a preliminary clinical and histologic evaluation *Dermatol. Surg.* **27** 111–6

Wat H, Wu D C and Chan H H 2017 Fractional resurfacing in the Asian patient: current state of the art *Lasers Surg. Med.* **49** 45–59

Wattanakrai P, Pootongkam S and Rojhirunsakool S 2012 Periorbital rejuvenation with fractional 1,550-nm ytterbium/erbium fiber laser and variable square pulse 2,940-nm erbium:YAG laser in Asians: a comparison study *Dermatol. Surg.* **38** 610–22

Weiss R A *et al* 2017 Safety and efficacy of a novel diffractive lens array using a picosecond 755 nm alexandrite laser for treatment of wrinkles *Lasers Surg. Med.* **49** 40–4

Weniger F G, Weidman A A and Barrero C C 2020 Full-field erbium:YAG laser resurfacing: complications and suggested safety parameters *Aesthet. Surg. J.* **40** P374–85

Wu D C *et al* 2021 A systematic review of picosecond laser in dermatology: evidence and recommendations *Lasers Surg. Med.* **53** 9–49

Wu D C *et al* 2021 A randomized, split-face, double-blind comparison trial between fractionated frequency-doubled 1064/532 nm picosecond Nd:YAG laser and fractionated 1927 nm thulium fiber laser for facial photorejuvenation *Lasers Surg. Med.* **53** 204–11

Xu X G *et al* 2011 Immunohistological evaluation of skin responses after treatment using a fractional ultrapulse carbon dioxide laser on back skin *Dermatol. Surg.* **37** 1141–9

Yim S *et al* 2020 Split-face comparison of the picosecond 1064-nm Nd:YAG laser using a microlens array and the quasi-long-pulsed 1064-nm Nd:YAG laser for treatment of photoaging facial wrinkles and pores in Asians *Lasers Med. Sci.* **35** 949–56

Yu W *et al* 2021 Three-year results of facial photoaging in asian patients after alexandrite 755 nm picosecond laser with diffractive lens array: a split-face, single-blinded, randomized controlled comparison *Lasers Surg. Med.* **53** 1065–72

Zhang M *et al* 2021 Comparison of 1064-nm and dual-wavelength (532/1064-nm) picosecond-domain Nd:YAG lasers in the treatment of facial photoaging: a randomized controlled split-face study *Lasers Surg. Med.* **53** 1158–65

IOP Publishing

Skin Photoaging (Second Edition)

Rui Yin, Yang Xu and Chengfeng Zhang

Chapter 11

Intense pulsed light therapy for photoaging

Xiang Wen and Li Xie

Intense pulsed light (IPL), also known as pulsed light and broadband light, is a non-laser light source. It has been widely used to improve cutaneous photoaging. IPL has a marked impact on reducing hyperpigmentation, lentignes and wrinkles, improving texture, rhytides, and pores for skin rejuvenation. The combination of wavelengths, fluences, pulse durations, and pulse delays facilitates the treatment efficacy.

11.1 Definition of IPL

Intense pulsed light (IPL) devices use flashlamps and bandpass filters to emit pulsed light with a wavelength range of 400–1200 nm. Different to lasers, IPL has the characteristics of polychromism, incoherence and non-parallelism (Stewart *et al* 2013). The features of laser and intense pulsed light are shown in figure 11.1. Intense pulsed light is generated by a high-intensity light source (such as a xenon lamp), which is focused and initially filtered to form a broadband light. A special filter is used in IPL devices to filter out light with a certain wavelength. The final emitted light is a special band of strong light suitable for the treatment of many skin diseases, such as freckles, melasma, rosacea, skin aging and photoaging.

In 1992, IPL therapy was first developed to treat telangiectasia of the legs. The first generation IPL devices, called Photoderm VL, produced by ESC-Sharplan (now Lumenis), was tested clinically in 1994 and officially approved to treat leg vascular diseases by the US FDA in 1995 (Goldman 1997). First generation IPL devices emitted single pulse light, including infrared light, which prevalently led to epidermis damage and a high incidence of side effects. In second generation IPL devices, water filters out the infrared portion, significantly reducing the risk of side effects (Babilas *et al* 2010).

The early IPL devices demonstrated a decrease in energy output due to unstable current, resulting in sudden high pulse energy emission. As the emission progressed, the rapid decline in current led to a weakening of the emitted energy, causing an inconsistent energy output. Each pulse created a spike wave with

doi:10.1088/978-0-7503-5112-6ch11

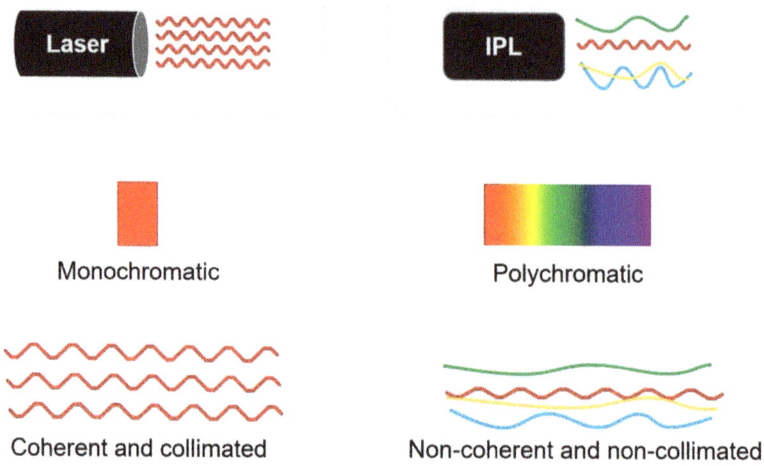

Figure 11.1. The features of laser and intense pulsed light.

gradually decreasing sub-pulse energy that could not be evenly distributed, leading to more adverse reactions such as epidermal burns or pigmentation.

In 2003, the Lumenis One (M22) of the new generation IPL devices enhanced the optimized pulse technology (OPT). By precisely regulating the current at both ends of the lamp tube through a computer, it effectively controls the entire process of intense pulse light generation and emission. Consequently, each pulse exhibits consistent pulse duration and balanced energy levels (Ruan *et al* 2019). With lower fluence and effective parameters only, it enables achieving an effective treatment. The recently advanced optimized pulse technology (AOPT) is based on multi-pulse emission, allowing independent adjustment of influence and pulse duration for each individual pulse (Chen *et al* 2022). Through the diversity of AOPT-IPL treatment parameters, it can target different layers and sizes of chromophores in the skin to solve a variety of skin problems.

In 1999, an American dermatologist named Bitter made the initial discovery that IPL can significantly improve skin photoaging and promptly applied and promoted this technology, later recognized as 'photo rejuvenation' (Bitter 2000). Numerous laser companies, including Lumenis, Candela, Syneron, Cutera, Alma, etc, have actively participated in market development by launching their own IPL devices. Each company's device possesses distinct characteristics and similarities. While similar indications can address certain clinical issues and achieve specific therapeutic effects, significant variations in clinical efficacy arise due to disparities in power control systems, filters, and pulse duration provided by different equipment manufacturers; thus rendering direct comparison of year-on-year efficacy unfeasible (Babilas *et al* 2010).

11.2 Mechanisms of intense pulsed light therapy for photoaging

The primary manifestations of skin photoaging consist of wrinkles, vascular lesions, hyperpigmentation, lentignes, texture, rhytides, and pores (Salminen *et al* 2022). Histologically, it is characterized by severe epidermal damage, aberrant keratinocyte proliferation, disorganized arrangement of elastic fibers within the dermis with partial clustering formation, modifications in collagen structure at the upper dermal layer (Shin *et al* 2023), changes of microvasculature networks (Chung and Eun 2007), as well as focal melanocyte proliferation. Intense pulsed light with a wavelength range of 400–1200 nm can be selectively targeted towards different chromophores in the skin such as melanin, hemoglobin and water through appropriate different filters, and simultaneously can address various skin concerns including pigmentation spots, dilated capillaries and texture to enable effective achievement of skin rejuvenation.

11.2.1 Selective photothermolysis

The theory of selective photothermolysis was first proposed by John Parrish and Rox Anderson in 1983 (Patel *et al* 2024). This theory suggests that specific wavelengths of light act specifically on a targeted tissues of the skin, causing the damage of targeted tissues without affecting other structures (Anderson and Parrish 1983). IPL is an incoherent light and essentially still belongs to ordinary light, which is similar to lasers in treatment mechanisms, according to the principle of selective photothermolysis.

Melanin, hemoglobin, and water are the three main chromophores of human skin. After being irradiated by a specific wavelength of IPL spectrum, hemoglobin selectively absorbs and produces photothermal effects, inducing denaturation and coagulation reactions within the hemoglobin molecules. This process damages the endothelial function of capillaries, causing them to lose their ability to stretch and relax, thereby achieving the effect of sealing blood vessels to treat telangiectasia by photoaging. The thermal relaxation time (TRT) of melanosomes is 250–1000 ns, and the lasers with a pulse width shorter than 50 ns are effective for the specific disruption of melanosomes. The pulse duration of IPL of milliseconds is much larger than the thermal relaxation time of melanosomes. The keratinocytes throughout the epidermis contain a large number of melanosomes and melanin granules.

The target tissue should be the pigment-containing cells and/or localized pigmented areas in epidermis for IPL therapy to epidermal pigmentation (Yamashita *et al* 2006), and the thermal relaxation time of the epidermis is about 3–10 ms. Thermal coagulation of melanin in the epidermis appears after IPL therapy. The thermal energy produced by the photothermal effect diffuses into the epidermal layer, promoting rapid differentiation of keratinocytes, and melanosomes also move up and fall off with the necrotic keratinocytes, which results in the elimination of melanosomes (Yamashita *et al* 2006). The target tissue of the near infrared band of IPL is water within the dermis. After the collagen within the dermis containing a lot of water absorbs heat power, the dermal temperature reaches 50–55 °C, leading to reversible thermal damage in the dermal collagen, causing contraction

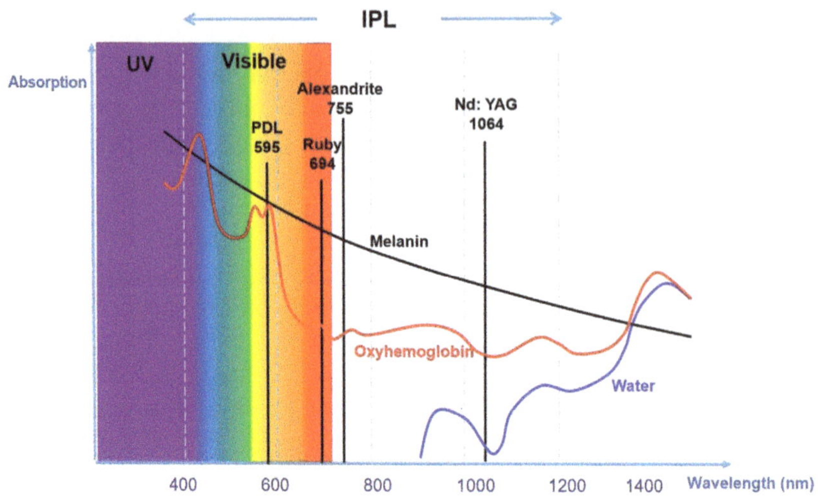

Figure 11.2. Intense pulsed light spectrum and absorption coefficient versus wavelength.

of the collagen fibers and temporarily skin firming. The intense pulsed light spectrum and the absorption coefficient versus wavelength of skin chromophores are shown in figure 11.2.

11.2.2 Biothermal stimulation effects

The histological manifestations of photoaging include a disordered arrangement of collagens in the dermis (Franco *et al* 2022), and abnormal aggregation of elastic fibers (Kohl *et al* 2011). These aging manifestations are improved by IPL therapy through multiple pathways. IPL-induced dermal thermal injury initiates damage repair, promotes fibroblast activation, and forms new collagen (Raulin *et al* 2003). Negishi *et al* found that intense pulsed light not only increases the synthesis of type I collagen but also increases the synthesis of type III collagen (Negishi *et al* 2002). Feng *et al* conducted a study on photo rejuvenation treatment in Asians. Pathological examination of four cases showed that both type I and type III collagen increased following IPL treatment, along with a decrease in elastin content. However, there was also evidence of more orderly arrangement of elastic fibers, as well as an increase number of active fibroblasts (Feng *et al* 2008). In an *in vitro* study of human fibroblasts following IPL irradiation, the expression of type III collagen and transforming growth factor $\beta 1$ (TGF-$\beta 1$) was upregulation (Wong *et al* 2009). It is worth noting that TGF-$\beta 1$ serves as a key regulatory factor in extracellular matrix biosynthesis within the skin, and stimulates the expression of type I collagen genes (Massagué and Sheppard 2023). Another study found that mRNA for collagen I and III of human skin fibroblasts were increased following IPL irradiation (Cuerda-Galindo *et al* 2015). The above studies indicate that IPL therapy can promote the regeneration and remodeling of collagen fibers, and is thus beneficial to skin rejuvenation.

The increased expression of matrix metalloproteinases (MMPs) is also associated with photoaging (Kohl *et al* 2011), as MMPs facilitate collagen and other extracellular matrix degradation (Mukherjee *et al* 2006). IPL irradiation on BALB/c mice resulted in histological examination showing dermal thickening accompanied by collagen proliferation, enhanced staining of type I and III collagen compared to the control group, increased mRNA expression levels of type I and III procollagen, and decreased mRNA expression levels of MMP-1 and MMP-2 (Luo *et al* 2009). It is believed that IPL therapy not only increases the generation of new collagen but also reduces collagen degradation in skin rejuvenation treatment.

11.3 The parameters of IPL

The fundamental mechanism of IPL treatment encompasses several crucial parameters: wavelength, pulse duration, energy density, pulse delay time, spot size, and cooling system.

11.3.1 Wavelength

In the concept of 'wavelength' of intense pulsed light, it is not a single wavelength, but a spectrum. By using different filters, specific wavelength ranges of light can be obtained. The filters used in clinical practice mainly include 420, 515, 560, 590, 615, 640, and 695 nm. When a 560 nm filter is used, the output wavelength of intense pulsed light is within the range of 560–1200 nm. The wavelength affects the absorption of light for chromophores, and also affects the depth of penetration into the skin. Specifically, long wavelengths tend to penetrate more deeply. This concept is particularly important when using intense pulsed light to treat patients with darker skin types, as melanin tends to absorb shorter wavelengths of light. It is important to choose longer wavelengths of intense pulsed light for patients with darker skin types to protect the epidermis and avoid dyschromia.

11.3.2 Pulse duration

Pulse duration, also known as pulse width, refers to the time that light energy is delivered to the tissue, usually measured in milliseconds (ms) in IPL. The length of the pulse width is based on the principle of selective photothermolysis and relies on the thermal relaxation time of the target chromophores. To minimize thermal damage to skin, it is important for the pulse duration to be shorter than the TRT of the chromophores.

11.3.3 Energy density

Energy density describes the power exerted by the light on tissue per square centimeter, also known as fluence, expressed in $\mathrm{J\,cm^{-2}}$ in IPL. Energy density is usually one of the most important treatment parameters of intense pulsed light, which not only determines the efficacy of treatment but also plays a crucial role in the occurrence of complications. In order to achieve therapeutic effects as soon as possible, a higher energy density is preferred. However, excessive fluence absorbed by the skin can lead to heat damage and potential side effects such as blisters, pigmentation, or scar formation. These side effects are more likely to occur in IPL treatment for individuals with higher fluences and patients with darker skin types (Halachmi and Lapidoth 2012). Therefore, it is important to consider maximizing the benefits while ensuring safety.

11.3.4 Pulse delay time

Some intense pulsed light devices are capable of producing only a single pulse, while others can produce multiple sequential pulses. The pulse delay time refers to the interval between two pulses, which also represents the time for cooling the target tissue. It is important for the pulse delay time to be shorter than the cooling time of the chromophore and longer than or equal to the cooling time of the epidermis, ensuring sufficient cooling time is provided for protecting the epidermis while maintaining high heat in the target tissue. For patients with darker skin types, it is recommended to increase the pulse delay time (Monib and Hussein 2020). Allowing an appropriate pulse delay time can prevent the temperature of the epidermis from exceeding 70 °C, while providing adequate thermal stimulation for dermal collagen remodeling, thus achieving skin rejuvenation and reducing adverse reactions. The optical parameters of IPL and their definitions are shown in table 11.1.

11.3.5 Spot size

The size of the spot refers to the diameter of the light emitted from the flashlamp, with small changes in spot size resulting in much larger changes in fluence (Bogdan Allemann and Kaufman 2011). Increased spot size allows for light to penetrate deeper, because scattering is minimized. The size of the spot is adjustable using interchangeable aperture pieces. When treating localized small lesions, it may be helpful to use interchangeable aperture pieces to match the lesions (Soon and Victor Ross 2008).

11.3.6 Cooling system

The IPL devices usually have a contact cooling system, which provides skin protection and allows greater energy to reach deeper target tissues. When treating

Table 11.1. Main parameters of IPL.

Parameter	Definition	Unit
Wavelength	Wavelength of IPL is not a single wavelength, but a spectrum range of 400 ~ 1200 nm.	Nm
Pulse duration	The length of time over which laser energy is delivered to the tissue. Also known as pulse width.	Ms
Energy density	Refers to power per square centimeter, also known as fluence.	J cm^{-2}
Pulse delay time	Refers to the cooling time of the target issue.	ms

vascular lesions, in order to avoid excessive constriction of capillaries, the cooling system needs to be turned off. In addition, the cold ultrasound gel applied before IPL treatment can also cool the skin.

11.4 Clinical applications of IPL

In recent years, there has been a continuous emergence of various new types of instruments and drugs, which have promoted the rapid development of non-surgical treatment for skin rejuvenation, such as retinoids, IPL, radiofrequency, and fractional laser. Intense pulsed light was first used to improve skin photoaging in 1999. With the continuous improvement and optimization of IPL technology, the fundamental mechanism of IPL has been expanded, and intense pulsed light acts as an important tool to treat photoaging. IPL is particularly advantageous, because IPL is non-invasive, safe, less painful, and readily available. Photoaging is the primary indication for IPL treatment, and it can be applied to all skin areas including the face, neck, chest and hands.

A substantial number of clinical trials have demonstrated that IPL can significantly enhance the qualitative characteristics of facial photoaging after several treatment sessions. Sales *et al* conducted a systematic review for the first time using digital photographs and self-evaluated efficacy (Sales *et al* 2022). A total of 16 studies were reviewed, involving 637 patients. They found that most studies results showed the efficacy of IPL treatment in telangiectasia, wrinkles, pore, erythema, rhytids, texture, lentigines, hyperpigmentation, and photoaging score. A randomized controlled split-face trial found that IPL therapy improved skin texture, telangiectasia, and irregular pigmentation (Hedelund *et al* 2006).

Twenty-four patients received five treatment sessions with an intense pulsed light at two week intervals, and improvements in skin elasticity, wrinkles and hyperpigmentations were noted (Kołodziejczak and Rotsztejn 2022). Broadband light (BBL), also an intense pulsed light, increases the total number of enhancer pulses while reducing pulse delay duration to achieve a total pulse width of 200 ms, thus achieving a nearly continuous photon output effect. After treatment, the patients exhibited noticeable improvements in zygomatic wrinkles, nasolabial folds, perioral expression wrinkles at the cheek, and marionette lines (In *et al* 2019). The newly added 800–1200 nm filter specifically targets the dermis. In the fast-sliding mode of the large light spot, near-infrared light continuously emits in seconds, providing more effective thermal stimulation to the dermal tissue for skin rejuvenation treatment.

The aging of the hands is typically characterized by wrinkles, skin thinning, and solar lentigines. 128 Japanese patients who had lentigines and wrinkles on the dorsa of both hands received appropriate IPL treatment, and improvement of lentigines and wrinkles was noticed. This result indicates that IPL is an effective choice for hand rejuvenation (Maruyama 2016). After four treatment sessions, lightening of the solar lentigines on patients' hands was observed, with no significant side effects (Goldman *et al* 2008).

Photodynamic therapy (PDT) combines a photosensitizer (PS), light, and oxygen to cause to destroy the target cells or tissues. The selection of light sources is an important component of PDT that can affect its effectiveness. Researches indicate its efficacy in skin rejuvenation. A split-face clinical trial demonstrated that PDT with topical 5-aminolevulinic acid (ALA-PDT) using IPL as a light source for the treatment of actinic keratosis (AKs) with photodamaged skin, brought a greater improvement in photoaging than IPL alone, including the appearance of crow's feet, tactile skin roughness, hyperpigmentation, and telangectasias (Gold *et al* 2006). Patients received IPL and blue light as light sources for PDT treatment of photodamage, and the results indicated that IPL + blue light-PDT demonstrated increased efficacy when compared to IPL-PDT, pulsed dye laser (PDL)-PDT, and PDL + blue light-PDT treatment groups (Abrouk *et al* 2022).

In addition to IPL monotherapy for skin rejuvenation, it can also be combined with treatments such as botulinum toxin, fillers, radiofrequency, and lasers to enhance treatment effectiveness. A recent study has demonstrated that the combination of IPL and an ablative fractional laser significantly reduced skin wrinkles and pore size and improved skin texture and elasticity in photoaging skin of Chinese people, and combined therapy is also effective in the treatment of photoage-induced hyperpigmentation, telangiectasia and erythema, without increasing the risk of adverse reactions (Mei and Wang 2018). Another study found that the combination therapy of Intense pulsed light, near infrared pulsed light, and fractional laser for skin rejuvenation is safe and more effective strategy than IPL monotherapy for skin rejuvenation (Tao *et al* 2015).

However, a study demonstrated that IPL was superior to radiofrequency therapy in the treatment of facial skin rejuvenation through clinical efficacy evaluation combined with immunohistochemical expression of MMP-1 (Nassar *et al* 2020).

IPL therapy is relatively safe and highly accepted by patients. IPL may not be suitable for patients such as those with a recent sunburn, pregnant or breast feeding women, those who take retinoids or photosensitive drugs, and those with photosensitive diseases. Patients with a history of herpes simplex require antiviral therapy before IPL treatment (Qiao *et al* 2021).

11.5 Procedure of IPL

IPL treatment is typically administered in outpatient settings and should be performed by trained doctors or relevant professionals following established operating procedures. In some regions, registered nurses are also permitted to provide IPL treatments.

Before each treatment, clinicians should fully address patient needs and any contraindications, and provide comprehensive information on the potential benefits and risks. Preprocedure informed consent is very important for IPL treatments. Photo documentation is mandatory prior to each treatment.

The treatment areas have to be free from makeup and shaved. It is necessary to use cold optical coupling gel or ultrasound gel for IPL treatment. This cold gel reduces friction between the handpiece and skin while dissipating surface heat released by the handpiece and decreasing refractive index between the air and skin for improved penetration and light absorption. Eye protection with appropriate goggles is mandatory for IPL treatment, both for the patients and clinicians.

Based in the skin Fitzpatrick type and manifestations of the patient's treatment area skin, appropriate parameters including filters, pulse width, pulse delay time, fluence, and pulse numbers should be taken into consideration. It is recommended to conduct a test whenever treatment parameters are introduced for the first time. The testing site is typically placed at an inconspicuous site in the treatment area, such as mandibular angle area. After illuminating 2–3 light spots, it is important to observe the skin reaction. If necessary, the treatment parameters should be adjusted, and followed by another test shot. Because the effect of IPL is cumulative, it usually requires 4–6 treatments or more treatments every 3–5 weeks in order to achieve improvements.

If necessary, the treatment area can be cooled down after IPL treatment by using a cooling gel pack or a medical cold compress for 10–30 min. After treatment, patients must stay away from sunlight or use sufficient UV protection measures for at least the next 8 weeks (Adamic *et al* 2007).

11.6 Adverse events of IPL

The most common adverse events are pain and erythema in the treatment regions. Generally, the feeling of pain during the treatment is minimal, and can typically be managed through local cooling (applied during or after treatment) or topical surface anesthesia if necessary. Additional reported side effects include edema, bullae, hypopigmentation, hyperpigmentation, atrophia, scarring, hypertrophic scarring, or keloid formation and infection (Radmanesh *et al* 2008). Given that the use of lasers and optical treatments is challenging in patients with darker skin types, IPL leads to promising results and is a safe treatment modality. Selecting appropriate parameters of intense pulsed light can effectively avoid adverse reactions. Furthermore, delayed skin responses may occur among those with darker skin within 24–48 h post-treatment. Therefore, prior to initiating an IPL treatment, it is crucial to thoroughly assess patients' skin types and characteristics.

11.7 Conclusion

IPL has been widely used in the field of dermatology to treat many skin diseases, and plays an irreplaceable role in the treatment of skin rejuvenation. It is non-ablative, and the biggest advantages of IPL over lasers are its variety, safety, and cost-effectiveness. IPL is an effective and well-tolerated treatment option with a good safety profile.

References

Abrouk M, Gianatasio C, Li Y and Waibel J S 2022 Prospective study of intense pulsed light versus pulsed dye laser with or without blue light in the activation of PDT for the treatment of actinic keratosis and photodamage *Lasers Surg. Med.* **54** 66–73

Adamic M, Troilius A, Adatto M, Drosner M and Dahmane R 2007 Vascular lasers and IPLS: guidelines for care from the European Society for laser dermatology (ESLD) *J. Cosmet. Laser Ther.* **9** 113–24

Anderson R R and Parrish J A 1983 Selective photothermolysis: precise microsurgery by selective absorption of pulsed radiation *Science* **220** 524–7

Babilas P, Schreml S, Szeimies R M and Landthaler M 2010 Intense pulsed light (IPL): a review *Lasers Surg. Med.* **42** 93–104

Bitter P H 2000 Noninvasive rejuvenation of photodamaged skin using serial, full-face intense pulsed light treatments *Dermatol. Surg.* **26** 835–42 discussion 843

Bogdan Allemann I and Kaufman J 2011 Laser principles *Curr. Prob. Dermatol.* **42** 7–23

Chen J, Liu J and Wu J 2022 Treatment of melasma by a combination of intense pulsed light with advanced optimal pulse technology and human-like collagen repair dressing: a case series study *Medicine* **101** e29492

Chung J H and Eun H C 2007 Angiogenesis in skin aging and photoaging *J. Dermatol.* **34** 593–600

Cuerda-Galindo E, Díaz-Gil G, Palomar-Gallego M A and Linares-GarcíaValdecasas R 2015 Intense pulsed light induces synthesis of dermal extracellular proteins *in vitro Lasers Med. Sci.* **30** 1931–9

Feng Y, Zhao J and Gold M H 2008 Skin rejuvenation in Asian skin: the analysis of clinical effects and basic mechanisms of intense pulsed light *J. Drugs Dermatol.* **7** 273–9

Franco A C, Aveleira C and Cavadas C 2022 Skin senescence: mechanisms and impact on whole-body aging *Trends Mol. Med.* **28** 97–109

Gold M H, Bradshaw V L, Boring M M, Bridges T M and Biron J A 2006 Split-face comparison of photodynamic therapy with 5-aminolevulinic acid and intense pulsed light versus intense pulsed light alone for photodamage *Dermatol. Surg.* **32** 795–801 discussion 801–793

Goldman A, Prati C and Rossato F 2008 Hand rejuvenation using intense pulsed light *J. Cutan. Med. Surg.* **12** 107–13

Goldman M P 1997 Treatment of benign vascular lesions with the Photoderm VL high-intensity pulsed light source *Adv. Dermatol.* **13** 503–21

Halachmi S and Lapidoth M 2012 Low-fluence vs standard fluence hair removal: a contralateral control non-inferiority study *J. Cosmet. Laser Ther.* **14** 2–6

Hedelund L, Due E, Bjerring P, Wulf H C and Haedersdal M 2006 Skin rejuvenation using intense pulsed light: a randomized controlled split-face trial with blinded response evaluation *Arch. Dermatol.* **142** 985–90

In S, Park H, Song H, Park J, Kim H and Cho S B 2019 Broadband light treatment using static operation and constant motion techniques for skin tightening in Asian patients *J. Cosmet. Laser Ther.* **21** 132–7

Kohl E, Steinbauer J, Landthaler M and Szeimies R M 2011 Skin ageing *J. Eur. Acad. Dermatol. Venereol.* **25** 873–84

Kołodziejczak A and Rotsztejn H 2022 Efficacy of fractional laser, radiofrequency and IPL rejuvenation of periorbital region *Lasers Med. Sci.* **37** 895–903

Luo D, Cao Y, Wu D, Xu Y, Chen B and Xue Z 2009 Impact of intense pulse light irradiation on BALB/c mouse skin-*in vivo* study on collagens, matrix metalloproteinases and vascular endothelial growth factor *Lasers Med. Sci.* **24** 101–8

Maruyama S 2016 Hand rejuvenation using standard intense pulsed light (IPL) in Asian patients *Laser Ther.* **25** 43–54

Massagué J and Sheppard D 2023 TGF-β signaling in health and disease *Cell* **186** 4007–37

Mei X L and Wang L 2018 Ablative fractional carbon dioxide laser combined with intense pulsed light for the treatment of photoaging skin in Chinese population: a split-face study *Medicine* **97** e9494

Monib K M E and Hussein M S 2020 Nd:YAG laser vs IPL in inflammatory and noninflammatory acne lesion treatment *J. Cosmet. Dermatol.* **19** 2325–32

Mukherjee S, Date A, Patravale V, Korting H C, Roeder A and Weindl G 2006 Retinoids in the treatment of skin aging: an overview of clinical efficacy and safety *Clin. Interv. Aging* **1** 327–48

Nassar S, Assem M, Mohamed D and Hassan G 2020 The efficacy of radiofrequency, intense pulsed light and carboxytherapy in facial rejuvenation *J. Cosmet. Laser Ther.* **22** 256–64

Negishi K, Wakamatsu S, Kushikata N, Tezuka Y, Kotani Y and Shiba K 2002 Full-face photorejuvenation of photodamaged skin by intense pulsed light with integrated contact cooling: initial experiences in Asian patients *Lasers Surg. Med.* **30** 298–305

Patel A D, Chopra R, Avram M, Sakamoto F H, Kilmer S, Anderson R R and Ibrahimi O A 2024 Updates on lasers in dermatology *Dermatol. Clin.* **42** 33–44

Qiao C, Li L, Wang H, Zhao C, Ke L, Sen D, Qi M, Li S, Wang M and Zeng Q 2021 Adverse events of intense pulsed light combined with meibomian gland expression versus meibomian gland expression in the treatment of meibomian gland dysfunction *Lasers Surg. Med.* **53** 664–70

Radmanesh M, Azar-Beig M, Abtahian A and Naderi A H 2008 Burning, paradoxical hypertrichosis, leukotrichia and folliculitis are four major complications of intense pulsed light hair removal therapy *J. Dermatolog. Treat.* **19** 360–3

Raulin C, Greve B and Grema H 2003 IPL technology: a review *Lasers Surg. Med.* **32** 78–87

Ruan F, Zang Y, Sella R, Lu H, Li S, Yang K, Jin T, Afshari N A, Pan Z and Jie Y 2019 Intense pulsed light therapy with optimal pulse technology as an adjunct therapy for moderate to severe blepharitis-associated keratoconjunctivitis *J. Ophthalmol.* **2019** 3143469

Sales A F S, Pandolfo I L, de Almeida Cruz M, Parisi J R, Garcia L A, Martignago C C S, Renno A C M and Vassão P G 2022 Intense pulsed light on skin rejuvenation: a systematic review *Arch. Dermatolog. Res.* **314** 823–38

Salminen A, Kaarniranta K and Kauppinen A 2022 Photoaging: UV radiation-induced inflammation and immunosuppression accelerate the aging process in the skin *Inflamm. Res.* **71** 817–31

Shin S H, Lee Y H, Rho N K and Park K Y 2023 Skin aging from mechanisms to interventions: focusing on dermal aging *Front. Physiol.* **14** 1195272

Soon S L and Victor Ross E 2008 Letter: use of a perforated plastic shield for precise application of intense pulsed light *Dermatol. Surg.* **34** 1149–50

Stewart N, Lim A C, Lowe P M and Goodman G 2013 Lasers and laser-like devices: part one *Australas J. Dermatol.* **54** 173–83

Tao L, Wu J, Qian H, Lu Z, Li Y, Wang W, Zhao X, Tu P, Yin R and Xiang L 2015 Intense pulsed light, near infrared pulsed light, and fractional laser combination therapy for skin rejuvenation in Asian subjects: a prospective multi-center study in China *Lasers Med. Sci.* **30** 1977–83

Wong W R, Shyu W L, Tsai J W, Hsu K H and Pang J H 2009 Intense pulsed light effects on the expression of extracellular matrix proteins and transforming growth factor beta-1 in skin dermal fibroblasts cultured within contracted collagen lattices *Dermatol. Surg.* **35** 816–25

Yamashita T, Negishi K, Hariya T, Kunizawa N, Ikuta K, Yanai M and Wakamatsu S 2006 Intense pulsed light therapy for superficial pigmented lesions evaluated by reflectance-mode confocal microscopy and optical coherence tomography *J. Invest. Dermatol.* **126** 2281–6

Chapter 12

Photodynamic therapy for photoaging

Rui Yin

Photodynamic therapy (PDT) is an emerging non-invasive treatment for photoaging and diverse skin conditions, including acne, rosacea, psoriasis, and non-melanoma skin cancer. By applying topical photosensitizers such as aminolevulinic acid (ALA) and methyl aminolevulinate (MAL), PDT generates reactive oxygen species, promoting collagen synthesis and alleviating aging signs. Clinical studies report enhancements in skin texture, fine lines, and pigmentation following PDT. The method's effectiveness is augmented when combined with intense pulsed light (IPL) or pulsed dye lasers (PDL), positioning it as a moderate approach between ablative and nonablative rejuvenation. Despite effective outcomes, the optimal PDT parameters for photorejuvenation are yet to be standardized, offering a promising avenue for future research.

12.1 Introduction

Photodynamic therapy (PDT) is recognized as a non-invasive procedure for managing a spectrum of dermatological conditions. It has shown clinical efficacy and is progressively being used for conditions such as acne, sebaceous gland enlargement, rosacea, viral warts, psoriasis, and more (Celli *et al* 2010, Ozog *et al* 2016). PDT has also proven to be a valuable treatment for non-melanoma skin cancer (NMSC), including precancerous lesions such as actinic keratosis (AK), basal cell carcinoma (BCC), and Bowen's disease (a form of *in situ* squamous cell carcinoma–SCC) (Queiros *et al* 2020, Shi *et al* 2021).

The topical application of PDT has shown promise in both treating and preventing NMSC. The innovative use of PDT for skin rejuvenation was highlighted by Bitter's pioneering clinical study published in 2000, which introduced the concept of photodynamic therapy for aesthetic enhancement (Gardner and Weiss 1990). In studies focused on AK, researchers observed an improvement in fine lines, uneven pigmentation, and other signs of sun damage, which were considered positive side effects (Gold 2002). This led to further investigations into PDT's potential for

doi:10.1088/978-0-7503-5112-6ch12

cosmetic skin enhancement. Numbers of clinical studies indicated that patients noted significant improvements in photoaging symptoms, including a reduction in fine lines, changes in pigmentation, a decrease in pore size and the improvement of skin texture (Kohl *et al* 2010). However, at present, standardized treatment protocols for this purpose have not yet been established.

12.2 Concept and mechanism of action

PDT process involves the application of visible light at specific wavelengths to activate non-toxic photosensitizer molecules in an oxygen-rich environment. Prodrugs such as MAL and ALA are converted into protoporphyrin IX (PpIX), the photosensitizer, within the cell's cytoplasm and mitochondria. Upon light exposure, these molecules produce cytotoxic singlet oxygen ($^1O_2$) and reactive oxygen species (ROS) that interact with cellular components such as RNA, DNA, proteins, and lipids, leading to effects such as lipid peroxidation and vascular changes (Jang *et al* 2013). While the exact mechanisms of PDT in combating skin aging are still being explored, it is believed that the activation of dermal fibroblast functions through pathways such as MAPK and ERK may play a role. This activation could enhance the proliferation and activity of dermal fibroblasts, resulting in increased production of matrix metalloproteinase (MMP)-3 and collagen type Iα RNA and protein (Jang *et al* 2013), which results in rejuvenated skin (Lubart *et al* 2007).

12.3 Clinical application

Topical PDT has demonstrated effectiveness in addressing multiple aspects of skin aging, with studies showing improvements in fine wrinkles, pigmentation irregularities, skin texture, and overall complexion. The choice of photosensitizer is critical to the success of PDT. Aminolevulinic acid (ALA) and its methyl ester (MAL) are commonly utilized for their ability to penetrate the skin and convert into protoporphyrin IX (PpIX), a potent photosensitizer. ALA is favored for its rapid metabolism, while MAL offers deeper penetration due to its lipophilic nature (Lee and Kloser 2013). Weiss (1990) found no significant difference in short-term efficacy and side effects between ALA and MAL, suggesting both can be recommended for PDT. Clinical studies have demonstrated that both ALA-PDT and MAL-PDT can improve skin texture, reduce fine lines, and alleviate pigmentary changes associated with photoaging. Clinical trials have shown that MAL may induce less pain and be better tolerated (Elson 1990).

Another critical aspect of PDT is the choice of light source, which has an optimal wavelength that is suitable for photoactivation of PpIX. The light source includes continuous-wave red or blue light, intense pulsed light (IPL), and pulsed dye laser (PDL) (Friedmann *et al* 2014). Each light source has its unique spectral characteristics that can influence the depth of penetration and the resultant clinical outcomes. LEDs are the primary light source for topical PDT due to their lower cost and easier maintenance compared to lasers (Moseley *et al* 2006, Brancaleon and Moseley 2002). For instance, LED red light (around 630 nm) is preferred for its deeper penetration. They are less painful and can be used for prolonged periods without causing significant thermal damage. Patients often report a high level of satisfaction with LED-PDT, citing minimal downtime and a comfortable treatment experience (Sanclemente *et al* 2011). Blue light (415–420 nm) is highly effective in activating PpIX, particularly in ALA-PDT. It is useful for treating superficial conditions but may not penetrate as deeply as red light. Patients treated with blue light often report a sensation of warmth during the procedure, with some experiencing mild to moderate discomfort (Tschen *et al* 2006). IPL sources, with their broad spectrum (410–1200 nm or 515–1200 nm), have been used in PDT for photorejuvenation. They target a range of chromophores, leading to an overall improvement in skin texture and pigmentation. Specific lasers, including pulsed dye laser (PDL) emitting at 585 or 595 nm, have been employed for targeted PDT. These lasers offer the advantage of deeper penetration and can stimulate collagenesis more effectively (Karrer *et al* 1999). Key (2005) evaluated PDL-PDT for facial photodamage, with patients reporting improved skin texture and clearance of roughness on the treated side. Orringer *et al* (Orringer *et al* 2008) found that PDL-PDT increased keratinocyte proliferation and induced collagen production without affecting p53 levels. However, PDL-PDT may be associated with a higher risk of adverse effects such as purpura and crusting, particularly in patients with darker skin types. The use of fractional resurfacing lasers in conjunction with PDT has also shown promising results by enhancing photosensitizer penetration and inducing collagen synthesis (Ruiz-Rodriguez *et al* 2008).

In clinical practice, the combination of PDT with light sources such as IPL, PDL, and fractional light exposure has been shown to be more effective for skin rejuvenation than PDT alone. Several randomized split-face trials using 20% ALA have been conducted for photodynamic rejuvenation, with results favoring the ALA-PDT side over IPL-only treatments, albeit with a higher incidence of side effects such as erythema, scaling, dryness, edema, crusting, and vesiculation (Alster *et al* 2005, Dover *et al* 2005, Gold *et al* 2006).

Following PDT, significant histological changes occur in sun-damaged skin. These include a reduction in epidermal thickness and atypical keratinocytes, as well as a decrease in p53 expression, a marker associated with early epidermal carcinogenesis (Sjerobabski and Situm 2014, Zhang *et al* 2014, Orringer *et al* 2008). In the dermis, there is a reduction in elastotic material and an increase in procollagen and collagen types I and III, along with a decrease in MMPs that break down collagen and elastin (MMP-1, MMP-3, and MMP-12) (Park *et al* 2010).

The frequency of PDT treatments is determined by the individual's skin condition and the desired aesthetic outcome. Typically, a series of treatments is required, ranging from three to six sessions with intervals of 2–4 weeks between each. Maintenance treatments may be recommended to sustain the rejuvenating effects. Each treatment protocol should be tailored to the patient's needs, taking into account the potential for phototoxicity and the necessity for post-treatment photoprotection.

Nearly all clinical trials have consistently reported the aesthetic benefits of PDT in facial rejuvenation. Improvements in fine wrinkles, skin laxity, and overall skin tone have been documented. Patient feedback has been predominantly positive, with many reporting enhanced skin texture and a reduction in the appearance of photodamage. However, the response to PDT can vary, and factors such as the patient's skin type, the severity of photoaging, and the specific treatment protocol can all influence outcomes.

12.4 Adverse reactions

Post-treatment, patients may experience a range of reactions, which are generally short-lived and manageable:

Erythema: A common transient reaction, erythema usually subsides within hours to a day after treatment. Patients are advised to use cold compresses and anti-inflammatory creams to alleviate this symptom.

Edema: Some patients may develop mild swelling, which can be managed with cold compresses and elevation of the treated area.

Crust formation and desquamation: These are more commonly seen with laser and intense light sources. Patients are instructed to keep the area clean and avoid picking at the crusts to prevent infection.

Pain: Variable reports of pain during and after PDT exist, with some patients describing a burning sensation. The use of topical anesthetics, cold air cooling, and analgesics can significantly reduce this discomfort.

Phototoxicity: The use of LED blue can easily lead to acute phototoxicity, usually occurring within 24 h after treatment. Second, if exposed to light within 24–48 h after treatment, phototoxic reactions are also likely to occur, so it is important to pay attention to sun protection after treatment.

Long-term adverse effects: Rarely, patients may report long-term changes in skin pigmentation or textural changes. Hyperpigmentation is prone to occur in dark skinned individuals (Fitzpatrick IV–V skin type), especially with LED light sources. These are typically managed with skin lightening agents or textural correction modalities.

12.5 Conclusion

PDT for photoaging uses nonablative light sources to activate the photosensitizer, resulting in modest clinical improvements. PDT is a promising tool for treating photodamage, offering better cosmetic results than nonablative rejuvenation techniques and with fewer and less severe side effects compared to ablative methods. The clinical outcomes are influenced by the light source, photosensitizer concentration, and incubation time. Its nonablative nature means it cannot resurface the skin but can target blood vessels, disperse pigment, and heat dermal collagen, leading to mild neocollagenesis. However, it may not provide the dramatic improvements in deep wrinkles and solar elastosis that ablative lasers or deep chemical peels can achieve. The ideal parameters for PDT in photo-rejuvenation are still being determined, and establishing standardized guidelines will require further research and time.

References

Alster T S, Tanzi E L and Welsh E C 2005 Photorejuvenation of facial skin with topical 20% 5-aminolevulinic acid and intense pulsed light treatment: a split-face comparison study *J. Drugs Dermatol* **4** 35–8

Brancaleon L and Moseley H 2002 Laser and non-laser light sources for photodynamic therapy *Lasers Med. Sci.* **17** 173–86

Celli J P *et al* 2010 Imaging and photodynamic therapy: mechanisms, monitoring, and optimization *Chem. Rev.* **110** 2795–838

Dover J S *et al* 2005 Topical 5-aminolevulinic acid combined with intense pulsed light in the treatment of photoaging *Arch. Dermatol.* **141** 1247–52

Elson M L 1990 Treatment of striae distensae with topical tretinoin *J. Dermatol. Surg. Oncol.* **16** 267–70

Friedmann D P *et al* 2014 The effect of multiple sequential light sources to activate aminolevulinic acid in the treatment of actinic keratoses: a retrospective study *J. Clin. Aesthet. Dermatol.* **7** 20–5

Gardner S S and Weiss J S 1990 Clinical features of photodamage and treatment with topical tretinoin *J. Dermatol. Surg. Oncol.* **16** 925–31

Gold M H 2002 The evolving role of aminolevulinic acid hydrochloride with photodynamic therapy in photoaging *Cutis* **69** 8–13

Gold M H *et al* 2006 Split-face comparison of photodynamic therapy with 5-aminolevulinic acid and intense pulsed light versus intense pulsed light alone for photodamage *Dermatol. Surg.* **32** 795–801 801–803

Jang Y H *et al* 2013 Prolonged activation of ERK contributes to the photorejuvenation effect in photodynamic therapy in human dermal fibroblasts *J. Invest. Dermatol.* **133** 2265–75

Karrer S *et al* 1999 Long-pulse dye laser for photodynamic therapy: investigations *in vitro* and *in vivo Lasers Surg. Med.* **25** 51–9

Key D J 2005 Aminolevulinic acid-pulsed dye laser photodynamic therapy for the treatment of photoaging *Cosmet. Dermatol.* **18** 31–6

Kohl E *et al* 2010 Aesthetic effects of topical photodynamic therapy *J. Eur. Acad. Dermatol. Venereol.* **24** 1261–9

Lee P K and Kloser A 2013 Current methods for photodynamic therapy in the US: comparison of MAL/PDT and ALA/PDT *J. Drugs Dermatol.* **12** 925–30

Lubart R *et al* 2007 A reasonable mechanism for visible light-induced skin rejuvenation *Lasers Med. Sci.* **22** 1–3

Moseley H *et al* 2006 Ambulatory photodynamic therapy: a new concept in delivering photodynamic therapy *Br. J. Dermatol.* **154** 747–50

Orringer J S *et al* 2008 Molecular effects of photodynamic therapy for photoaging *Arch. Dermatol.* **144** 1296–302

Ozog D M *et al* 2016 Photodynamic therapy: a clinical consensus guide *Dermatol. Surg.* **42** 804–27

Park M Y *et al* 2010 Photorejuvenation induced by 5-aminolevulinic acid photodynamic therapy in patients with actinic keratosis: a histologic analysis *J. Am. Acad. Dermatol.* **62** 85–95

Queiros C *et al* 2020 Photodynamic therapy in dermatology: beyond current indications *Dermatol. Ther.* **33** e13997

Ruiz-Rodriguez R *et al* 2008 Photorejuvenation using topical 5-methyl aminolevulinate and red light *J. Drugs Dermatol.* **7** 633–7

Sanclemente G, Medina L, Villa J F, Barrera L M and Garcia H I 2011 A prospective split-face double-blind randomized placebo-controlled trial to assess the efficacy of methyl amino-levulinate + red-light in patients with facial photodamage *J. Eur. Acad. Dermatol. Venereol.* **25** 49–58

Shi L *et al* 2021 Chinese guidelines on the clinical application of 5-aminolevulinic acid-based photodynamic therapy in dermatology (2021 edition) *Photodiagnosis Photodyn. Ther.* **35** 102340

Sjerobabski M I and Situm M 2014 Photorejuvenation--topical photodynamic therapy as therapeutic opportunity for skin rejuvenation *Coll. Antropol.* **38** 1245–8

Tschen E H *et al* 2006 Photodynamic therapy using aminolaevulinic acid for patients with nonhyperkeratotic actinic keratoses of the face and scalp: phase IV multicentre clinical trial with 12-month follow up *Br. J. Dermatol.* **155** 1262–9

Weiss J S 1990 Topical tretinoin in the treatment of photodamaged skin. Current status and future prospects *Int. J. Dermatol.* **29** 183–4

Zhang H Y *et al* 2014 Evaluation of 5-aminolevulinic acid-mediated photorejuvenation of neck skin *Photodiagnosis Photodyn. Ther.* **11** 498–509

Chapter 13

Low level light therapy for photoaging

Shanglin Jin and Chengfeng Zhang

Low level light therapy (LLLT) is a promising approach for treating photodamaged skin. It utilizes specific wavelengths of light to stimulate cellular processes, such as collagen production and tissue repair, leading to improvements in skin texture, tone, and elasticity. LLLT is non-invasive, painless, and generally well-tolerated, making it an attractive option for addressing signs of skin aging caused by chronic sun exposure. Research indicates its efficacy in reducing wrinkles, fine lines, and pigmentation irregularities associated with photoaging, offering patients a non-surgical means of rejuvenating their skin.

13.1 Definition of LLLT

Low-level light therapy (LLLT), also known as cold laser therapy, uses light-emitting diodes (LEDs), lasers or other light sources to regulate cell activity and thus achieve therapeutic effects (Caruso-Davis *et al* 2011). When these light sources are used, they usually take parameters of low energy, low power density, narrow spectrum, specific pulse mode and pulse width, and will not cause significant temperature increase of the treated tissue during the treatment, so that the tissue structure will not be damaged. The underlying mechanism is mainly based on photobiomodulation (PBM), which also becomes biostimulation, a photobiological phenomenon that stimulates intracellular photobiochemical reactions and regulates cell activity through non-photothermal action.

In 1967, Hungary's Endre Mester *et al* first proposed the biological effects of low-energy laser therapy on tissue (McGuff *et al* 1965). Mester used a ruby laser with lower energy to treat tumors in order to replicate McGuff's experiment in Boston and found that the low-energy ruby laser (694 nm) had no effect on cancer, but promoted hair growth on the backs of animals shaved for experimental purposes. This was the first demonstration of photobiomodulation using low-energy laser therapy, opening up a new field of medicine.

doi:10.1088/978-0-7503-5112-6ch13

The photobiomodulation effect of LLLT can be divided into two approaches: one is using cold lasers (such as LEDs, etc) to produce non-thermal effects and the other is using traditional ablative lasers to produce photothermal effects and non-photothermal effects on skin tissue, the latter of which belongs to photobiomodulation. LLLT usually selects coherent light sources dominated by lasers or incoherent light sources dominated by LEDs. Compare with lasers, LED light sources are relatively inexpensive and also safe and effective. They can be combined with multiple wavelengths and have gradually replaced traditional laser equipment, becoming the preferred light source for LLLT.

Currently, LLLT is widely used in the field of dermatology, including preventing scars, alleviating inflammation, promoting repair, photorejuvenation, treating hair loss and pigmented diseases, etc. This chapter summarizes the latest findings on low level light therapy and its clinical application in the field of dermatology.

13.2 Mechanisms of LLLT

The first law of photobiology states that the effect of light on biological systems must depend on the absorption of photons by photoreceptors on cells, resulting in conformational changes, thereby converting light energy into chemical energy, or absorbing photons to cause biological macromolecules and then energy increases and becomes an excited state (Sutherland 2002). Photoreceptors usually contain chromophores, each of which has its own specific absorption spectrum.

The most well-known chromophore is plant chlorophyll. Common chromophores in the human body include retinal, porphyrin, pteroline, flavin, bile pigment, melanin, uric acid, tryptophan, vitamin B6, vitamin K, etc. These chromophores can form ligand complexes with some metals and auxochromes, which eventually become photoreceptors. Under the action of light at a specific wavelength, the internal chromophore and complex undergo a conformational change, causing changes in the second messenger and downstream pathways, thereby exerting its biological effects (Kottke *et al* 2018). Important chromophores that initiate light modulation include cytochrome oxidase C, opsin, and intracellular water, porphyrin, hemoglobin, and flavoprotein that act at specific sites in cells (de Freitas and Hamblin 2016).

Currently known photoreceptors include cytochrome C oxidase (CCO), NADPH oxidase (NOX), cryptochromes (CRY), opsin (OPN), etc (Mignon *et al* 2016).

CCO is the most studied cellular photoreceptor. It is an oxidoreductase, which exists on the mitochondrial membrane of cells and is the fourth enzyme complex in the electron transport chain of respiration. The structure of CCO includes the heme formed by iron-protoporphyrin, so its two spectral absorption peaks are determined in the red 600–700 nm and near-infrared 760–900 nm bands, respectively. Karu (2014) found that after illumination, CCO becomes an excited state, which on the one hand accelerates the electron transfer rate in the respiratory chain, increases the mitochondrial membrane potential to generate more ATP, and then changes the level of cAMP to directly regulate cell biology behavior. On the other hand, excited CCO can also exert photobiomodulation effect through reactive oxygen species (ROSs).

ROSs are a by-product of the respiratory chain. After light exposure, the activity of the electron transport chain in mitochondrial respiration is intensified. The balance between the ROSs generated by the respiratory chain and the antioxidant systems is broken, thereby increasing the content of intracellular ROSs. ROSs activate nuclear factor κB (NF-κB) and regulate DNA transcription, cytokine production, and cell proliferation. In addition to the ATP pathway and the ROSs pathway, Poyton and Ball(2011) proposed that a new mechanism for CCO to exert PBM is to increase the content of nitric oxide (NO). Excited CCO can catalyze the reduction of nitrites to generate NO or photoinduced CCO. The complex dissociates NO, which acts as an intracellular second messenger.

Intracellular redox is not only affected by the respiratory chain, but also exists in the redox chain to regulate cellular metabolism, in which NOX can play a role in photobiomodulation. NOX is a membrane protein of cells that catalyzes the reduction of oxygen molecules to form reactive oxygen species. Tafur and Mills (2008) found that LLLT can activate the redox chain through NOX in phagocytes, and finally catalyze the production of ROSs. This has an important role in immune defense, where phagocytes produce large amounts of ROSs, an oxygen burst, that kills pathogens that have been phagocytosed into the cell. In addition, ROSs can also act as second messengers and participate in cell signaling regulation. Gizinger *et al* (2016) used 635 nm LLLT to irradiate patient peripheral blood neutrophils *in vitro*. When the irradiation dose reached 10.8–12 J cm^{-2}, they found that the intracellular NOX activity increased and the release of lysosomal particles was promoted.

CRY is a conserved blue light receptor protein that exists widely in animals and plants and is mainly related to biological circadian rhythms. It contains two chromophores, flavin and pterin, which determine its absorption peak at 350–420 nm (Rajagopalan and Handler 1964). In recent years, studies have revealed that in the plant *Arabidopsis thaliana*, blue light can activate CRY to generate ROSs and hydrogen peroxide (H_2O_2) to promote cell death (Consentino *et al* 2015). Humans and mammals express two proteins, CRY1 and CRY2, which are closely related to cellular circadian rhythm genes (such as BMAL1 and CLOCK). Mice with mutations in CRY1 or CRY2 lack the circadian rhythm in response to external light (Nakao 2014). LLLT-activated CRY can also inhibit the NF-κB and PKA pathways, down-regulate cAMP, and ultimately inhibit the production of inflammatory factors to relieve inflammation (Narasimamurthy *et al* 2012).

OPN is a class of transmembrane G protein-coupled receptors on cells with photoreceptor effects. In humans, OPN contains seven isoforms: OPN1, OPN2 (rhodopsin), RRH (peropsin), retinal G protein-coupled receptor (RGR), OPN3 (encephalopsin), OPN4 (melanopsin), and OPN5 (neuropsin). The spectral absorption peak of opsin is 380–570 nm (Terakita 2005). Among them, OPN1, OPN2, RRH and RGR are vision-related opsins, which mainly exist in retinal photoreceptors and participate in the visual signal transduction pathway, and are the most widely studied. Some studies have pointed out that OPN2 is expressed in human skin cells, and 380–420 nm LED blue-violet light can inhibit the differentiation of human keratinocytes through OPN2 (Kim *et al* 2013). A similar study by Denda and Fuziwara reported that 430–510 nm blue light inhibits skin barrier repair, while

550–679 nm red light can promote it (Denda and Fuziwara 2008). OPN2 plays an important role in visible light-mediated skin barrier homeostasis. In addition to vision-related opsins, in recent years, non-vision-related OPN3, OPN4, and OPN5 have gradually attracted attention as photoreceptors. Among them, OPN3 is highly expressed in keratinocytes and melanocytes, and is closely related to pigmentation. Ozdeslik *et al* found that OPN3 is a light-independent inhibitor of melanogenesis (Ozdeslik *et al* 2019). Regazzetti *et al* found that in dark-skinned individuals, the action of blue light on OPN3 can lead to hyperpigmentation (Regazzetti *et al* 2018). Castellano-Pellicena *et al* found that OPN3 and OPN5 can be expressed in human facial and abdominal skin, where keratinocytes and fibroblasts highly express OPN3, and both blue and red light can enhance cell activity through OPN3, thereby promoting wound healing and repairing the skin barrier (Castellano-Pellicena *et al* 2018). OPN4, also known as melanopsin, has a λ_{max} of 490 nm and is mostly expressed in the hypothalamus to participate in the regulation of the biological clock. Bertolesi and McFarlane clarified in a review that OPN4 plays an important role in the regulation of skin pigmentation (Bertolesi and McFarlane 2018). First, light can affect the PI3K pathway through OPN4 to mediate melanosome movement (Isoldi *et al* 2005), and second, light stimulates OPN4 through the neuroendocrine system which regulates the production of α-melanocyte-stimulating hormone (α-MSH) (Bertolesi *et al* 2015) and melatonin (Alkozi *et al* 2017), thereby regulating skin pigment synthesis. OPN5, also known as neuropsin, has a λ_{max} of 380 nm and is mostly distributed in neural tissues to participate in circadian rhythm perception and regulation. The discovery of opsins provides a new idea photobiomodulation.

In addition to the above receptors, there are some new photoreceptors and their functions have yet to be discovered. In recent years, Santana-Blank *et al* proposed that water can also be used as a special photoreceptor (Santana-Blank *et al* 2016). As the most important medium in the cellular microenvironment, water is a polar molecule, and the water oscillator paradox produced by near-infrared (NIR) can transfer energy to cells like a rechargeable battery, thereby affecting their biological activities.

13.3 Biological effects of LLLT

At present, LLLT has been widely used in pain relief, anti-inflammatory repair, etc. How does it exert such biological effects? After photoreceptors are activated in the cells, how do they act? Although the mechanism of action of LLLT has not been fully elucidated, it been been shown to be closely related to mitochondria. Karu proposed the theory of a mitochondrial retrograde signaling pathway (Karu 2008). They found that LLLT of red light and near-infrared light (600–1000 nm) can change the mitochondrial membrane potential through the action of photoreceptors such as mitochondrial $\Delta\Psi$m, ATP, ROS, NO, Ca^{2+}, and these molecules can affect the expression of genes in the nucleus, so as to achieve light regulation to change the biological activities of cells.

Gavish *et al* irradiated human keratinocytes at 780 nm and 2 J cm^{-2}, and found that $\Delta\Psi$m changed immediately, while the expressions of interleukin IL-1α, IL-6, and keratinocyte growth factor (KGF) were increased and IL-1β was inhibited

(Gavish *et al* 2004). The change of $\Delta\Psi$m can be accompanied by the increase of ATP. Hu *et al* found that accumulation of ATP in melanoma cells irradiated by He-Ne laser can promote cell proliferation through the JNK/AP-1 pathway (Hu *et al* 2007). Photoreceptors CCO and NOX can generate large amounts of ROSs and NO when activated by LLLT. When the level of cellular redox changes, NF-κB pathway can be induced (Morgan and Liu 2011). Usually in the cell membrane, the binding expression level of NF-κB and its inhibitory protein IκB is inhibited. ROSs can activate the phosphorylation of IκB, which results in the dissociation from NF-κB. NF-κB is finally activated and then enter cell nucleus, regulating cellular activity. Schroeder *et al* found that 760–1440 nm, 30 J cm^{-2} near-infrared light can cause ROSs accumulation in human fibroblasts and eventually upregulate matrix metalloproteinase-1 (MMP-1) (Schroeder *et al* 2007). In addition to ROSs, NO also plays a complex role in PBM. The classical downstream pathway is NO-cyclic guanosine monophosphate (cGMP)-protein kinase G (PKG) (NO/cGMP/PKGs). PKGs can enter the nucleus to regulate the expression of a series of downstream genes and proteins (Martínez-Ruiz *et al* 2011). In addition, NO may be associated with the cyclooxygenase COX pathway, upregulating prostaglandin expression and promoting inflammation (Sorokin 2016). NO can also reversely regulate the inhibition of CCO and the ROSs generated (Erusalimsky and Moncada 2007).

In addition to the above signaling pathways, it has been found in recent years that after OPN is activated by light, the down-regulated ion channels are opened, and the transient receptor potential cation channel (TRP) can regulate the intracellular calcium ion Ca^{2+} level. As early as the last century, Young *et al* found that light at 660 nm and 4–8 J cm^{-2} can change intracellular Ca^{2+}, which can affect cell membrane permeability, intracellular pH, etc (Young *et al* 1990). More and more studies on intracellular calcium levels have been performed. This notion was revived in 2017 on the 50th anniversary of the discovery of the role of photobiomodualtion, and has received renewed attention in recent years with more in-depth studies on the roles of new photoreceptors such as OPN and ion channels such as TRP (Amaroli *et al* 2018). Lan *et al* irradiated melanoblasts with a 632.8 nm He-Ne laser and found that it could promote cell differentiation, increase mitochondrial DNA copy number and synthesis, and enhance CCO activity dependent on Ca^{2+} (Lan *et al* 2012). Wang *et al* proposed that the TRP ion channel is the target of LLLT treatment (Wang *et al* 2014). On mast cells, they found that TRPV1 can be activated by 637 nm (4.3 J cm^{-2}) and 532 nm (4.8 J cm^{-2}), and TRPV2 can be activated by 640 nm (57.6 J cm^{-2}) and TRPV4 can be activated by light at 405 and 532 nm. The activation of these ion channels ultimately leads to up-regulation of intracellular Ca^{2+} levels, increased ATP synthesis, and mast cell degranulation. Collectively, intracellular Ca^{2+} might serve as an important mediator of LLLT photobiomodulation.

13.4 Clinical applications of LLLT

13.4.1 The parameter selection of LLLT

The classification of LLLT depends on the parameter selection of LLLT. The biological activity of LLLT photobiomodulation is not only related to the target cells, but also tightly associated with the illumination parameters of the light source. LLLT usually chooses a light source with a wavelength of 390–1100 nm, an energy density of 0.04–50 J cm^{-2}, and a power density of less than 100 mW cm^{-2}. The wavelength determines the depth of penetration of the light source. 390–600 nm is usually used to treat superficial lesions, while 600–1100 nm is used to treat deeper lesions. Light of 700–770 nm has limited effect on skin and is less applied in clinics (Avci *et al* 2013).

The selection of LLLT light source has always been controversial. LLLT can use a coherent light source dominated by lasers or an incoherent light source dominated by LEDs. LEDs are semiconductors that convert electrical current into light, emitting incoherent light of the same wavelength as a laser, but with lower energy. The output power of a laser is in watts, and the output power of an LED is usually in milliwatts. The LEDs consist of panels that can treat large areas in a relatively short period of time (seconds to minutes). Laser is the abbreviation of 'light amplification by stimulated emission of radiation', which is characterized by monochromaticity, coherence, parallelism and high energy.

Coherence refers to the property of waves in order to produce significant interference phenomena. As an electromagnetic wave, laser can interact with each other and cause an interference phenomenon, which can be divided into spatial coherence and temporal coherence. LEDs, on the other hand, have poor light coherence and are generally considered to be incoherent light. The most commonly used coherent light sources in LLLT are He-Ne lasers and GaAs lasers (Chung *et al* 2012). It was once thought that coherence is an important property of light to exert its effects, but with the generation of LEDs, researchers have found that the efficacy of LEDs and traditional lasers as LLLT light sources in the treatment of diseases is controversial (Heiskanen and Hamblin 2018). Professor Hode believed that for superficial lesions, the photon energy flow is more concentrated, and when other optical parameters are consistent, LED light can achieve the same effect as laser light, while for deeper lesions, the laser energy is more concentrated and the effect is better (Hode *et al* 2009). However, some studies have also revealed that LEDs are more effective in tissue repair when 640 nm LED light and a 660 nm laser are used to treat wounds in diabetic mice at the same dose density (6 J cm^{-2}) (Dall Agnol *et al* 2009). Brochetti *et al* proposed that the coherence of light is not the only factor that determines the effect of disease treatment (Brochetti *et al* 2017). Although LED light is different from laser light, they are similar in terms of treatment effect. Seemingly it is meaningless to compare the efficacy between incoherent LEDs and coherent laser without specifying optical parameters and disease types. However, in the field of dermatology, due to the differences in the parameters of laser energy and individual tolerance to laser energy, even low level lasers may cause adverse reactions (such as pain, erythema, desquamation, post-inflammatory pigmentation, etc). However, as a cold laser, LED light is relatively mild, safe, inexpensive, broadly effective with a wide range of wavelengths so that it has been more and more widely used for clinical treatment.

Like drug treatment, LLLT also has concepts such as 'dose' and 'treatment window'. The critical parameters include irradiation density (mW cm^{-2}), irradiation time (s) and energy density (J cm^{-2}) of the light source itself and treatment interval. In recent years, Huang *et al* proposed a new 3D Arndt Schulz model based on the original biphasic dose curve of LLLT photobiomodulation effect (Huang *et al* 2011), which incorporates the two variables of irradiation density and irradiation time together into the photobiomodulation effect. It is believed that neither a too low irradiation density nor a too short irradiation time could exert the PBM effect, while a too high irradiation density or a too long irradiation time would inhibit the photobiological effect, as well. For different tissues, specific optical parameters should be applied.

In addition, the illumination mode is also worth discussing. Generally speaking, the illumination mode of the light source can be divided into pulsed light and continuous light. The effect and mechanism of these two modes on cell biological activities is still inconclusive. Some studies reported that the pulsed mode is superior to the continuous mode. In an *in vitro* study, Barolet *et al* irradiated human fibroblasts with 630 nm LED red light (8 J $cm^{-1\ 2}$) continuous light or microsecond pulsed light and, after 72 h, the pulsed light group showed more collagen type I synthesis, and the mechanism may be related to the dissociation of more NO by CCO (Barolet *et al* 2010). In an *in vivo* experiment, Barolet *et al* again irradiated the bilateral hands of skin systemic sclerosis patients with a 940 nm laser, using continuous light mode on one side and pulsed light mode on the other side (Barolet 2014). After 13 weeks, pulsed light mode showed higher efficacy and the mechanism may be related to the down-regulation of TGF-β expression level. However, some studies reported comparable therapeutic effect between pulsed light and continuous light. Al-Watban and Zhang irradiated rat back wounds with 635 nm, 1 J cm^{-2} continuous or pulsed light, and revealed similar effects of the two on wound healing (Al-Watban and Zhang 2004). Therefore, the effects of pulsed light and continuous light on biological activities need to be further studied.

The optical parameters of LLLT and their definitions are shown in tables 13.1 and 13.2 (de Freitas and Hamblin 2016).

With the application of LED light in LLLT, the clinical research of its effect on skin diseases has greatly increased. For some diseases where topical therapy or system therapy does not work, laser therapy may play an exceptional role. However, conventional ablative lasers may cause adverse reactions such as post-inflammatory

Table 13.1. Main light source parameters of LLLT.

Parameter	Unit	Description
Wavelength	Nm	Light, as a type of energy or electromagnetic wave, has the characteristics of a wave.
Power density	mW cm^{-2}	Refers to the power per unit area.
Pulse structure	Peak power (W) Pulse frequency (Hz) Pulse width (s) Duty cycle (%)	If a beam of light is pulsed light, its energy power can be calculated by the following formula: average energy power (W) = peak power (W) × pulse width (s) × pulse frequency (Hz)

Table 13.2. Main dose parameters of LLLT.

Parameter	Unit	Description
Energy	J	Calculation formula: energy (J) = power (W) × time (s)
Energy density	J cm^{-2}	Refers to the energy per unit area, as the main parameter to measure the dose of light modulation.
Irradiation time	S	Irradiation time and power density determine the energy density.

hyperpigmentation, pain, redness and swelling even at low energy levels. As a type of cold laser, LEDs can reduce the side reactions caused by photothermal action with superior curative effect, safety and low cost, and thus become the preferred light source for LLLT. Currently, it is mainly used in the treatment of acne vulgaris, post-herpetic neuralgia, photoaging, tissue repair, hyperpigmentation diseases, etc.

13.4.2 Application of LLLT in the treatment of photoaging

Skin aging begins in the late 20 s and early 30 s, including endogenous aging (natural aging) and extrinsic aging (mostly photoaging). UV damage is the main factor for exogenous aging, which is manifested by skin color dullness and unevenness, rough and wrinkled texture, and loose skin without firmness. The most popular methods for skin rejuvenation include retinoic acid, dermabrasion, chemical peels, CO_2 and Er:YAG lasers. The main principle is to exfoliate the epidermis and induce well-controlled skin damage in order to promote collagen synthesis and dermal matrix reshaping. However, exfoliative treatment requires the original epidermal cells to be shed and replaced by new ones. Therefore, delicate postoperative care and a longer recovery period are needed. In addition, adverse reactions such as edema, infection, pain, and post-inflammatory hyperpigmentation may occur. The painful feeling and the high cost limited its application, making non-exfoliative skin rejuvenation a better option, where the application of LLLT has been further investigated.

Weiss *et al* conducted a clinical study in which 300 patients were treated with 590 nm (0.1 J cm^{-2}) pulsed LEDs (twice a week for 60 s for 8 weeks) (Weiss *et al* 2005). The results showed that the skin texture of 90% patients was softened and the roughness and texture were visibly reduced. Wunsch and Matuschka conducted a prospective, randomized controlled study in which 136 volunteers were randomly divided into three groups: 611–650 nm, 570–850 nm and the control group, irradiated twice a week for a total of 30 times (Wunsch and Matuschka 2014). Compared with the control group, the skin appearance of the light group was significantly improved and the collagen density was increased. The two light source groups showed similar efficacy and no serious adverse effects. In another prospective half-face controlled RCT study in 2007, 112 patients were randomly allocated into four groups (control group, LED near-infrared light group, LED red light group, and combined treatment group), among which for the LED near-infrared light group 830 nm, 66 J cm^{-2} was applied and for the LED red light group 633 nm,

126 J cm^{-2} was applied (Lee *et al* 2007). The treatment was executed every 3–4 d, twice a week, for four consecutive weeks and then followed up for another 12 weeks. The results revealed that patients in the combination treatment group had the biggest reduction in skin wrinkles (up to 36%), increase of skin elasticity (up to 19%) (Lee *et al* 2007). The tissue immunohistochemistry further suggested the expression of MMP inhibitory proteins 1 and 2 (TIMP-1 and TIMP-2), IL-1, TNF-α and ICAM-1 were increased, while the level of IL-6 decreased.

13.5 Conclusion

Low-level light therapy has been widely used in the field of dermatology, mainly through photobiomodulation. In LLLT, the emerging LED light as a cold laser has gained much attention. Compared with traditional lasers, LED light is safer and more effective, and cheaper. In the near future, it may replace traditional lasers for the first choice for LLLT light sources. However, parameters of LLLT such as energy density, irradiation density, irradiation time, and treatment interval are still heterogeneous and require standardization in further clinical studies. Nevertheless, there is no doubt that the photobiomodulation effect of LLLT has broad application prospects as light therapy in the treatment of numerous skin diseases and conditions, especially photoaging.

References

Alkozi H A, Wang X, Perez de Lara M J and Pintor J 2017 Presence of melanopsin in human crystalline lens epithelial cells and its role in melatonin synthesis *Exp. Eye. Re.* **154** 168–76

Al-Watban F A H and Zhang X Y 2004 The comparison of effects between pulsed and CW lasers on wound healing *J. Clin. Laser Med. Surg.* **22** 15–8

Amaroli A, Ferrando S and Benedicenti S 2018 Photobiomodulation Affects Key Cellular Pathways of all Life-forms: Considerations on Old and New Laser Light-targets and the Calcium Issue *Photochem. Photobiol.* **95** 455–59

Avci P, Gupta A, Sadasivam M, Vecchio D, Pam Z, Pam N and Hamblin M R 2013 Low-level laser (light) therapy (LLLT) in skin: Stimulating, healing, restoring *Semin. Cutan. Med. Surg.* **32** 41–52

Barolet D 2014 Pulsed versus continuous wave low-level light therapy on osteoarticular signs and symptoms in limited scleroderma (CREST syndrome): A case report *J. Biomed. Opt.* **19** 118001

Barolet D, Duplay P, Jacomy H and Auclair M 2010 Importance of pulsing illumination parameters in low-level-light therapy *J. Biomed. Opt.* **15** 048005

Bertolesi G E, Hehr C L and McFarlane S 2015 Melanopsin photoreception in the eye regulates light-induced skin colour changes through the production of α-MSH in the pituitary gland *Pigment Cell Melanoma Res.* **28** 559–71

Bertolesi G E and McFarlane S 2018 Seeing the light to change colour: An evolutionary perspective on the role of melanopsin in neuroendocrine circuits regulating light-mediated skin pigmentation *Pigment Cell Melanoma Res.* **31** 354–73

Brochetti R A *et al* 2017 Photobiomodulation therapy improves both inflammatory and fibrotic parameters in experimental model of lung fibrosis in mice *J. Lasers Med. Sci.* **32** 1825–34

Caruso-Davis M K, Guillot T S, Podichetty V K, Mashtalir N, Dhurandhar N V, Dubuisson O, Yu Y and Greenway F L 2011 Efficacy of low-level laser therapy for body contouring and spot fat reduction *Obes. Surg.* **21** 722–9

Castellano-Pellicena I, Uzunbajakava N E, Mignon C, Raafs B, Botchkarev V A and Thornton M J 2018 Does blue light restore human epidermal barrier function via activation of Opsin during cutaneous wound healing? *Lasers Surg. Med.* **51** 370–82

Chung H, Dai T, Sharma S K, Huang Y-Y, Carroll J D and Hamblin M R 2012 The Nuts and Bolts of Low-level Laser (Light) Therapy *Annals of Biomedical Engineering* **40** 516–33

Consentino L *et al* 2015 Blue-light dependent reactive oxygen species formation by Arabidopsis cryptochrome may define a novel evolutionarily conserved signaling mechanism *New Phytol.* **206** 1450–62

Dall Agnol M A, Nicolau R A, de Lima C J and Munin E 2009 Comparative analysis of coherent light action (laser) versus non-coherent light (light-emitting diode) for tissue repair in diabetic rats *Lasers Med. Sci.* **24** 909–16

de Freitas L F and Hamblin M R 2016 Proposed Mechanisms of Photobiomodulation or Low-Level Light Therapy. *IEEE Journal of Selected Topics in Quantum Electronics : A Publication of the IEEE Lasers and Electro-Optics Society* **22** 348–64

Denda M and Fuziwara S 2008 Visible Radiation Affects Epidermal Permeability Barrier Recovery: Selective Effects of Red and Blue Light *J. Invest. Dermatol.* **128** 1335–6

Erusalimsky J D and Moncada S 2007 Nitric oxide and mitochondrial signaling: From physiology to pathophysiology *Arterioscler. Thromb. Vasc. Biol.* **27** 2524–31

Gavish L, Asher Y, Becker Y and Kleinman Y 2004 Low level laser irradiation stimulates mitochondrial membrane potential and disperses subnuclear promyelocytic leukemia protein *Lasers Surg. Med.* **35** 369–76

Gizinger O A, Moskvin S V, Ziganshin O R and Shemetova M A 2016 The effect of continuous low-intensity laser irradiation of the red spectrum on the changes in the functional activity and speed of NADPH-oxidase response of human peripheral blood neutrophils *Vopr. Kurortol. Fizioter Lech. Fiz. Kult.* **93** 28–33

Heiskanen V and Hamblin M R 2018 Photobiomodulation: Lasers vs. light emitting diodes? *Photochem. Photobiol. Sci.* **17** 1003–17

Hode T, Duncan D, Kirkpatrick S, Jenkins P and Hode L 2009 The importance of coherence in phototherapy *Mechanisms for Low-Light Therapy IV Proc. SPIE* **7165** 716507

Hu W-P, Wang J-J, Yu C-L, Lan C-C E, Chen G-S and Yu H-S 2007 Helium-neon laser irradiation stimulates cell proliferation through photostimulatory effects in mitochondria *J. Invest. Dermatol.* **127** 2048–57

Huang Y-Y, Chen A C-H, Carroll J D and Hamblin M R 2009 Biphasic dose response in low level light therapy *Dose-Responsey* **7** 358–83

Huang Y-Y, Sharma S K, Carroll J and Hamblin M R 2011 Biphasic dose response in low level light therapy—An update *Dose-Response* **9** 602–18

Isoldi M C, Rollag M D, Castrucci A M, de L and Provencio I 2005 Rhabdomeric phototransduction initiated by the vertebrate photopigment melanopsin *Proc. Natl. Acad. Sci. USA* **102** 1217–21

Karu T I 2008 Mitochondrial signaling in mammalian cells activated by red and near-IR radiation *Photochem. Photobiol.* **84** 1091–99

Karu T I 2014 Cellular and Molecular Mechanisms of Photobiomodulation (Low-Power Laser Therapy) *IEEE Journal of Selected Topics in Quantum Electronics* **20** 143–8

Kim H-J, Son E D, Jung J-Y, Choi H, Lee T R and Shin D W 2013 Violet Light Down-Regulates the Expression of Specific Differentiation Markers through Rhodopsin in Normal Human Epidermal Keratinocytes *PLOS ONE* **8** e73678

Kottke T, Xie A, Larsen D S and Hoff W D 2018 Photoreceptors Take Charge: Emerging Principles for Light Sensing *Annu. Rev. Biophys.* **47** 291–313

Lan C-C E, Wu S-B, Wu C-S, Shen Y-C, Chiang T-Y, Wei Y-H and Yu H-S 2012 Induction of primitive pigment cell differentiation by visible light (helium-neon laser): A photoacceptor-specific response not replicable by UVB irradiation *J. Mol. Med. (Berl)* **90** 321–30

Lee S Y, Park K-H, Choi J-W, Kwon J-K, Lee D R, Shin M S, Lee J S, You C E and Park M Y 2007 A prospective, randomized, placebo-controlled, double-blinded, and split-face clinical study on LED phototherapy for skin rejuvenation: Clinical, profilometric, histologic, ultrastructural, and biochemical evaluations and comparison of three different treatment settings *J. Photochem. Photobiol. B.* **88** 51–67

Martinez-Ruiz A, Cadenas S and Lamas S 2011 Nitric oxide signaling: Classical, less classical, and nonclassical mechanisms *Free Radic. Biol. Med.* **51** 17–29

McGuff P E, Deterling R A and Gottlieb L S 1965 Tumoricidal effect of laser energy on experimental and human malignant tumors *N. Engl. J. Med.* **273** 490–2

Mignon C, Botchkareva N V, Uzunbajakava N E and Tobin D J 2016 Photobiomodulation devices for hair regrowth and wound healing: A therapy full of promise but a literature full of confusion *Exp. Dermatol.* **25** 745–9

Morgan M J and Liu Z 2011 Crosstalk of reactive oxygen species and NF-κB signaling *Cell Res.* **21** 103–15

Nakao A 2014 Temporal regulation of cytokines by the circadian clock *J. Immunol. Res.* **2014** 614529

Narasimamurthy R, Hatori M, Nayak S K, Liu F, Panda S and Verma I M 2012 Circadian clock protein cryptochrome regulates the expression of proinflammatory cytokines *Proc. Natl. Acad. Sci. USA* **109** 12662–7

Ozdeslik R N, Olinski L E, Trieu M M, Oprian D D and Oancea E 2019 Human nonvisual opsin 3 regulates pigmentation of epidermal melanocytes through functional interaction with melanocortin 1 receptor *Proc. Natl. Acad. Sci. USA* **116** 11508–17

Poyton R O and Ball K A 2011 Therapeutic photobiomodulation: Nitric oxide and a novel function of mitochondrial cytochrome c oxidase. *Discov. Med.* **11** 154–159

Rajagopalan K V and Handler P 1964 The absorption spectra of iron-flavoproteins *J. Biol. Chem.* **239** 1509–14

Regazzetti C, Sormani L, Debayle D, Bernerd F, Tulic M K, De Donatis G M, Chignon-Sicard B, Rocchi S and Passeron T 2018 Melanocytes Sense Blue Light and Regulate Pigmentation through Opsin-3 *J. Invest. Dermatol.* **138** 171–8

Santana-Blank L, Rodríguez-Santana E A, Santana-Rodríguez J, Santana-Rodriguez K and Reyes H 2016 Water as a photoacceptor, energy transducer and rechargeable electrolytic biobattery in photobiomodulation Handbook of Low-Level Laser Therapy eds. M R Hamblin, M de Sousa and T Agrawal (Singapore: Pan Stanford Publishing)

Schroeder P, Pohl C, Calles C, Marks C, Wild S and Krutmann J 2007 Cellular response to infrared radiation involves retrograde mitochondrial signaling *Free Radic. Biol. Med.* **43** 128–35

Sorokin A 2016 Nitric Oxide Synthase and Cyclooxygenase Pathways: A Complex Interplay in Cellular Signaling *Curr. Med. Chem.* **23** 2559–78

Sutherland J C 2002 Biological effects of polychromatic light *Photochem. Photobiol.* **76** 164–170

Tafur J and Mills P J 2008 Low-intensity light therapy: Exploring the role of redox mechanisms *Photomed. Laser Surg.* **26** 323–8

Terakita A 2005 The opsins *Genome Biology* **6** 213

Wang L, Zhang D and Schwarz W 2014 TRPV Channels in Mast Cells as a Target for Low-Level-Laser Therapy *Cells* **3** 662–73

Weiss R A, McDaniel D H, Geronemus R G, Weiss M A, Beasley K L, Munavalli G M and Bellew S G 2005 Clinical experience with light-emitting diode (LED) photomodulation. *Dermatol. Surg.* **31** 1199–205

Wunsch A and Matuschka K 2014 A controlled trial to determine the efficacy of red and near-infrared light treatment in patient satisfaction, reduction of fine lines, wrinkles, skin roughness, and intradermal collagen density increase *Photomed. Laser Surg.* **32** 93–100

Young S R, Dyson M and Bolton P 1990 Effect of light on calcium uptake by macrophages *Laser Therapy* **2** 53–7

Chapter 14

Radiofrequency technology for treating photoaging

Rui Yin

Radiofrequency (RF) technology offers versatile applications in facial rejuvenation. By delivering controlled heat energy to targeted layers of the skin, RF stimulates collagen production and tissue remodeling, resulting in improved skin texture, firmness, and reduction of wrinkles. RF treatments are non-invasive, allowing for minimal downtime and side effects. Moreover, RF can be combined with other modalities such as IPL, LED therapy or topical agents to enhance clinical outcomes, making it a valuable tool in comprehensive facial rejuvenation strategies.

14.1 Definition of radiofrequency

Radiofrequency (RF) is an electromagnetic wave characterized by alternating amplitude and frequency modulations, with its energy existing and propagating in space in the form of electricity or magnetism (wave). Its frequency range is extensive, spanning from hundreds of KHz to hundreds of MHz. RF serves various medical purposes such as cutting, excising, electrocauterizing, ablating, and coagulating tissue through electrothermal action, effectively removing lesions and treating diseases. In the medical equipment realm, the narrow band of this frequency range (from 200 kHz to 40 MHz) has distinct applications due to reduced nerve and muscle stimulation within this range. Consequently, energy application can be gentle, allowing for tissue heating at different levels (Lapidoth and Halachmi 2015).

RF used on the skin relies on its ability to deliver volumetric heat in depth. It acts on the dermis and even the subcutaneous tissue, producing a columnar thermal damage zone. This action causes an immediate contraction of collagen fibers and subsequent post-traumatic repair response, which forms the basis for RF application in the field of skin beauty and treatment (Zelickson *et al* 2004). High-frequency currents heat cutaneous tissues as a whole, irrespective of skin type, making it safe for dark skin types and effective for clear chromophores. Unlike lasers, there are no

doi:10.1088/978-0-7503-5112-6ch14

losses by reflection or scattering with RF. Energy diffusion depends only on tissue conductivity.

Similar to light/laser treatments, the effect of RF on tissue is controlled by the concentration (fluency) or power density of the device. High power applied over a large area using large electrodes causes mild heating. However, when concentrated in small areas, such as an electrode in the form of a needle, it causes tissue ablation.

The penetration of RF energy into the tissue, or the attenuation of the energy as it penetrates the tissue, depends on the fluency used, the configuration of the electrodes (monopolar, bipolar, or unipolar), the anatomy of the treated area, and the tissue's conductivity characteristics.

14.2 Basic principles and parameters of RF

Electromagnetic fields interacting with human tissues can induce heating effects. The mechanism of RF heating primarily depends on the working frequency of RF. There are two heating mechanisms: one generates ion current by displacing charged particles in the alternating electromagnetic field, while the other involves the rotation of polar water molecules in AC electromagnetic fields.

These mechanisms interact with affected particles and biological tissues, resulting in the dissipation of electromagnetic energy, thereby heating and raising the temperature of biological tissues. At lower frequencies, ion current is the main heating mechanism, while frequencies above 10 MHz see the rotation mechanism of water molecules gradually taking precedence. Frequencies above 30–40 MHz generate more heat through this mechanism than ion current. The eyes and testes are particularly sensitive to RF (Maria and Tamura 2018).

RF can selectively transfer heat energy to the dermis and subcutaneous tissues. Following treatment, short-term primary collagen contraction occurs immediately. Subsequently, reversible thermal injury initiates the skin repair mechanism, upregulating the expression of collagen I mRNA, promoting new collagen fiber synthesis, and facilitating collagen remodeling. This process typically occurs 2–6 months after RF treatment and persists for an extended period (Zelickson *et al* 2004).

RF exhibits two primary characteristics: (i) it demonstrates little correlation with skin pigment, making it advantageous for treating individuals with dark skin; (ii) RF penetrates deeply and can heat the deep dermis and even subcutaneous fat, promoting fiber diaphragm contraction and effectively addressing skin laxity (Fisher *et al* 2005).

The RF field comprises both an electric field and a magnetic field. Typically, the electric field is expressed in watts per square meter ($W\ m^{-2}$), while the magnetic field is expressed in amperes per square meter ($A\ m^{-2}$). Furthermore, power density can accurately measure the energy in an area distant from the transmission source. Experimental findings indicate that the absorption frequency of RF energy at non-radiation sources ranges from 80 to 100 MHz. (Maria and Tamura 2018).

14.3 RF classification

14.3.1 Single pole RF

14.3.1.1 Monopolar RF

Monopolar RF circuitry utilizes a skin contact electrode and ground electrode to conduct current (Alster and Lupton 2007). When the current is transmitted to the skin, it induces volumetric heating in the dermis, improving skin laxity through immediate and long-term thermal effects (Elsaie 2009). The typical instrument frequency is 6.78 MHz, capable of generating a treatment area temperature ranging from 65 °C to 75 °C at a controllable depth of 3 to 6 mm within the skin, while safeguarding the epidermis with a cooling device (Elsaie 2009, Sukal and Geronemus 2008).

14.3.1.2 Loop free unipolar RF

The loop-free unipolar RF lacks a grounding electrode, with a typical instrument frequency of 40.68 MHz, facilitating volumetric heating of the dermis, fat, and other tissues (Lolis and Goldberg 2007).

14.3.1.3 D-focused RF

Focused monopole RF, based on an internal adjustment frequency of 40.68 MHz, enhances the external modulation frequency, compresses the RF sine wave waveform, and adjusts the propagation direction of the sine wave via wave phase matching technology. This concentrated thermal effect of the RF wave shock center increases energy focus, enabling the target tissue temperature to reach 55 °C–65 °C. High-frequency RF phase shifters achieve precise phase adjustments of RF waves, controlling phase movement at depths of 1.5, 2.5, 3.5, and 4.5 mm subcutaneously. This addresses the issue of uncontrollable penetration depth encountered with traditional RF methods (Lolis and Goldberg 2007).

14.3.2 Bipolar RF

This configuration comprises two symmetrical positive and negative electrodes. The penetration depth is approximately half the distance between the two electrodes, resulting in shallow penetration (Elsaie 2009). Due to its limited effectiveness, it is seldom used alone in clinical practice.

14.3.3 Multipolar RF

Multipolar RF consists of three or more electrodes, with all electrodes exchanging positive and negative charges alternately. This setup forms a current loop between any two electrodes, enabling multiple current loops to gather treatment energy. Therefore, using relatively low power can yield sufficient energy, rapidly increasing skin tissue temperature in the treatment area (Sadick and Rothaus 2016). Additionally, patients do not require electrode plates or pads during multi-level radiofrequency treatment. The alternating interchange of positive and negative electrodes among multiple electrodes not only expands the effective treatment area but also heats the skin tissue evenly, effectively saving treatment time and enhancing treatment safety and comfort.

14.3.4 Fractional RF

Fractional RF generates current through an array arrangement of bipolar RF electrodes. In fractional mode, it heats the local dermis to create a small thermal injury zone, offering the benefits of reduced skin damage and faster wound healing. Based on the mechanism of action, fractional RF is categorized into non-invasive fractional RF and invasive fractional RF, such as microneedle RF (Delgado and Chapas 2022).

14.3.4.1 Non-invasive fractional RF

This configuration contacts the skin directly through a matrix array of positive/negative electrodes to emit energy. Micro-exfoliation occurs in the epidermis of the electrode contact area, while discontinuous heating areas are observable in the dermis. Within a single heating area, continuous changes from shallow to deep can be observed, including exfoliation, coagulation, necrosis, and sub-necrosis tissues, forming a narrow upper and wide lower water drop pattern. Higher energy levels correspond to increased exfoliation, while lower energy levels result in a higher proportion of coagulation, necrosis, and sub-necrosis. During non-invasive electrode treatment, it is essential to maintain dry skin in the treatment area. This is because wet skin has extremely low impedance, allowing a circuit to form directly on the skin surface between the positive and negative electrodes during energy emission, potentially causing damage to the epidermis.

14.3.4.2 Invasive fractional RF

The apparatus, also known as microneedle fractional RF, directly transfers RF energy to the target tissue by employing insulated or uninsulated microneedles distributed in an array. This configuration serves dual functions: RF heating and mechanical damage via microneedles (Delgado and Chapas 2022). This type of radiofrequency microneedle can be categorized into two groups. The first is insulated microneedles, which penetrate the dermal papilla layer through the epidermis, with the needle tip emitting radiofrequency. Because the needle body is insulated, the thermal effect can be concentrated effectively on the dermis. The other type is semi-insulated microneedles. After penetrating the dermis, the needle body emits RF energy. Although it induces a larger area of thermal damage, it primarily capitalizes on the impedance difference between the epidermis and dermis to facilitate the passage of RF current through the dermis, resulting in minimal damage to the epidermis. Generally, adjustments can be made to the size of radiofrequency energy, the duration of action, and the depth of microneedles. The depth of microneedles and the duration of radiofrequency action mainly influence the histological changes of thermal injury caused by radiofrequency, while the size of energy may be associated with the density of dermal thermal injury.

14.3.5 Hybrid system

Radiofrequency technology plays a significant role in addressing skin sagging. However, in addition to sagging, skin photoaging encompasses changes in skin texture, wrinkles, fine lines, hyperpigmentation, and capillary dilation. To address these issues, a hybrid system emerged, integrating radiofrequency with other photonic methods, including intense pulsed light (IPL), vacuum systems, pulsed magnetic fields, infrared (IR), and more (Jiang *et al* 2017).

14.3.5.1 Electro optical synergy (ELOS)

The ELOS system leverages the synergistic effect of RF and optical equipment. The optical component encompasses broadband light (typically ranging from 400–980 nm; 580–980 nm; 680–980 nm) or single-wavelength lasers, spanning from visible light to near-infrared light wavelengths above 400 nm. RF and light energy are simultaneously emitted, enabling the simultaneous resolution of multiple skin concerns in the treated area. Additionally, the low impedance reduces the energy threshold required for RF to achieve the desired effect, enhancing comfort and minimizing adverse reactions (Jiang *et al* 2017).

14.3.5.2 Over-frequency system

The over-frequency system combines monopolar RF with an ultrasound component (e.g. BTL Exilis system, BTL Industries). The ultrasound component aims to modify tissue impedance, increase cell permeability, and enhance penetration of RF energy into deeper layers. A clinical study by Chilukuri *et al* demonstrated the system's versatility in evaluating efficacy across multiple body parts. Thirty-four patients were divided based on their indication and treated for laxity on the face, arms, as well as for fat in the thighs and abdomen. Four 30 min treatments were administered. Independent evaluators identified patient baseline photographs from the 3 month follow-up in 92% of cases, with all groups scoring above 90% (the highest on facial photographs at 93%, the lowest on arm photographs at 90.5%). Patients reported an average satisfaction rating of 4.15 on a scale of 1–5 (5 indicating strong satisfaction, 1 indicating strong dissatisfaction), and they agreed that the treatment was comfortable, with an average score of 4.06 (5 indicating strong agreement, 1 indicating strong disagreement). There were no reports of post-treatment pain or skin damage (Chilukuri *et al* 2017, Chilukuri and Lupton 2017).

14.3.5.3 Combined system of radiofrequency and pulsed magnetic field

The combined system of radiofrequency and pulsed magnetic field releases a pulsed magnetic field (PMF) while transmitting radiofrequency (e.g. Thermage ThermaCool, Solta Medical, San Francisco, CA). This system comprises several components enabling the delivery of electromagnetic energy to the skin via a single-use treatment tip, which cools its surface during RF energy transmission. Energy settings are tailored based on the anatomy of the treated area (Chilukuri and Lupton 2017). The pulsed magnetic field itself does not generate heat; rather, through a coil positioned above the skin, it induces the passage of the magnetic field through the skin, forming a load current (eddy current). This current alters the potential of charged receptors on the dermal cell membrane, affecting receptor distribution on fibroblast surfaces and promoting collagen production by modulating cAMP metabolism. This synergizes with RF treatment (Chilukuri and Lupton 2017).

A multicenter study evaluated a low-fluence algorithm for treating facial laxity. At 4 months, 95% of patients showed improvement, with most (65%) reporting good (26%–50%) or very good (51%–75%) improvement. Five percent of patients showed no improvement. Results were consistent at 6 months, with the proportion of patients showing no improvement increasing to 8%. Subjective satisfaction was 78% at 4 months and decreased to 70% at 6 months post-treatment (Bogle *et al* 2007).

14.4 Clinical application of radiofrequency

14.4.1 Aesthetic rejuvenation

Currently, monopolar RF is considered the gold standard for non-invasive treatment of skin laxity (Abraham and Mashkevich 2007). It boasts deep penetration and strong efficacy. Follow-up examinations conducted 1–10 months post-treatment reveal sustained improvement in skin laxity for 80% of patients, with 55% experiencing enhanced skin texture, indicating enduring effects (Edwards *et al* 2013). No severe complications such as scarring, pain, or fat atrophy were reported.

Bipolar RF, however, exhibits a shallower effective penetration depth compared to monopolar RF, necessitating repeated short-term treatments to ensure efficacy due to insufficient energy delivery to deep tissues. Dot array RF can effectively heat the deep skin layers while inducing a peeling effect on the epidermis. Dayan *et al* reported a 93% satisfaction rate among 247 patients treated for mandibular and cervical skin laxity, with notable improvements in skin texture (Dayan *et al* 2019).

Microneedle radiofrequency combines microneedle and radiofrequency treatment modalities, offering a relatively safe and controllable method for heating the deep skin layers. Matteo *et al* observed 33 patients with mild skin laxity in the lower face and neck. Six months post-treatment, they noted reductions of 28.58° and 16.68° in the chin and submandibular angles, respectively, demonstrating advantages in treating lower face and neck skin laxity and tightening (Clementoni and Munavalli 2016).

14.4.2 RF lipolysis

RF technology has a significant impact on adipocytes and fat metabolism. Utilizing alternating current, radiofrequency technology generates ion currents and local heat within fat cells, thereby improving the appearance of fat and fat mass to a certain extent. Unipolar volumetric radiofrequency devices can achieve more extensive and deeper heating, making them recommended for fat and cellulite treatment (Kennedy *et al* 2015). Kennedy *et al* conducted a review of 31 clinical studies involving 2937 patients who underwent radiofrequency and other technologies for subcutaneous fat reduction or body contouring. The majority of patients achieved satisfactory results, with radiofrequency studies reporting satisfaction rates ranging from 71% to 97% (Kennedy *et al* 2015). While radiofrequency lipolysis technology demonstrates efficacy comparable to other lipolysis technologies, there is currently no consensus research on treatment cycles, energy levels, or treatment methods.

14.4.3 Scar

While lasers are typically the first-line choice for scar treatment, RF offers unique advantages and plays a significant role in scar management (Seago *et al* 2020). RF targets polar water molecules and charged particles, independent of skin color, and can achieve volumetric or lattice mode heating. Through the electrothermal effect, RF stimulates, damages, or destroys scar tissue, initiating the body's post-traumatic self-repair mechanism, promoting the proliferation and rearrangement of dermal collagen fibers and elastic fibers, thus aiding in scar repair (Dai *et al* 2017, Taub 2019).

Microplasma RF employs monopolar RF emission energy to excite inert gases (such as nitrogen, argon, etc) into a plasma state, characterized by high temperature and energy. Upon contact with the skin, this energy is rapidly transferred to the treatment area, inducing micro-exfoliation on the skin surface and transmitting heat to the dermis, leading to improved elasticity, smoothness, and color uniformity (Lan *et al* 2018, Wang *et al* 2017). Fractional RF and RF microneedles are also viable options for scar treatment (Lan *et al* 2021).

Recently, low-temperature plasma (NTP) has emerged as a scar treatment modality. In animal experiments, NTP promotes wound healing, possibly by inhibiting the TGF β1/Smad2/Smad3 signaling pathway and adjusting α-SMA and type I collagen levels, thereby reducing scar formation (Wang *et al* 2020). For keloids, NTP can downregulate the expression of EGFR and STAT3, inhibiting fibroblast migration (Kang *et al* 2017).

14.5 Combined applications of radiofrequency and other medical cosmetic treatments

When combining radiofrequency therapy with other medical cosmetic treatments, attention should be given to potential overlapping adverse reactions and conflicts between them. Therefore, the combination therapy should be arranged in a rational sequence and interval.

14.5.1 Combined application with photoelectric equipment

The general principle is to prioritize non-invasive treatments, followed by invasive procedures after a short and reasonable interval. When combining non-invasive radiofrequency with invasive laser equipment or vice versa, it is advisable to administer the non-invasive treatment first, followed by the invasive one after a 2 week interval.

If non-invasive RF is combined with other non-invasive laser equipment, there is no prescribed order. As long as there are no noticeable adverse reactions, successive treatments can be performed on the same day. The interval between treatments can vary from minutes to hours, depending on the skin's reaction post-treatment.

For combinations involving invasive RF and other invasive laser equipment, treatment order is typically based on treatment requirements, with intervals between different treatments ranging from 1 to 3 months. If all treatments are invasive and aim to stimulate collagen proliferation, the interval should exceed 3 months.

14.5.2 Combined application with microplastic surgery

Microplastic surgery often necessitates the use of preparations or materials such as type A botulinum toxin, hyaluronic acid, absorbable wire, nano-fat, or fat glue. When combined with radiofrequency therapy, the sequence of treatments should be carefully considered. Although current research confirms that radiofrequency therapy does not affect the diffusion of botulinum toxin type A, if both radiofrequency therapy and botulinum toxin injection are chosen to be administered at the same site on the same day, priority should be given to radiofrequency therapy. There should be an interval of several minutes to several hours before botulinum toxin injection, allowing the skin tissue to cool down. Subsequently, botulinum toxin injection should be performed based on the patient's needs and tolerance. If botulinum toxin is injected first, radiofrequency treatment at the same site is recommended after two weeks. Hyaluronic acid injection can be administered the day after radiofrequency treatment. Patients who have undergone hyaluronic acid injection should be managed differently based on the molecular weight and degree of cross-linking of the hyaluronic acid. Radiofrequency therapy should be conducted approximately 2 weeks to 3 months later (Hsu *et al* 2019).

Following radiofrequency treatment, the skin temperature typically returns to normal on the same day, allowing for thread lifting treatment. If thread embedding has been performed, radiofrequency therapy is generally required at intervals exceeding 3 months. For fat filling, RF treatment can precede fat filling. However, if fat filling has already been performed, RF treatment is recommended three to six months later.

14.5.3 Combined application with surgery

When combining radiofrequency therapy with surgery, invasive surgery is typically conducted first. Subsequently, adequate recovery time should be allocated, generally necessitating a waiting period of 1–2 months after surgery. Once the surgical wound has healed, radiofrequency treatment can be initiated.

14.5.4 Combined treatment with micro-needling

According to patients' specific conditions, if combined treatment with radiofrequency (excluding fractional microneedling radiofrequency) is required, roller microneedling treatment can be administered first, followed by radiofrequency treatment after a 2 week interval. Alternatively, non-invasive radiofrequency treatment can precede roller microneedling. On the same day as the non-invasive radiofrequency treatment, roller microneedling can be performed immediately after the skin temperature returns to normal.

14.5.5 Combined treatment with focused ultrasound

Focused ultrasound can be clinically employed for facial firming and lifting, as well as for fat reduction and shaping. It creates high-energy focus in subcutaneous tissue and generates thermal freezing points to achieve facial firming effects. When combined with radiofrequency treatment, it can reduce fat and enhance the synergistic effect of firming and upgrading. Depending on the individual's circumstances, different sequential schemes can be chosen, such as starting with radiofrequency treatment and performing focused ultrasound treatment on the same day once the skin temperature returns to normal. If focused ultrasound treatment precedes radiofrequency treatment, it should be administered at least two weeks later. However, to minimize the risk of adverse reactions such as accumulation of thermal coagulation zones and excessive thermal effects, allowing for a certain interval between treatments is safer than continuous sequential treatment.

For fat reduction and shaping, high-energy focused ultrasound includes focused non-thermal ultrasound and focused thermal ultrasound. When combined with radiofrequency therapy, radiofrequency treatment can be utilized to increase local blood circulation for 15–20 min, potentially enhancing the mechanical effect of focused ultrasound (Chang *et al* 2014). Subsequently, focused ultrasound therapy can be employed to destroy fat cells. If focused ultrasound treatment is conducted first, it is advisable to schedule RF treatment one week later to expedite fat metabolism and skin tightening.

14.6 Adverse reactions and clinical precautions of RF

In general, RF is safer and elicits fewer adverse reactions compared to other energy-based beauty technologies. Most adverse reactions are temporary and transient. Serious adverse events are often associated with high treatment parameter settings, the use of counterfeit equipment, and/or treatment heads. Pain is the most common adverse reaction, but it's usually proportional to efficacy to some extent and can be alleviated by adjusting the treatment level or discontinuing treatment (Paasch *et al* 2009). Transient erythema, mostly subsiding within 24 h, is common. Edema may appear immediately post-treatment and typically resolves spontaneously within 1–3 days. Secondary burns may manifest as persistent erythema resembling the treatment head's contact area, followed by clear scabbing or small blisters that vanish within 6 or 7 days after treatment initiation. Transient skin depression is rare but may occur in thin-skinned areas due to high treatment energy, repeated pulse superposition, and excessive deep tissue heating. Fat atrophy and excessive fibrous septa contraction can occur due to these factors but generally resolve within 1–3 months (Ekelem *et al* 2019). Fat necrosis and atrophy are very rare and may result from local over-operation leading to fat liquefaction and degeneration. Transient pigmentation, occasionally observed in lattice RF, plasma RF, etc, correlates with operating energy and density (Ekelem *et al* 2019). Subcutaneous nodules and hematomas are rare and typically stem from overly high energy density during treatment and a lack of timely application of appropriate cold gel.

Skin numbness in the treatment area is rare and usually subsides without special treatment, typically following the nerve distribution pattern. Scar formation is sporadic, often due to incomplete treatment head contact or improper management post-blister occurrence.

Medical RF falls under the medical treatment category and should be administered by a qualified physician. Patients should remove metal ornaments before treatment, and the use of sedatives, local blockers, or narcotic analgesics during treatment is not recommended as they may interfere with thermal sensory feedback and increase the risk of adverse events.

During treatment, the treatment head should fit the skin snugly, and some technologies may require the use of standard grid paper to prevent pulse superposition. Employing proper low energy and multiple coverage can yield favorable outcomes while mitigating the risk of adverse reactions caused by excessively high energy. Energy should be reduced in thinner subcutaneous tissue areas (zygoma, mandible, temporal part, forehead) (Chapas *et al* 2020).

In recent years, a new treatment model has emerged, involving lower energy usage, repeated treatments, and treatment endpoint determination based on patient thermal sensory feedback. This approach can eliminate or reduce unacceptable adverse reactions and significantly alleviate treatment-related pain, enabling most treatments to be conducted without anesthesia.

References

Abraham M T and Mashkevich G 2007 Monopolar radiofrequency skin tightening *Facial Plast. Surg. Clin. North Am.* **15** 169–77

Alster T S and Lupton J R 2007 Nonablative cutaneous remodeling using radiofrequency devices *Clin. Dermatol.* **25** 487–91

Bogle M A *et al* 2007 Evaluation of the multiple pass, low fluence algorithm for radiofrequency tightening of the lower face *Lasers Surg. Med.* **39** 210–7

Chang S L *et al* 2014 Combination therapy of focused ultrasound and radio-frequency for noninvasive body contouring in Asians with MRI photographic documentation *Lasers Med. Sci.* **29** 165–72

Chapas A, Biesman B S, Chan H H L, Kaminer M S, Kilmer S L, Lupo M P, Marmur E and Van Dyke S 2020 Consensus recommendations for 4th generation non-microneedling monopolar radiofrequency for skin tightening: a Delphi consensus panel *J. Drugs Dermatol.* **19** 20–6

Chilukuri S, Denjean D and Fouque L 2017 Treating multiple body parts for skin laxity and fat deposits using a novel focused radiofrequency device with an ultrasound component: safety and efficacy study *J. Cosmet. Dermatol.* **16** 476–9

Chilukuri S and Lupton J 2017 'Deep heating' noninvasive skin tightening devices: review of effectiveness and patient satisfaction *J. Drugs Dermatol.* **16** 1262–6

Clementoni M T and Munavalli G S 2016 Fractional high intensity focused radiofrequency in the treatment of mild to moderate laxity of the lower face and neck: a pilot study *Lasers Surg. Med.* **48** 461–70

Dai R *et al* 2017 The efficacy and safety of the fractional radiofrequency technique for the treatment of atrophic acne scar in Asians: a meta-analysis *J. Cosmet. Laser Ther.* **19** 337–44

Dayan E *et al* 2019 Adjustable depth fractional radiofrequency combined with bipolar radiofrequency: a minimally invasive combination treatment for skin laxity *Aesthet. Surg. J.* **39** S112–9

Delgado A R and Chapas A 2022 Introduction and overview of radiofrequency treatments in aesthetic dermatology *J. Cosmet. Dermatol.* **21** S1–S10

Edwards A F *et al* 2013 Clinical efficacy and safety evaluation of a monopolar radiofrequency device with a new vibration handpiece for the treatment of facial skin laxity: a 10-month experience with 64 patients *Dermatol. Surg.* **39** 104–10

Ekelem C *et al* 2019 Radiofrequency therapy and noncosmetic cutaneous conditions *Dermatol. Surg.* **45** 908–30

Elsaie M L 2009 Cutaneous remodeling and photorejuvenation using radiofrequency devices *Indian J. Dermatol.* **54** 201–5

Fisher G H *et al* 2005 Nonablative radiofrequency treatment of facial laxity *Dermatol. Surg.* **31** 1237–41

Hsu S H, Chung H J and Weiss R A 2019 Histologic effects of fractional laser and radiofrequency devices on hyaluronic acid filler *Dermatol. Surg.* **45** 552–6

Jiang M *et al* 2017 A prospective study of the safety and efficacy of a combined bipolar radiofrequency, intense pulsed light, and infrared diode laser treatment for global facial photoaging *Lasers Med. Sci.* **32** 1051–61

Kang S U *et al* 2017 Opposite effects of non-thermal plasma on cell migration and collagen production in keloid and normal fibroblasts *PLoS One* **12** e187978

Kennedy J *et al* 2015 Non-invasive subcutaneous fat reduction: a review *J. Eur. Acad. Dermatol. Venereol.* **29** 1679–88

Lan T *et al* 2021 Comparison of fractional micro-plasma radiofrequency and fractional microneedle radiofrequency for the treatment of atrophic acne scars: a pilot randomized split-face clinical study in China *Lasers Surg. Med* **53** 906–13

Lan T *et al* 2018 Treatment of atrophic acne scarring with fractional micro-plasma radiofrequency in Chinese patients: a prospective study *Lasers Surg. Med.* **50** 844–50

Lapidoth M and Halachmi S 2015 *Radiofrequency in Cosmetic Dermatology* **vol 2** (Basel: Karger) pp 1–6

Lolis M S and Goldberg D J 2007 Radiofrequency in cosmetic dermatology: a review *Dermatol. Surg.* **38** 1765–76

Maria C A I and Tamura B L 2018 *Lights and Other Technologies. Clinical Approaches and Procedures in Cosmetic Dermatology* (Cham: Springer International) pp 193–7

Paasch U *et al* 2009 Skin rejuvenation by radiofrequency therapy: methods, effects and risks *J. Deutsch. Dermatol. Ges.* **7** 196–203

Sadick N and Rothaus K O 2016 Aesthetic applications of radiofrequency devices *Clin. Plast. Surg.* **43** 557–65

Seago M *et al* 2020 Laser treatment of traumatic scars and contractures: 2020 international consensus recommendations *Lasers Surg. Med.* **52** 96–116

Sukal S A and Geronemus R G 2008 Thermage: the nonablative radiofrequency for rejuvenation *Clin. Dermatol.* **26** 602–7

Taub A F 2019 The treatment of acne scars, a 30-year journey *Am. J. Clin. Dermatol.* **20** 683–90

Wang S *et al* 2017 Fractional microplasma radiofrequency technology for non-hypertrophic post-burn scars in Asians: a prospective study of 95 patients *Lasers Surg. Med.* **49** 563–9

Wang X F *et al* 2020 Potential effect of non-thermal plasma for the inhibition of scar formation: a preliminary report *Sci Rep.* **10** 1064

Zelickson B D *et al* 2004 Histological and ultrastructural evaluation of the effects of a radiofrequency-based nonablative dermal remodeling device: a pilot study *Arch. Dermatol.* **140** 204–9

IOP Publishing

Skin Photoaging (Second Edition)

Rui Yin, Yang Xu and Chengfeng Zhang

Chapter 15

Intense focused ultrasounds for facial rejuvenation

Huaxu Liu

Intense focused ultrasound (IFU) finds diverse applications in facial rejuvenation. By targeting specific layers of skin, such as the SMAS and dermis, IFU induces collagen remodeling and skin tightening, resulting in the reduction of wrinkles and improved skin elasticity. Additionally, IFU stimulates the production of growth factors and hyaluronic acid, promoting tissue regeneration and enhancing overall skin texture and tone. Its non-invasive nature and ability to precisely deliver energy make IFU a valuable tool for addressing signs of aging and achieving natural-looking facial rejuvenation.

15.1 Introduction

Intense focused ultrasound (IFU) is an emerging non-invasive treatment that utilizes low-energy ultrasound directed outside the body onto a specific area, forming a high-energy target. This targeted approach induces irreversible coagulative necrosis in the tissues of the target area while preserving surrounding tissues. IFU has demonstrated both safety and efficacy in the clinical treatment of various cancers, including liver, breast, kidney, and bone tumors. Its non-invasive and non-radiative nature has also led to its application in guiding plastic surgery procedures. IFU-guided plastic surgery now joins the array of aesthetic technologies, alongside traditional methods such as liposuction, laser therapy, and radiofrequency ablation (Xiao *et al* 2022).

15.2 Definition of intense focused ultrasound

The mechanism of IFU involves concentrating ultrasound waves, which can be directed, penetrate, and focus on tissues, releasing a large amount of energy at the lesion site until the diseased tissue is damaged.

doi:10.1088/978-0-7503-5112-6ch15

15.3 Mechanisms of IFU in facial rejuvenation

As for the mechanism of IFU-guided cosmetic surgery, ultrasound energy is focused on subcutaneous tissues, such as the superficial musculoaponeurotic system (SMAS), rapidly heating them to 55 °C–65 °C. This temperature range disrupts and deforms the peptide bonds between collagen molecules in the triple helix, leading to immediate skin contraction in loose areas. When applied to collagen-rich tissues, IFU can also compact adipose tissue by tightening adipose fiber septa or lifting skin lengthwise by shrinking the SMAS layer. IFU stimulates hyaluronic acid and growth factor production in the extracellular matrix, promoting fibroblast regeneration, matrix production, collagen synthesis, and skin tissue growth. This process induces only diffuse and limited thermal damage in the target tissue, with ultrasound energy outside the thermal zone insufficient to cause tissue damage. Unlike lasers or microwaves, IFU can penetrate deep tissues without high-energy exposure to the epidermis, thus avoiding epidermal damage. Skin color does not affect the targeted delivery of ultrasound energy. Ultimately, the non-invasive tightening and lifting effects flatten wrinkles on the face and neck.

A transducer and a driver constitute the core components of the IFU system. In cosmetic surgery utilizing focused ultrasound (FU), low-energy ultrasound from outside the body is focused on the dermis or SMAS to induce coagulative necrosis without harming surrounding tissue in the treated area. Unlike HIFU, which employs long focal lengths, large calibers, and low frequencies (around 1 MHz) to treat deep tumors, the ultrasound used in cosmetic procedures operates within a much smaller range. Therefore, typically in MFU the transducer features a short focal length, a large caliber, and a high frequency.

Compared to laser and microwave technologies, IFU offers the advantage of focusing energy to create a high-energy zone at the target tissue. The transducer can converge the acoustic beam within a specific range, concentrating energy and enhancing ultrasound transmission.

15.4 Biological effects of IFU

During transmission, ultrasonic waves inevitably interact with the medium, resulting in phenomena such as reflection, refraction, diffraction, scattering, absorption, and attenuation. Ultrasonic waves also induce mechanical, thermal, and cavitation effects on the medium.

15.4.1 Mechanical effects

Ultrasonic energy causes high-frequency vibrations in the medium, leading to effects such as displacement, velocity, acceleration, sound pressure, and intensity. For instance, high sound pressure can disrupt the mechanical structure of the medium, facilitating cohesion, breaking, grinding, cutting, drilling, stirring, welding, and other material processing tasks.

15.4.2 Thermal effect

Absorption of ultrasonic energy by the medium generates heat due to internal friction between molecules, raising the medium's temperature. Higher ultrasound frequencies result in greater absorption and stronger thermal effects.

15.4.3 Cavitation effect

Ultrasonic waves in liquid induce pressure changes, leading to compression or stretching of the liquid. Excessive stretching can cause cavitation, forming cavities near vacuum conditions. Upon cavity collapse, the liquid generates shock waves, high pressure, temperature, and even ionization, useful for cleaning, atomization, emulsification, oxidation, catalysis, polymerization, and liquid processing.

15.4.4 Thixotropic effect

Ultrasound alters the binding state in biological tissues, decreasing viscosity and causing thinning of plasma and precipitation of blood cells, known as the thixotropic effect. At low sound intensities, this effect may be reversible, but high intensities can induce irreversible tissue changes.

Mechanical, heating, cavitation, and other effects collectively alter the properties of biological tissue, termed the biological effect of ultrasound. Ultrasound, a mechanical vibration wave, can penetrate living tissue to treat diseases without harming surrounding tissue. The ultrasound beam focuses energy on the lesion, inducing coagulative thermal necrosis and irreversible cell death, primarily through thermal and mechanical effects (Zhou 2011, Fu *et al* 2021).

15.5 Classification of IFU

Clinically, FU is generally divided into two types. One is high intensity focused ultrasound (HIFU) that vibrates the molecules in the target area to generate strong shear forces and cause coagulative necrosis through frictional heat generation, without causing damage to other surrounding tissues. It is often used for ablation of tumors *in vivo* (Zhou 2011). The other is microfocused ultrasound (MFU), which focuses low energy to produce small-site coagulative necrosis in the dermal reticular layer and subcutaneous tissues. The energy can be focused on a depth of 1–5 mm to stimulate the denaturation and contraction of collagen fibers in the dermal reticular layer, as well as the production of new collagen and elastin, thus achieving a cosmetic effect (Fabi 2015). Currently, IFU is commonly used for lipolysis, skin tightening, skin rejuvenation and wrinkle removal.

15.6 Clinical applications

15.6.1 FU for lipolysis

For conventional plastic surgery, dissection, laser and radiofrequency can remove subcutaneous fat in large areas, but may bring complications. Due to its non-invasive and non-radiative characteristics, IFU provides a new lipolytic therapy, with a high safety preliminarily verified by relevant studies. A study has observed the effect of IFU on the fat under the skin of live pigs (Li *et al* 2012). IFU was used to irradiate the subcutaneous fat on the back of pigs with a frequency of 1.9 MHz and an acoustic power of 550 W. The specimens were immediately prepared and stained to observe the immediate histological manifestations. The results showed that a large number of punctured adipose cells appeared in the adipose layer, and the skin and muscle layers were not damaged. Moravvej *et al* (2015) have performed a total of 194 IFU lipolytic treatments on the abdomen of 28 subjects, and found that the average circumference of the abdomen decreased to varying degrees immediately after treatment and at the 3 month follow-up. IFU may serve as an effective non-invasive treatment for subcutaneous liposculpture.

15.6.2 Focused ultrasound for skin rejuvenation

IFU is also frequently adopted for skin tightening. As the skin ages, the skill tissue becomes significantly less elastic, and prone to facial sagging and laxity. Surgical procedures for skin laxity have proven to be effective, but leave surgical scars. To ensure safety and provide a good treatment experience, Shenzhen Peninsula Medical Co. Ltd (Shenzhen, China) developed large-focal area high-energy IFU technology in 2010, which was approved by the Chinese NMPA in 2021.

The Peninsula Ultrasound System is designed for the treatment of skin tissue, subcutaneous fat layer and SMAS fascia layer, including an energy generation system, a computing control system, interactive system, two handpieces and multiple cartridges. The two handpieces are the MFUS and MFUS-D.

The MFUS handpiece controls a high-precision stepper motor by electronic pulses, which in turn drives the ultrasound transducer to mechanically slide over the skin surface, releasing ultrasound energy at 1.2–2.0 mm intervals, with a maximum range of 25 mm. The MFUS handpiece, in combination with the 2.0, 3.0, and 4.5 mm cartridges, focuses ultrasonic energy onto the derma, fat, and SMAS fascia layers.

The MFUS-D handpiece is more flexible than MFUS and can be applied to some uneven and difficult-to-treat areas. MFUS-D combined with 2.0, 3.0 and 4.5 mm cartridges can deliver energy to the superficial dermis, deep dermis, fat layer and SMAS fascial layer. The ultrasound frequency and transducer can be adjusted to achieve a large focus on the target area. Combined with sliding scanning technology, the tissue temperature can rise to 60 °C–70 °C without creating coagulation denaturation or forming scar adhesions, which ensures a safe and comfortable treatment. The sliding scanning handpiece can be easily manipulated to treat narrow areas around the eyes, lips, lines, neck, etc.

Each cartridge generates directional ultrasound energy with a high accuracy, releasing precise energy into the depth of the target. Ultrasonic energy causes intermolecular vibration, friction and a small amount of cavitation in human tissues. This mechanical energy causes a thermal effect through heating the local tissue to 55 °C–65 °C. At this temperature, the peptide bonds between the triple helix collagen molecules are broken and deformed, resulting in immediate contraction and tightening of the originally sagging skin, and when applied to collagen-rich tissues, it can also cause significant contraction, such as tightening fatty fiber septa to compact fatty tissue, or contracting the SMAS layer to lift up the skin longitudinally. IFU increases hyaluronic acid and growth factors in the extracellular matrix, which promotes the regeneration of fibroblasts and the production of matrix and collagen, as well as the growth of skin tissue. Only diffuse and limited thermal damage is produced in the target tissue area, and the ultrasound energy outside the thermal damage area is not sufficient to cause tissue damage. The mechanism of FU is different from that of lasers or microwaves. For a laser to reach deep tissue, the epidermis needs to be subjected to a high laser energy, whereas ultrasound can pass precisely through the epidermis to the dermis or below, avoiding damage to the epidermis. The targeted delivery of ultrasound energy is not affected by the color base of the skin tissue.

Two handpieces and cartridges at three depths of 2.0, 3.0 and 4.5 mm enable layered anti-aging in the dermis, fat layer, superficial fascia and SMAS fascial layer. The focal plane distance, which is the distance from the target tissue to the skin surface, can be decided by the transducer and the ultrasound frequency. When treating superficial tissues, the focal plane distance can be set at 2.0 mm, where a cartridge with a higher frequency can produce thermal damage while avoiding the spread of heat to the skin surface. At a depth of 2.0 mm that reaches the dermis, ultrasound stimulates continuous collagen regeneration to reduce static lines. When treating deeper tissues, such as fat and SMAS, the 3.0 and 4.5 mm depths can be chosen with lower frequencies and larger thermal dispersion range. At a depth of 3.0 mm that reaches the subcutaneous fat layer, the ultrasound can contract the skin and compress fat cells, thus improving the contour curve. At a depth of 4.5 mm that reaches the fascia layer of SMAS, IFU can lift the skin base and tighten the skin.

In a study by Li *et al* (Li *et al* 2022), an MFU therapy instrument (model MFUS One, cartridges M3.0, M4.5, D3.0, and D4.5) was used to treat the face and neck of 70 patients, and the results showed that the total effective rate was 97.14%, and none showed any surgical scars or other forms of injury.

15.7 Limitations

Currently, the treatment equipment has not been equipped with a temperature control system, and the treatment may rely heavily on experience. Different devices on the market are similar, and their long-term clinical effects still need to be confirmed further through strictly designed clinical studies.

15.8 Conclusion

HIFU, as a novel, minimally invasive treatment, has become one potential alternative treatment for a variety of benign and malignant diseases. MFU, another non-invasive, painless and highly effective IFU, can transmit ultrasound energy through the epidermis to the dermis and subcutaneous tissues, and stimulate collagen renewal and reorganization to achieve skin rejuvenation. In the future, research into MFU can be shifted to its effects on deeper tissue layers. Higher resolution ultrasound will also provide better visualization of facial tissue layers, thus facilitating precise treatment. The safety and effectiveness of IFU will be further improved in the field of aesthetics.

References

Fabi S G 2015 Noninvasive skin tightening: focus on new ultrasound techniques *Clin. Cosmet. Investig. Dermatol.* **8** 47–52

Fu X, Si X and Zhu J 2021 Research progress in clinical application of high-intensity focused ultrasound technology *China Clin. New Med.* **14** 1044–8

Li H, Sun Y and Tang G 2022 Clinical application of microfocused ultrasound technology in facial and neck rejuvenation *China Med. Aesthet.* **12** 21–5

Li W *et al* 2012 Histological observation of non-invasive focused ultrasound lipolysis quantitative efficacy in animal experiments *Chin. Cosmetol.* **21** 1156–8

Moravvej H *et al* 2015 Focused ultrasound lipolysis in the treatment of abdominal cellulite: an open-label study *J. Lasers Med. Sci.* **6** 102–5

Xiao S, Li H and Li F 2022 Focus on the application of ultrasound in cosmetic surgery and the progress of research on key issues *China Med. Aesthet.* **12** 74–8

Zhou Y F 2011 High intensity focused ultrasound in clinical tumor ablation *World J. Clin. Oncol* **2** 8–27

Chapter 16

Home-based modalities for photoaging

Lei Li and Rui Yin

Home-based modalities for photoaging offer a convenient and noninvasive approach to skin rejuvenation. These devices, while not as potent as professional-grade equipment, provide consumers with the flexibility of conducting treatments at home. They are particularly beneficial for individuals seeking mild improvements in skin texture, tone, and fine lines. However, it is essential to recognize their limitations in efficacy compared to clinic-based treatments. Home-based modalities can complement clinical interventions but should not replace them entirely for more pronounced anti-photoaging effects.

16.1 Introduction to home-based modalities

In the pursuit of enhancing their appearance, many individuals have turned to medical aesthetic services and treatments. Home-use modalities, serving as a complement to anti-photoaging strategies, are medical devices designed for use outside of professional healthcare facilities. They offer beauty enthusiasts a cost-effective, safe, highly portable, and relatively efficient solution for home use, following provided instructions.

As public awareness of these devices grows, individuals increasingly seek their dermatologist's opinion on their safety and efficacy. Despite this, the market for home-use modalities continues to expand rapidly, alongside professional medical treatments, which remain the primary choice for most individuals (Brown 2011).

The majority of home-use modalities are low-powered and available over the counter, eliminating the need for a physician's prescription. While their therapeutic power may be lower compared to medical devices, the benefits of accessibility, convenience, and privacy protection are highly desirable. Technologies such as radiofrequency, lasers, intense pulsed light (IPL), and light-emitting diode (LED) are commonly employed in home-use modalities. With proper use, these devices hold significant promise for anti-photoaging therapy with minimal side effects (Kim *et al* 2020).

doi:10.1088/978-0-7503-5112-6ch16

16.2 Classification

The increasing availability of anti-photoaging home-use modalities on the market can be categorized into three main types: LED systems, nonablative lasers, and radiofrequency devices (Town *et al* 2014).

LED systems are non-laser and nonablative light systems used for therapeutic purposes. The wavelength of light emitted by the LED determines its penetration depth into tissues, available in blue (400–470 nm), yellow (570–590 nm), red (630–700 nm), or near-infrared (800–1200 nm). Through photo-biochemical reactions within cells, LED can impact cellular metabolism, reduce oxidative stress, promote collagen synthesis, and stimulate blood flow. Home-use LED modalities can complement professional medical devices in improving photodamaged facial skin (Opel *et al* 2015, Gold *et al* 2017).

Nonablative fractional lasers, utilizing fractional photothermolysis, create isolated microscopic skin columns without harming surrounding tissues. Home-use nonablative lasers offer multiple reduced-coverage treatments with lower energy settings for enhanced skin safety and minimal reactions, making them more convenient and accessible. Thermal zones of injury trigger a natural wound-healing response, leading to quick recovery, low risk of allergic reactions, and increased collagen production (Leyden *et al* 2012).

Radiofrequency technology has been utilized for several years in clinical settings, demonstrating efficacy in diminishing wrinkles not only on the face and neck but also on other body areas. This technology functions by generating an electrical current, inducing heat production through resistance build-up within the dermis and subcutaneous tissues. Consequently, thermal stimulation triggers collagen contraction and initiates a wound-healing response, facilitating the remodeling of dermal collagen and skin tightening. While home-use RF modalities emitting low-intensity energy have exhibited modest improvements in wrinkles and skin texture compared to office-based counterparts, this may be attributed to their superficial penetration and relative energy deficit (Shemer *et al* 2014).

Despite the proven effectiveness of these devices in treating photoaging, a growing body of scientific evidence suggests that combining unique technologies or integrating them with topical agents could yield more comprehensive and complementary clinical outcomes (Geronemus *et al* 2016). Guermonprez *et al* propose that, in comparison to LED therapy or topical serums alone, the combination of LED therapy with a topical serum yields superior effects in reducing wrinkles (Guermonprez *et al* 2020).

16.3 Application

Photodamaged skin manifests with fine lines, wrinkles, mottled pigmentation, and other undesired changes, attributed to reduced collagen and diminished epidermal water content. Home-based modalities offer the advantage of multiple treatments at lower energy levels, enhancing skin safety and minimizing adverse reactions, thus rendering them more consumer-friendly and suitable. These modalities operate on principles akin to professional treatments. Within the subcutaneous layer, devices emitting low-intensity light, lasers, or radiofrequency have demonstrated dermal alterations, such as stimulating fibroblast growth, collagen synthesis, and enzymatic activities. Consequently, these effects contribute to the reduction of hyperpigmentation, lentigines, and overall facial wrinkles, alongside an improvement in skin texture, tone, and appearance (Shemer *et al* 2014, Gold *et al* 2017).

16.4 Objective evaluation home-based modalities for photoaging

The demand for nonsurgical, noninvasive, and minimal-downtime procedures for wrinkle and laxity reduction has seen a significant surge in the past decade. This heightened demand has spurred the development of new technologies aimed at delivering these noninvasive and nonablative anti-photoaging effects.

Clinic-based treatments come with several drawbacks, including limited availability, inconvenience in terms of travel time and distance to the clinic, higher treatment costs, discomfort from professional devices with high fluency, and risks of pigmentation issues and scarring. These drawbacks can be mitigated by utilizing small, low-energy home-use systems. Home-based modalities serve as a complement rather than competition to clinic-based procedures, aiding physicians in maintaining long-term patient relationships (Gold *et al* 2017).

With proper patient education and instruction, home-based modalities are safe, pleasant, and user-friendly, allowing for long-term and repeated treatments directly at home. Patients can undergo procedures in the privacy and comfort of their homes. Additionally, home-based treatments are potentially more cost-effective, appealing to consumers who may not be willing to pay higher fees for in-office treatments (Metelitsa and Green 2011).

Although home-based modalities have gained popularity due to their safety and convenience, they come with limitations. Typically designed for 'nonprofessional' users with a focus on low cost and absolute safety, these devices are often underpowered compared to their in-clinic counterparts, resulting in reduced efficacy and necessitating more treatment sessions to achieve visible clinical effects (Metelitsa and Green 2011).

16.5 Conclusion

Home-based modalities provide a noninvasive, effective, safe, and comfortable treatment option for photodamaged facial skin. With these modalities, consumers can undergo multiple treatments with reduced coverage and lower energy settings, resulting in fewer skin reactions and in a more convenient environment. Both patients and dermatologists benefit from this technology, as it enhances expected clinical outcomes with daily use. Consequently, dermatologists can extend their therapeutic plans by incorporating home-based modalities. However, it is important to note that while these devices are available, they cannot replace visits to a dermatologist's office, nor are they as effective as medical-grade devices found in clinics. Therefore, we recommend patients receive treatment at clinics equipped with medical-grade devices, with home-based devices serving as adjunctive treatments to improve facial rejuvenation.

References

Brown A S 2011 At-home laser and light-based devices *Curr. Probl. Dermatol.* **42** 160–5

Geronemus R, Du A, Yatskayer M, Lynch S, Krol Y and Oresajo C 2016 Enhanced efficacy of a topical antioxidants regimen in conjunction with a home-use non-ablative fractional diode laser in photodamaged facial skin *J. Cosmet. Laser Ther.* **18** 154–61

Gold M H, Biron J, Levi L and Sensing W 2017 Safety, efficacy, and usage compliance of home-use device utilizing RF and light energies for treating periorbital wrinkles *J. Cosmet. Dermatol.* **16** 95–102

Guermonprez C, Declercq L, Decaux G and Grimaud J A 2020 Safety and efficacy of a novel home-use device for light-potentiated (LED) skin treatment *J. Biophoton.* **13** e202000230

Kim D S, Song K U, Lee H K, Park J H, Kim B J, Yoo K H and Shin J H 2020 Synergistic effects of using novel home-use 660- and 850-nm light-emitting diode mask in combination with hyaluronic acid ampoule on photoaged Asian skin: a prospective, controlled study *J. Cosmet. Dermatol.* **19** 2606–15

Leyden J, Stephens T J, Herndon J H and JR 2012 Multicenter clinical trial of a home-use nonablative fractional laser device for wrinkle reduction *J. Am. Acad. Dermatol.* **67** 975–84

Metelitsa A I and Green J B 2011 Home-use laser and light devices for the skin: an update *Semin. Cutan. Med. Surg.* **30** 144–7

Opel D R, Hagstrom E, Pace A K, Sisto K, Hirano-Ali S A, Desai S and Swan J 2015 Light-emitting diodes: a brief review and clinical experience *J. Clin. Aesthet. Dermatol.* **8** 36–44

Shemer A, Levy H, Sadick N S, Harth Y and Dorizas A S 2014 Home-based wrinkle reduction using a novel handheld multisource phase-controlled radiofrequency device *J. Drugs Dermatol.* **13** 1342–7

Town G, Petersen R and Du Crest D 2014 The recent rapid development of the directed-energy, home-use device sector *Eur. Med. J.* **2** 50–5

IOP Publishing

Skin Photoaging (Second Edition)

Rui Yin, Yang Xu and Chengfeng Zhang

Chapter 17

Conclusions and future directions

As we move into the twenty-first century it is becoming clear that solutions to the perennial problems of skin aging are becoming increasingly sought after. The reasons for this increased activity are many and diverse. We can highlight the following. First, there is the relentless emphasis in the new digital world on visual appearance, and the need to conform to an ideal image of what a face should be. This pressure is compounded by advertisers who push images of super models on film, television, and the ubiquitous Internet. Second there is the phenomenon of increased life expectancy, especially among women, who are the main consumers of beauty products. Women are expecting to remain attractive to a much greater age than they once did. Third, there is the phenomenon in the developed world of a much greater disposable income, especially at the age at which solutions to skin aging become a primary concern. Fourth there is the fact that a wide variety of innovative, relatively non-invasive treatment approaches are now available, that were undreamed of a few years ago.

This book has attempted to review the progress on etiopathology and treatment for skin aging and photoaging. Public information has been growing on the dangers of excessive sun exposure, and advances in sunscreens that protect against UVB, UVA, blue light and NIR light, and more education about behavior modification and especially avoidance of tanning salons has led to an overall drop in dangerous sunlight exposure compared to a few decades ago. Photoaging research provides scientific evidence regarding the damaging effects of UV radiation on the skin and methods for protection. Based on this research, more effective sun protection products such as sunscreens, sprays, and sun-protective clothing can be developed to prevent photoaging and skin cancer.

Research on photoaging can uncover the mechanisms through which ultraviolet radiation and other light sources contribute to skin aging, leading to the development of anti-aging skincare products. For example, skincare products based on photoaging research may contain antioxidants, photorepair agents, anti-inflammatory agents, and

doi:10.1088/978-0-7503-5112-6ch17

other ingredients to reduce signs of skin aging. Antioxidants can be classified as either primary or secondary, depending on whether the substance itself quenches ROSs, or whether it indirectly increases the host anti-oxidant system. There are a large number of topical preparations containing vitamins, retinoids, plant-based extracts, flavonoids, polyphenols, tannins and anthocyanidins. Even metal-chelating compounds have been tested.

The rapid rise of botox (botulinum toxin type A, which actually is the most poisonous substance known on the planet) has astonished many in the field. The global botox market was valued at USD 6.4 billion in 2022 and is expected to reach USD 15.2 billion by 2030. More and more dermal fillers are being approved for clinical use by the FDA. The global dermal filler market size was worth around USD 4.8 billion in 2021 and is predicted to grow to around USD 11.7 billion by 2030 with a compound annual growth rate (CAGR) of roughly 10.5% between 2022 and 2030.

A wide array of intradermal fillers composed of preparations of hyaluronic acid, collagen from different species, fat, calcium hydroxyapatite, and polymers made of polylactic acid, PMMA and silicone have all been tried.

Various agents such as α-hydroxy-acids (glycolic acid), salicylic acid, trichloro-acetic acid, Jessner's solution, pyruvic acid and phenol-based formulas have been employed to give a chemical peel that can be classified as superficial, medium, and deep. The idea here is that when the epidermis grows back it will have a new 'spring in its step' and fine lines and wrinkles will be less pronounced.

Phototherapy utilizes specific wavelengths of light for skin treatments, such as laser or photorejuvenation. In-depth research on photoaging helps optimize the parameters and procedures of phototherapy, enhancing treatment effectiveness and safety. Many lasers and light sources have been employed to treat skin aging. The original ablative laser systems were effective, but had a long 'down-time' while the superficial wounds to the skin healed. Fractional ablative laser system, however, takes advantage of the finding that healing is dramatically quicker when islands of undamaged epidermis are left between the zones of thermal damage. Non-ablative lasers systems and low-level light devices have much lower incidence of side-effects, but are not as effective, needing to be repeated several times. Photodynamic therapy mediated by aminolevulinic acid and intense pulsed light (IPL)/red light has also been widely used for fine lines and wrinkles. The use of LLLT devices for photorejuvenation has expanded apace. The non-invasive nature of these devices and their suitability for home use suggests that they will continue to expand in the future. The advent of inexpensive LEDs with surprisingly high-power levels, means that numerous devices can be specifically constructed to allow therapeutic illumination of different parts of the face. The rise of Internet marketing using outlets such as eBay and Ali Baba have enabled small companies to easily build up a consumer base for these types of products. LED devices are accepted as being of low to zero-risk and so can be used at home with minimal regulation.

Energy can also be delivered into the skin using radiofrequency technology that can be monopolar, bipolar or fractional. Microneedling radiofrequency is suitable for patients with different skin types and age groups, providing a minimal-invasive treatment option for varying degrees of photoaging symptoms. Intense Focused Ultrasounds has also been applied for photoaging in clinical practice. Hybrid systems contain the integration of radiofrequency and other photoelectric means, including intense pulsed light (IPL), vacuum system, pulsed magnetic field, infrared (IR), and so on, and have been applied in clinical practice and have proven to have satisfying effects on rejuvenation.

Mesotherapy is another therapeutic direction for photoaging, which induces an optimal physiologic environment to stimulate the fibroblast activity and collageno-genesis by intradermal (ID) or subcutaneous (SQ) microinjection of beneficial compounds, such as pharmaceutical and homeopathic medications, plant extracts, vitamins, hyaluronic acid (HA), and other bioactive substances into the dermis and/or subcutaneous fat. The delivery of active ingredients with mesotherapy can exactly reverse the photoaging process such as the degeneration of elastin and the continuous trans-epidermal water loss.

In the future, more and more novel modalities (including laser, light and energy based modalities) and mesotherapeutic relative ingredient will be available. However, the safety and efficacy of the treatment will still be a great concern in the clinical treatment of anti-photoaging. Tailored skincare products and treatment plans can be developed based on different skin types, colors, ages, and other characteristics, meeting the specific needs of individuals.